March 25–28, 2012
Napa, California, USA

Association for Computing Machinery

Advancing Computing as a Science & Profession

WELCOME to this world famous wine growing region

NAPA VALLEY

ISPD'12

Proceedings of the 2012
International Symposium on Physical Design

Sponsored by:
ACM SIGDA

Technical Co-Sponsor:
IEEE CAS

Supported by:
ATopTech, Cadence, IBM Research, Intel, Mentor Graphics, Oracle, SpringSoft, Synopsys, & TSMC

**Association for
Computing Machinery**

Advancing Computing as a Science & Profession

The Association for Computing Machinery
2 Penn Plaza, Suite 701
New York, New York 10121-0701

Notice to Past Authors of ACM-Published Articles

ISBN: 978-1-4503-1167-0

Additional copies may be ordered prepaid from:

ACM Order Department
PO Box 30777
New York, NY 10087-0777, USA

Phone: 1-800-342-6626 (USA and Canada)
+1-212-626-0500 (Global)
Fax: +1-212-944-1318
E-mail: acmhelp@acm.org
Hours of Operation: 8:30 am – 4:30 pm ET

Printed in the USA

Foreword

On behalf of the organizing committee for the 2012 *ACM International Symposium on Physical Design* (ISPD), we welcome you with great pleasure to Napa Valley, California, for the symposium. Continuing the fine tradition established by its twenty predecessors, a series of five ACM/SIGDA Physical Design Workshops held intermittently in 1987-1996 and fifteen editions of ISPD in the current form since 1997, the 2012 ISPD provides a premier forum to exchange ideas and promote research on critical areas related to the physical design of VLSI systems and other related systems.

We received over 60 submissions from all around the world. After a rigorous, month-long, double-blind review process, the Technical Program Committee (TPC) met to select papers to be included in the technical program based on over 290 reviews provided by 21 TPC members and 41 external reviewers. Only 20 papers were selected to be presented in this symposium. These papers present advances in floorplanning, placement, and routing, address challenges in advanced processes, identify opportunities in emerging technologies, offer new cooling and clocking solutions, and propose novel gate sizing techniques.

In addition, the program is complemented by one keynote speech and twelve invited talks delivered by distinguished researchers in both industry and academia. Dr. Burn Lin from TSMC, a member of the United States National Academy of Engineering, will present in the keynote speech candidate solutions to push lithography beyond the 20nm node. A commemorative session on Monday afternoon allows us to pay tributes to Professor Chung-Laung (Dave) Liu, an outstanding teacher, a pioneer in EDA, and a Phil Kaufman Award recipient. His former students will share with us how Professor Liu's research transformed *ad hoc* EDA to algorithmic EDA and the far reaching impacts of his work, as illustrated in today's nanometer-era routing solutions and the wide-spread use of simulated annealing in EDA. Professor Liu will also grace the symposium with a delightful recount of his 30 years in EDA. An invited session on Wednesday morning will focus on research topics related to congestion-driven logic and physical synthesis. Other invited talks will be interspersed with the presentations of accepted papers, covering the following topics: design-aware lithography, challenges in the 3D integration of CMOS-memristors, synthesis of minimal functional skew clock trees, and analysis of power grids. An invited talk will showcase the latest edition of the ISPD contest.

Since 2005, ISPD has organized highly competitive contests to promote and advance research in placement, global routing, and clock network synthesis. This year, we feature a new contest topic, namely, discrete gate sizing. This year's contest continues to attract a large number of participants from all over the world. The results of this contest will be announced by the ISPD Contest Chair. Continuing with the tradition of the past contests, a new large-scale real-world benchmark suite for gate sizing will also be released.

We owe the success of 2012 ISPD to many people. We would like to thank the authors, the presenters, and the keynote/invited speakers for contributing to the high-quality program. We also thank the session chairs for moderating the sessions. We would like to express our gratitude to the TPC members and external reviewers, who provided constructive comments and detailed reviews to the authors. We greatly appreciate the wonderful set of invited talks put together by the Steering Committee, which is chaired by Yao-Wen Chang. We also thank the Steering Committee for carefully selecting the best paper. Special thanks go to the Publications Chair Cliff Sze and the Publicity Chair Azadeh Davoodi for their tremendous services. We would like to acknowledge the Intel team led by the Contest Chair Mustafa Ozdal for organizing the inaugural gate sizing contest. We are also grateful to the sponsors for financial assistance. The symposium is sponsored by the ACM SIGDA (Special Interest Group on Design Automation) with technical co-sponsorship from the IEEE Circuits and Systems Society. Generous financial contributions have also been provided by (in alphabetical order): ATopTech, Cadence, IBM Research, Intel Corporation, Mentor Graphics, Oracle, SpringSoft, Synopsys, and TSMC. Last but not least, we thank Lisa Tolles of Sheridan Printing Company for her expertise and enormous patience during the production of the proceedings.

We sincerely hope that you enjoy ISPD 2012 and look forward to your participation in future editions of ISPD.

Jiang Hu
ISPD 2012 General Chair

Cheng-Kok Koh
Technical Program Chair

Table of Contents

Keynote Address
Session Chair: Jiang Hu *(Texas A&M University)*

Session 1: Advanced Processes
Session Chair: Markus Olbrich *(University of Hannover)*

Session 2: Emerging Challenges and Technologies
Session Chair: David Pan *(University of Texas - Austin)*

Session 3: Commemoration for Professor C.-L. Liu
Session Chair: Yao-Wen Chang *(National Taiwan University)*

Session 4: Analog, Datapath, and Detailed Placement
Session Chair: Ismail Bustany *(Mentor Graphics)*

Session 5: Power and Thermal Modeling and Optimization
Session Chair: Charles Liu *(TSMC)*

Session 6: Clocking and Routing
Session Chair: Yiyu Shi *(Missouri University of Science and Technology)*

Session 7: Gate Sizing
Session Chair: Ozdal Mustafa *(Intel)*

Session 8: Congestion-Driven Logic and Physical Synthesis
Session Chair: Zhuo Li *(IBM)*

Session 9: Floorplanning and Mixed-Size Placement
Session Chair: Ting-Chi Wang *(National Tsing Hua University)*

ISPD 2012 Symposium Organization

General Chair: Jiang Hu *(Texas A&M University)*

Program Chair: Cheng-Kok Koh *(Purdue University)*

Past Chair: Yao-Wen Chang *(National Taiwan University)*

Steering Committee Chair: Yao-Wen Chang *(National Taiwan University)*

Steering Committee: Yao-Wen Chang *(National Taiwan University)*
Jason Cong *(UCLA)*
Patrick Groeneveld *(Magma)*
Gi-Joon Nam *(IBM Research)*
Sachin Sapatnekar *(University of Minnesota)*
Prashant Saxena *(Synopsys)*
PV Srinivas *(Mentor Graphics)*

Program Committee: Saurabh Adya *(Magma Design Automation)*
Laleh Behjat *(University of Calgary)*
Ismail Bustany *(Mentor Graphics)*
Hung-Ming Chen *(National Chiao Tung University)*
Tung-Chieh Chen *(SpringSoft)*
Salim Chowdhury *(Oracle)*
Azadeh Davoodi *(University of Wisconsin-Madison)*
Masanori Hashimoto *(Osaka University)*
Shiyan Hu *(Michigan Technological University)*
Marcelo de Oliveira Johann *(Universidade Federal do Rio Grande do Sul)*
Vishal Khandelwal *(Synopsys)*
Cheng-Kok Koh *(Purdue University)*
Charles C. C. Liu *(TSMC)*
Malgorzata Marek-Sadowska *(University of California, Santa Barbara)*
Markus Olbrich *(University of Hannover)*
Mustafa Ozdal *(Intel)*
Seungweon Paek *(Samsung)*
Yiyu Shi *(Missouri University of Science and Technology)*
Cliff Sze *(IBM Research)*
Chin-Chi Teng *(Cadence)*
Rasit Topaloglu *(GLOBALFOUNDRIES)*
Ting-Chi Wang *(National Tsing Hua University)*

ISPD 2012 Sponsors & Supporters

Sponsor:

Technical
Co-sponsor:

Supporters:

IBM Research

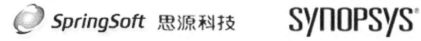

Lithography Till the End of Moore's Law

Burn J. Lin

TSMC, Ltd.

burnlin@tsmc.com

ABSTRACT

Soon after the realization of Moore's law, the suspicion on this law reaching an end persisted. Now, the concern is as serious as it can get. Not only the feature size has to be scaled down, the manufacturing cost also has to be contained. This presentation covers the lithography tools and processes to sustain Moore's law of scaling and Moore's law of economy. It includes the three viable candidates to push lithography beyond the 20nm logic node. (1) Extension of immersion lithography beyond the optical resolution limit by multiple patterning. (2) Reduction of the ArF immersion wavelength by an order of magnitude to the 13.5 nm EUV wavelength. (3) Use of massive parallelism on e-beam direct write.

Categories and Subject Descriptors

B.7.0 [Integrated Circuits]: General

General Terms

Design, Performance, Reliability.

Keywords

Lithography, Moore's Law.

Design-Aware Lithography

Shayak Banerjee
IBM Semiconductor R&D Center
East Fishkill, NY 12533, USA
banerjsh@us.ibm.com

Kanak B. Agarwal
IBM Austin Research Laboratory
Austin, TX – 78758, USA.
kba@us.ibm.com

Sani R. Nassif
IBM Austin Research Laboratory
Austin, TX 78758, USA
nassif@us.ibm.com

ABSTRACT

In the face of continued technology scaling with limited lithographic capabilities, there has been a push towards increased co-optimization of design and process. A key enabler is enhancing the design-manufacturing interface to allow more information than traditional layout shapes to propagate to lithography. We describe a method to generate this additional information in the form of shape tolerances on layout polygons. We further develop two different manufacturing methods to utilize these tolerances during mask optimization. One is a tolerance-driven optical proximity correction algorithm to limit on-wafer lithographic hotspots by constraining process window contours to lie within tolerances. The second is a layout optimization approach that modifies layout shapes during OPC to make them more robust to process variations. Our experiments show that this increased level of interaction between design and lithography leads to fewer process hotspots on-wafer compared to conventional design-oblivious methods.

Categories and Subject Descriptors: B.7 [Integrated Circuits]: Design Aids, Lithography.

General Terms: Algorithms, Design, Reliability.

1. INTRODUCTION

The interface between design and manufacturing has traditionally been very rigid. Design has relied heavily on the contract of ground rules from the foundry to outlaw layout constructs which are not manufacturable. In return, foundry obtains a layout consisting of a set of polygons on multiple levels, which has been designed adhering to the ground rules (Fig. 1a). While such a methodology proved extremely effective in older technology generations, the nanometer regime has introduced fresh challenges in following the old path. The primary challenge for manufacturing lies in the process of lithography, where tools have failed to keep pace with Moore's Law. In particular, the wavelength of light used for lithography has remained constant at 193nm for several technology generations, while feature sizes being printed are a small fraction of that. Despite the use of techniques such as immersion lithography, the lithographic $k1$ factor has been steadily approaching its theoretical single exposure limit of 0.25 [1].

Such sub-wavelength printing makes features highly susceptible to any variations in the lithographic process conditions. This has led to poor yield due to a spiraling number of lithographic hotspots – ground-rule clean structures which fail to print reliably [2,3,4] across the range of process conditions. Two different approaches have been proposed to deal with this patterning challenge: (i) the

use of restricted design rules (RDR) [5] to severely constrain the design space (ii) co-optimizing design and process to attain better yield. The former has the advantage of not perturbing existing design and manufacturing methodologies, but is often counteracted by increase in design cost such as area or via density. The latter requires a rethink of our traditional approaches, but has the capability to minimize the impact on design cost, while simultaneously managing manufacturing risks down to acceptable levels. In this paper, we focus on the second method to achieve continued technology scaling despite limited lithographic capabilities. In particular we elaborate on methods to enhance the flexibility of the design-manufacturing interface to enable more design-aware patterning optimizations (Fig. 1b). To this end, we show how to utilize the knowledge of electrical criticality to communicate the available flexibility of design shapes in the form of geometric shape tolerances. We further describe two manufacturing techniques to utilize such tolerances while generating a lithographic mask for the given layout, in order to obtain a more optimal patterning solution.

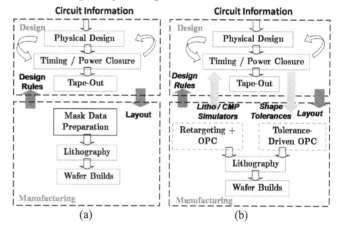

Figure 1: (a) Traditional design-manufacturing interface (b) Enhanced interface with shape tolerances

Traditionally, lithography has treated all layout structures equally by considering them as a set of polygons to be faithfully reproduced on wafer. Such layout structures can be broadly partitioned into devices and wires, with the latter further distributed between various metal levels starting from 1x (narrow, short wires) to 8x (wide, tall wires). The main device layer with tightest patterning requirement is the polysilicon layer. In current technologies, the critical polysilicon features are placed on a regular grid at constant pitch and single orientation [5]. This design restriction allows us to tune the optics and resolution enhancement techniques (RET) for better control over critical dimension (CD). The restriction also results in a rather predictable process variability-band behavior for different devices in the circuit. For this layer, we can generate shape tolerances by converting electrical slack of a design into equivalent shape slack for different polygons [6]. The methodology assigns different shape tolerances to different devices depending on their electrical sensitivity. However, this scheme requires very tight coupling with the timing tools while providing limited benefits during manufacturing due to inherent regularity imposed by the restricted polysilicon design rules.

The 1x metal layer, on the other hand, is primarily used for *local* interconnects and has parasitic resistances (R) and capacitances (C) which are several orders of magnitude smaller than the device loads and impedances in the circuit. The result is that a significant fraction of M1 target shapes have little or no impact on electrical performance and these shapes can be safely modified within acceptable bounds without impacting designer intent. This flexibility in printing 1x metal layer provides an opportunity to significantly improve manufacturing yield. This is made even more important due to higher susceptibility of 1x metal layer to lithographic variability, making it the primary source of hotspots.

In this paper, we propose to quantify the amount of modification possible for 1x metal layer by utilizing the amount of flexibility available between the designed value and ground rules, which we call *ground rule slack*. We use this to generate tolerance bands, which indicate how much shape perturbation is "safe" from a manufacturing point of view. These tolerance bands are subsequently used in the optical proximity correction (OPC) flow for converting the layout to a mask for lithography. We develop a novel OPC technique called tolerance-driven OPC (TD-OPC). This method uses our ground rule slack-based tolerance bands coupled with process window (PW) simulation [2,3] to find a mask solution, which ensures that contours across the lithographic process window lie within the specified tolerances. Since ground rule slack indicates the amount of leeway available in shapes without causing hotspots, the use of this methodology reduces the number of lithographic defects, a result which is highlighted in our experiments in Section 5. Further details of ground rule slack generation and usage are detailed in Sections 2 and 3 respectively.

We then note that using process window simulations during OPC incurs a high runtime penalty, due to the need to simulate contours at multiple process conditions. Further, we observe that in electrically non-critical shapes like local wires, the exact target shape is not a fixed value, but can be modified, which increases the feasibility region for the OPC tool, allowing it to reach a more optimal solution. The process of modifying the target layout during dataprep is called *retargeting* [7,8,9]. We propose to utilize the notion of tolerances to drive retargeting, to enable us to reach a more robust patterning solution without perturbing electrical properties. To overcome the runtime challenge, we use image slope at the wafer plane as a measure of process window [8] and perform retargeting to improve this quantity, while constraining total target movement using the tolerances. We refer to this technique as simultaneous mask and target optimization (SMATO) (Section 4).

Our experiments on the M1 layer of sample industrial testcases show that on-wafer lithographic hotspots can be reduced by 47% on an average using TD-OPC in conjunction with ground rule slack based tolerance bands, compared to traditional process window optical proximity correction (PWOPC). Using SMATO, we were able to reduce the runtime by 2.6X, albeit at a possibly lower improvement in hotspot reduction.

2. GENERATING SHAPE TOLERANCES BASED ON GROUND RULE SLACK

We performed experiments on sample cells from an advanced technology library, where we perturbed all the M1 wires in the layout between +/- 10% of their nominal designed width. We then ran an extraction tool to calculate parasitics, followed by SPICE simulation to obtain the cell delay post-perturbation. Table 1 shows the results of our experiments. It can be seen that metal width

modification had less than 1% impact on cell delay. The result is that a significant fraction of M1 target shapes have little or no impact on electrical performance and these shapes can be safely modified without impacting designer intent.

Figure 2: M1 target shapes and corresponding inner and outer bands based on available ground rule slack

Table 1.   Delay Dependence on Metal1 Width change

Cell	No. of M1 Lines	Max. Delay Var. (%)	Min. Delay Var. (%)
INV	3	0.06	0.03
NAND2	4	0.07	0.90
NOR2	4	0.35	0.05
NAND3	8	0.10	0.37
NOR3	8	0.41	0.10
Average		**0.20**	**0.29**

It follows from Table 1 that trying to reproduce the exact target for M1 makes the problem unnecessarily hard for lithography as there exist a range of alternative target shapes that can provide the same electrical functionality. Unfortunately, designers do not have the tools and the bandwidth needed to select the lithographically optimal target from the set of possible target layouts. For local wires, designers arbitrarily produce *one* design rule clean layout that enables all required electrical connectivity without sacrificing layout density. It is practically infeasible and also undesirable to make major changes in the layout after the design is completed but we can easily make provisions for small layout changes without altering electrical behavior. The layout flexibility or *shape slack* [6] information can be communicated to manufacturing through target tolerance bands [6,11]. Fortunately, most M1 shapes are used for local interconnects and are electrically non-critical (Table 1). This simplifies the tolerance-band generation process as we do not need to run parasitic extraction and circuit simulation to quantify the sensitivity of each metal shape on circuit performance. Under this assumption, we can generate tolerance bands for M1 by simply considering the electrical connectivity and ground rule requirements. The electrical connectivity checks ensure that the circuit functionality is maintained while the ground rule checks ensure that the target bands do not violate manufacturing constraints such as minimum spacing, minimum contact overlap, etc. Figure 2 shows a sample M1 layout from a real design. The figure shows the original target shapes along with the outer and the inner target tolerance bands obtained by considering the available ground rule slack. The tolerance bands shown in the figure indicate that for this layout, very few shapes are at minimum ground rule. In other words, there is a significant amount of ground rule slack available in real designs and this slack can be utilized during RET to improve lithographic yield without impacting circuit yield.

The traditional RET method for generating tolerances is to use constant symmetric bands on each feature [2]. A common practice in lithography is to aim to control a feature to within +/-10% of its nominal drawn size. The primary advantage of using ground rule slack is that the tolerances generated are indicators of possible manufacturing defects. In particular, regions with small inner tolerance are possible pinching regions, while bridging is likely to occur in regions with small outer tolerance. Similarly, contact overlap regions with small tolerance define regions of potential poor contact overlap. In the following section, we will describe a manufacturing methodology that utilizes this ground rule slack information to find a mask solution with minimum number of lithographic hotspots [4].

3. TOLERANCE DRIVEN OPTICAL PROXIMITY CORRECTION (TD-OPC)

Each layer of a target layout passes through a series of processing steps known as dataprep, in order to be converted to a mask for use in manufacturing. Optical proximity correction (OPC) is the primary method in use for mask generation. This technique uses models of the lithographic process to predict the image of the mask on the wafer, and uses an iterative optimization algorithm to converge to a mask solution that matches this image to the target accurately. The error metric optimized for is the geometric difference between the wafer contour and the target feature, which is also called edge placement error (EPE) [12]. This technique was popular in the past due to low runtime overheads in generating a mask solution, even for very large full-chip layouts. However, the major drawback of OPC is that it only simulates the image under nominal process conditions and hence cannot account for variations in the manufacturing process. As the industry has progressed to printing smaller feature sizes with essentially the same imaging system, the sensitivity to manufacturing variations has increased. In particular, one observes greater susceptibility to dose errors, defocus and mask/overlay errors [1], which define the lithographic process window. In the recent past, an improvement in simulation capability has led to the popularity of a new technique called process-window optical proximity correction (PW-OPC) [2,3]. This method simulates the wafer contours at multiple process corners and optimizes for the sum of weighted EPE across these contours. This has the impact of making the final mask solution more robust to process variation as compared to OPC. In this section we describe a tolerance-driven OPC (TD-OPC) algorithm that utilizes our previously generated tolerance band information in conjunction with process window simulations to produce a robust mask solution [13].

The first step in the process of OPC or PW-OPC is to fracture a given layout into a large number of movable edges. Each of these edges is called a *fragment* [10,12] and can be moved independently of the other fragments. Figure 3 shows a fragment at which the contours at various process conditions have been simulated and compiled in the form of a process variability band (PV-band) [14]. The inner and outer tolerances for this fragment are also defined at distances t_{out} and $-t_{in}$ from the target respectively. The edge placement error between the target and the outer and inner edges of the PV band are depicted as e_{out} and $-e_{in}$ respectively. Finally, the edge placement error for the contour at nominal process conditions is shown as e_{nom}. In standard OPC, the cost function (C_{OPC}) optimized for is the sum of the nominal EPE across the fragments in the design:

$$C_{OPC} = \sum_{i=1}^{N} e_{nom,i} \quad (1)$$

In current algorithms for process-window OPC, one adds robustness to process variations by optimizing for a weighted sum of EPE across the process window. For a simple three contour scheme, the cost function for PW-OPC (C_{PWOPC}) is the weighted sum of nominal, inner and outer EPE (Eq. 2), where the individual weights are user-specified.

$$C_{PWOPC} = \sum_{i=1}^{N} \left(w_0 e_{nom,i} + w_1 e_{out,i} + w_2 e_{in,i} \right) \quad (2)$$

For TD-OPC, we also adopt a simple three contour approach. We assume that one can prune the process space to a few combinations of dose, focus and mask error [1] values that gives us worst-case behavior. This allows us to generate inner and outer contours in addition to the nominal contour. We wish to place a high penalty on any exceeding of the inner or outer tolerances. We further require this cost function to be easily differentiable since we will adopt a gradient-descent based minimization scheme. For this purpose, we place an exponential cost on tolerance band violation. We also desire high fidelity to the drawn target whenever the tolerance bands are not violated, thus motivating the need to optimize for the nominal EPE as well. These factors combined lead us to define a new cost function (C_{TDOPC}) for TD-OPC which involves a weighted sum of the nominal EPE and violation costs (Eq. 3). The weights are user specified and may be optimized for the particular lithographic process used.

$$C_{TDOPC} = \sum_{i=1}^{N} \left[w_0 e_{nom,i} + w_1 \exp\left(e_{out,i} - t_{out}\right) + w_2 \exp\left(e_{in,i} - t_{in}\right) \right] \quad (3)$$

With the cost function for TD-OPC in place, we now describe a gradient descent method for minimizing the cost function. For this purpose, we need to relate the change in the cost function to the movement of a fragment at a particular site. We make a simplifying assumption that the EPE at a particular site is primarily affected by the movement of the site itself. This assumption is, in fact, widely used in OPC literature, especially when binary or attenuated-PSM masks are used [1]. This allows us to approximate the sensitivity of the cost function to the fragment movement at a site using the gradient information:

$$\frac{\partial C_{TDOPC}}{\partial d_i} = w_0 \frac{\partial e_{nom,i}}{\partial d_i} + w_1 \frac{\partial e_{out,i}}{\partial d_i} \exp\left(e_{out,i} - t_{out,i}\right)$$
$$+ w_2 \frac{\partial e_{in,i}}{\partial d_i} \exp\left(e_{in,i} - t_{in,i}\right) \quad (4)$$

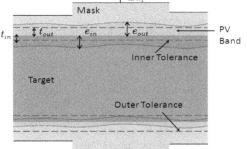

Figure 3: Mask and target layer with tolerances and EPE

The sensitivity of EPE at a site to the fragment movement at the site ($\partial e/\partial d$) can be analytically computed during image simulation and is called the mask error enhancement factor (MEEF) [12]. This adds a small amount of overhead to the image simulation time. Alternatively, one can use a constant approximation of the MEEF (Eq. 5). Since the proximity correction process involves multiple iterations between image simulation and fragment movement, a reasonable approximation of the MEEF still leads to quick convergence to a final mask solution [12]. With the knowledge of the sensitivity, we then move the fragment by a fraction (η) of this value. This fraction is also called the feedback factor and can be tuned dynamically, at each iteration, to improve the speed of convergence.

$$\frac{\partial e_{nom,i}}{\partial d_i} = \alpha_{nom}, \quad \frac{\partial e_{out,i}}{\partial d_i} = \alpha_{out}, \quad \frac{\partial e_{in,i}}{\partial d_i} = \alpha_{in} \quad (5)$$

$$\Delta d_i = -\eta \frac{\partial C_{TDOPC}}{\partial d_i} \quad (6)$$

$$= -\eta \left(w_0 \alpha_{nom} + w_1 \alpha_{out} \, exp\left(e_{out,i} - t_{out,i}\right) + w_2 \alpha_{in} \, exp\left(e_{in,i} - t_{in,i}\right) \right)$$

The result of this gradient descent based minimization is to move the mask towards a solution where the final wafer contours have the lowest number of tolerance violations. In case there cannot be a solution without violations at a particular site, the exponential nature of the cost function drives the optimization engine to find a solution where the *magnitude* of the violation will be minimized. The entire process runs for a pre-specified number of iterations of image simulation and fragment movement, similar to OPC and PW-OPC. We also constrain the maximum amount of fragment movement based on mask rule checks (MRC) to maintain manufacturability of the mask itself [10]. This is implemented by writing MRC constraints into the OPC flow such that if a particular fragment movement violates the MRC, then it will not be moved.

4. SIMULTANEOUS MASK AND TARGET OPTIMIZATION (SMATO)

The above TD-OPC method suffers from the drawback of having to simulate multiple contours at different process conditions, which leads to high algorithm runtime. The printability of a layout can also be improved by modifying the target layout itself to eliminate bad patterns from the design. This process of modifying the target layout during manufacturing is called *retargeting* [7,9]. Current retargeting methods are rules-based which are quite narrow in scope because of the exhaustive search required to cover the entire space of possible two dimensional patterns. A few alternative non-rule based retargeting and hotspot fixing [9,15] methods have also been but like PWOPC, these methods also suffer from high runtime problems due to the number of image simulations involved.

In this section we describe a simultaneous mask and target optimization (SMATO) method [18] with the dual purpose of ensuring the wafer contour matches the target and is robust in the presence of variability. Consider a target structure and its corresponding mask in one dimension as shown in Fig. 4. The intensity profile from the mask is also depicted. We assume a constant threshold resist model (CTR) such that any intensity above a certain level (I_{th}) can be assumed to print on the wafer. Image slope is the gradient of the image ($\partial I/\partial x$) as measured at the target edge. The edge placement error (Δe) corresponding to this particular image is also shown in Fig. 4. It can be seen that for the CTR, minimizing edge placement error is equivalent to minimizing the error (ΔI) between the intensity at the target edge (I_{tgt}) and the threshold intensity (Eq. 7). We henceforth refer to this as the *intensity error*. It is defined for a particular evaluation site and is used to develop a cost function to drive OPC [17]. Given N sites in a layout, the cost function is defined as the mean square error (MSE) of intensity across these points (Eq. 8). The MSE form

allows the use of tractable optimization techniques to minimize the cost function [17].

$$\Delta I = I_{tgt} - I_{th} \quad (7)$$

$$C_{OPC} = \sum_{i=1}^{N} \left(\Delta I_i\right)^2 \quad (8)$$

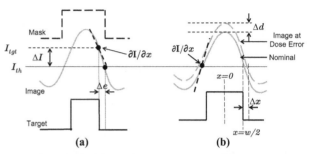

Figure 4: (a) Definition of intensity error and image slope (b) Image profile in presence of dose errors

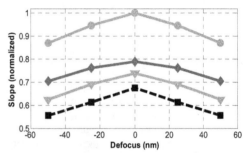

Figure 5: Variation of image slope with defocus

Minimizing the above cost function improves fidelity of the wafer contour to the target at nominal process conditions. However, it cannot guarantee adequate process window for features in the presence of variability. This is because the intensity is only computed at a single process point and does not account for dose errors or defocus. Dose errors (Δd) are modeled as variation in the intensity of incident light. This causes a shift in the position of the edge of the image (Δx), which manifests itself in the form of feature width variations. For a CTR model, Δx can be approximated in terms of the image slope (Eq. 9) [16]. This equation shows that sensitivity of feature size variation to dose can be reduced if the image slope is high.

$$\Delta x = \Delta d \left[\frac{1}{I} \frac{\partial I}{\partial x} \Big|_{x=\frac{CD}{2}, f_{nom}} \right]^{-1} = \frac{I \, \Delta d}{\nabla I} \quad (9)$$

Further, we know that slope degrades with variation in focus (Δf) from nominal conditions [8]. Fig. 5 shows variation of the gradient with defocus for four different features from a sample layout. An observation from Fig. 5 is that higher image slope (∇I) at nominal focus (f_{nom}) translates to higher slope at defocus i.e. $\nabla I_1\left(f_{nom} + \Delta f\right) \geq \nabla I_2\left(f_{nom} + \Delta f\right)$ if $\nabla I_1\left(f_{nom}\right) \geq \nabla I_2\left(f_{nom}\right)$ This means reduced sensitivity to dose errors not only at nominal focus, but also at defocus - which translates to improved process window (PW) of features. The added advantage of using image slope over full PW simulation is that it is readily available during image simulation at no extra computational cost.

We therefore propose a modified cost function for SMATO which satisfies the dual objective of minimizing mean square intensity error for maintaining fidelity and maximizing slope for improving process window of layout features (Eq. 10).

$$C^j_{SMATO} = \sum_{i=1}^{N}\left[w_j \left(I_{tgt,i} - I_{th} \right)^2 + \frac{\left(1-w_j\right)}{|\nabla I_i|} \right] \quad (10)$$

Due to complex optical interactions, there may exist situations where moving mask/target improves slope but degrades fidelity or vice versa [8]. We introduce an iteration (j) dependent weighting function ($0 \le w_j \le 1$) that may be utilized to prioritize one over the other. We further propose to minimize this cost function by using a local search algorithm. For this purpose we utilize the fact that the intensity profile at a site (for binary/att-PSM masks) is most impacted by the movement of the fragment at that site [8,12]. This allows us to write the cost function at each site (i) purely in terms of mask movement (λ) and target movement (η) at that site:

$$C^j_{SMATO,i}\left(\lambda_j,\eta_j\right) = w_j \left(I_{tgt,i}\left(\lambda_j,\eta_j\right) - I_{th} \right)^2 + \frac{\left(1-w_j\right)}{\left|\nabla I_i\left(\lambda_j,\eta_j\right)\right|} \quad (11)$$

The above function can be evaluated analytically for small modifications in mask and target [18]. Given this capability, we describe the local search algorithm used to minimize the cost function. At each iteration we pick three movements of the target {-λ_j, 0, λ_j} and three movements of the mask {-η_j, 0, η_j}. The values of target movement are chosen such that the maximum amount of target movement, summed across the iterations, is limited. This is a necessary condition to avoid large perturbations of the wire shapes in order to prevent deviations in their electrical properties. For each mask/target combination, we analytically evaluate the cost function. We then pick the pair of target and mask movement that provides the local minimum of the cost function. There are two key reasons for using local search to solve the problem: (a) The non-linear nature of optical interactions makes it difficult to predict *apriori* what will be the best direction of mask and target movement to minimize the cost function. The degree of difficulty is increased because in some cases reducing intensity error degrades slope and vice versa [8]. Also note that in OPC we move only mask fragments to achieve fidelity to target, but in SMATO we may move the target itself to obtain lower intensity error if it achieves better image slope; (b) Only small perturbations are allowed in the target, in order to not disturb the electrical properties. Such modifications are also limited to be in steps of the mask grid (typically 1nm), consequently limiting the search space. The algorithm for SMATO is summarized as follows:

Algorithm: Simultaneous Mask and Target Optimization

Inputs: L – input layout
Output: R – retargeted layout
          M – mask to print retargeted layout
1. $F \leftarrow$ **Fragment_Layout**(L)
2.   **set** $M \leftarrow L$ and $R \leftarrow L$
3.   **for** $j \leftarrow 1$ to *num_of_iterations*
4.   **Image_Simulate**(M)
5. $T \leftarrow \{ -\eta_j, 0, \eta_j \}$
6. $D \leftarrow \{ -\lambda_j, 0, \lambda_j \}$
7.   **for** $i \leftarrow 1$ to *num_of_fragments*
8. $C_{min} \leftarrow Max\_Cost$
9.   **foreach** $\eta \, \varepsilon \, T$ and **foreach** $\lambda \, \varepsilon \, D$
10.   **Evaluate_Cost_Function**($C^j_{SMATO,i}(\lambda,\eta)$) (Eq. 11)
11.   **if** $C^j_{SMATO,i}(\lambda,\eta) < C_{min}$
12. $C_{min} \leftarrow C^j_{SMATO,i}(\lambda,\eta)$
13.   Set_Mask_Movement(F_i) = λ
14.   Set_Target_Movement(F_i) = η
15.   **next** i
16.   **Move_Fragments**(F)
17.   **Update_Mask**(M, F)
18.   **Update_Target**(R, F)
19. **next** j

Table 2. Comparison of PW-OPC to TD-OPC with Ground Rule Slack Based Tolerance Bands

Clip	PW-OPC		TD-OPC		% Red. In PV Area	% Red. In ORC Marker Area
	PV Band Area	ORC Marker Area	PV Band Area	ORC Marker Area		
1	7.2	17.1	8.6	18.3	-18.8	-6.9
2	5.8	12.3	5.3	11.7	9.5	11.1
3	6.0	8.8	5.5	8.4	8.1	6.7
4	4.0	8.0	4.9	8.3	-24.0	-4.1
5	8.0	14.2	5.7	12.0	29.2	15.1
6	83.4	368.8	69.7	161.8	16.5	56.1
Average					12.2	47.9

5. EXPERIMENTAL RESULTS

The tolerance band generation was performed using the *Calibre SVRF* kit in conjunction with a set of design rules which included spacing, width and contact/via overlap checks. We further used *Calibre DenseOPC* to implement our TD-OPC algorithm. For lithography simulation, we utilized high-NA immersion lithography optical and resist models. *Calibre DenseOPC* functionality was exploited to generate MEEF values as well as perform tagging commands, which identify the different fragment movement values for different regions. Experiments were performed on the 1x metal layer of several designs. For each testcase, we performed a conventional PWOPC [10] – assuming equal weighting of EPE at the inner, nominal and outer contours. We then used our tolerance band generation coupled with our TD-OPC algorithm on each of these testcases. For each of the output masks, we ran a full process window simulation. We assumed dose errors of +/-2% and defocus of +/-100nm. Using models calibrated at these different process corners, we compiled the process variability bands (PV-bands) [14]. The area of the PV band is indicative of the degree of robustness of a layout to process variability [14]. Lower PV band area indicates lower sensitivity to variations in dose and focus and vice versa. Table 2 shows a compilation of the normalized PV band area for the different designs. We further ran optical rule checks (ORC) [4] to check for areas of potential pinching or bridging. We used *Calibre OPCVerify* to perform these checks and generated a marker layer for each violation encountered. The total area of the ORC markers is a measure of the printability of a particular layout. Larger ORC marker area indicates higher possibility of lithographic defects and vice versa.

Table 2 shows that the total PV band area can be reduced by 12.2% on an average by using our ground rule derived tolerances in conjunction with TD-OPC, compared to the use of regular PWOPC. This indicates higher robustness to process variation. We further observed an average of 47.9% reduction in the total ORC marker area, indicating lower possibilities of pinching and bridging in metal lines. This result is expected because our ground rule slack based tolerance bands contain information about possible areas of pinching and bridging and they constrain the OPC tool in these regions so as to avoid such potential hotspots. Figure 6 shows an example testcase of metal lines at minimum width and spacing, where the use of regular OPC causes an area of potential pinching (indicated by the circled region) (Fig. 6a). The use of regular PWOPC tries to fix the pinching problem, but ends up creating a bridging problem between the two neighboring metal lines (Fig. 6b). This problem is solved by propagating the design aware tolerance bands to our TD-OPC algorithm, as shown in Fig. 6c.

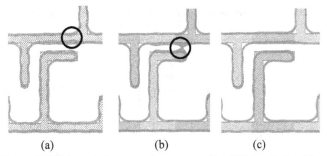

(a)   (b)   (c)

Figure 6: M1 target shapes and corresponding PV-bands for (a) OPC, (b) regular PW-OPC, and (c) ground rule slack + TD-OPC

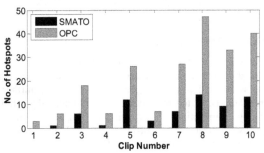

Figure 7: Comparison of hotspots for OPC and SMATO

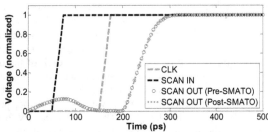

Figure 8: CLK to Out Delay comparison for latch pre- and post-SMATO shows negligible electrical impact of retargeting

We further implemented the SMATO algorithm to compare against OPC and PWOPC. The simulation corners were selected to be at (i) nominal dose/focus (ii) +2% dose and nominal focus (iii) -2% dose and +25nm defocus. All algorithms utilized the same fragmentation scheme and also ran for the same number of iterations. The contours from multiple process points were used to compile the PV band again. Results demonstrate that the use of SMATO reduces the PV Band area by 15.4%, on average, over OPC. Further, this improvement is achieved at a low (5.5%) average runtime overhead. Compared to PWOPC, we obtained a runtime reduction of 2.6X. To enable further comparison we implemented a set of optical rule checks (ORC) on the PV bands. Such checks determine the presence of hotspots leading to shorting of metal lines or opening of a wire at different process conditions. Results are shown in Fig. 7 where it is clear that SMATO reduces the number of ORC violations over OPC. On an average, SMATO

was able to reduce the number of hotspots by 69% compared to OPC.

As an electrical experiment, we took the 1x metal level of a latch in 32nm technology. We first ran extraction on the original layout to determine the parasitic resistance and capacitances of the metal wires. These parasitics were subsequently used in a SPICE simulation to determine the latch delay from the input clock signal (CLK) and the input data (D) to the output (Q). We then updated the latch layout with the modified layout produced by SMATO and re-evaluated the parasitics. This was once again used in SPICE to determine the updated latch delay values. Figure 8 shows the comparison of clock to output (CLK-Q) delays before and after retargeting. We observed 0.11% change in CLK-Q delay and 0.06% change in D-Q delay. This level of perturbation is well below the expected noise from other sources of cell delay variation, such as transistor threshold voltage and gate dimension variations. This proves that small perturbations of local wire dimensions, as produced by TD-OPC and SMATO does not significantly alter the electrical properties of the design, while significantly improving printability.

6. REFERENCES

[1] C. Mack, "Fundamental Principles of Optical Lithography: The Science of Microfabrication," *Wiley-Interscience*, 2008.

[2] A. Krasnoperova, J. Culp, I. Graur and S. Mansfield, "Process Window OPC for Reduced Process Variability and Enhanced Yield", *Proc. SPIE*, vol. 6154, 61543L-1, 2006.

[3] J. T. Azpiroz, *et al*, "Improving Yield through the Application of Process Window OPC", *Proc. SPIE*, vol. 7274, 2009.

[4] J. A. Bruce *et. al.*, "Model-based Verification for First Time Right Manufacturing", *Proc. of SPIE*, vol. 5756, 2005.

[5] L. Pileggi, A. J. Strojwas, "Regular Fabrics for Nano-Scaled CMOS Technologies", *Proc. of ISSCC*, 2006.

[6] S. Banerjee, K.B. Agarwal, C. Sze, S. Nassif, M. Orshansky, "A Methodology for Propagating Design Tolerances to Shape Tolerances for use in Manufacturing", *Proc. of DATE*, 2010.

[7] S. Mansfield, G. Han and L. Liebmann, "Through-Process Modeling for Design-for-Manufacturability Applications", *J. Micro/Nanolith. MEMS MOEMS*, vol. 6(3), 031007-1, 2007.

[8] N. Cobb, Y. Granik, "Using OPC to Optimize for Image Slope and Improve Process Window", *Proc. of SPIE*, vol. 5130, 2003.

[9] E. Yang, C.H. Li, X.H. Kang and E. Guo, "Model-Based Retarget for 45nm Node and Beyond", *Proc. of SPIE*, vol. 7274, 727428-1, 2009.

[10] Mentor Graphics, "Calibre nmOPC Manual", 2009.

[11] S. Mansfield, L. Liebmann, A. Krasnoperova and I. Graur, "Designer's Intent Tolerance Bands for Proximity Correction and Checking", US Patent: 7266798.

[12] N. Cobb, Y. Granik, "Model-Based OPC using the MEEF Matrix," *Proc. of SPIE*, vol. 4889, pp. 1281-92, 2002.

[13] S. Banerjee, K.B. Agarwal, M. Orshansky, "Ground Rule Slack Aware Tolerance Driven Optical Proximity Correction for Local Metal Interconnects", *Proc. of CICC*, 2010.

[14] J.A.Torres, C.N.Berglund, "Integrated Circuit DFM Framework for Deep Subwavelength Processes", *Proc. of SPIE*, vol. 5756, 2005.

[15] S. Kobayashi, *et al.*, "Automated Hot-Spot fixing System Applied for Metal Layers of 65 nm Logic Devices", *Proc. of SPIE*, vol. 6283, 62830R-1, 2006.

[16] H. J. Levinson, W. H. Arnold, "Focus: the Critical Parameter for Submicron Lithography", *Journal of Vacuum Science and Technology B*, vol. 5, Issue 1, pp. 293-298, 1987.

[17] Y. Liu and A. Zachor, "Optimal Binary Image Design for Optical Lithography," *Proc. of SPIE*, vol. 1264, pp. 401–412, 1990.

[18] S. Banerjee, K.B. Agarwal, M. Orshansky, "SMATO: Simultaneous Mask and Target Optimization for Improving Lithographic Process Window", *Proc. of ICCAD*, 2010.

Graph-Based Subfield Scheduling for Electron-Beam Photomask Fabrication *

Shao-Yun Fang
Graduate Institute of Electronics Engineering
National Taiwan University
Taipei 10617, Taiwan
yuko703@eda.ee.ntu.edu.tw

Wei-Yu Chen
Graduate Institute of Electronics Engineering
National Taiwan University
Taipei 10617, Taiwan
cweiyu@eda.ee.ntu.edu.tw

Yao-Wen Chang
Department of Electrical Engineering
National Taiwan University
Taipei 10617, Taiwan
ywchang@cc.ee.ntu.edu.tw

ABSTRACT

Electron beam lithography (EBL) has shown great promise for photomask fabrication; however, its successive heating process centralizing in a small region may cause a severe problem of critical dimension (CD) distortion. Consequently, subfield scheduling which reorders the sequence of the writing process is needed to avoid successive writing of neighboring subfields. In addition, the writing process of a subfield raises the temperature of neighboring regions and may block other subfields for writing. This paper presents the first work to solve the subfield scheduling problem while taking into account blocked regions by formulating the problem into a *constrained maximum scatter travelling salesman problem* (constrained MSTSP). To tackle the constrained MSTSP which can be shown to be NP-complete in general, we identify a special case thereof with points on two parallel lines and solve it optimally in linear time. We then decompose the constrained MSTSP into subproblems conforming to the special case, solve each subproblem optimally and efficiently by a graph-based algorithm, and then merge the sub-solutions into a complete scheduling solution. Experimental results show that our algorithm is effective and efficient in finding good subfield scheduling solutions that can alleviate the successive heating problem (and thus reduce CD distortion) for e-beam photomask fabrication.

Categories and Subject Descriptors

B.7.2 [**Integrated Circuits**]: Design Aids

General Terms

Algorithms, Design, Performance

Keywords

Electron Beam Lithography, Subfield Scheduling, Manufacturability

*This work was partially supported by SpringSoft, Synopsys, TSMC and NSC of Taiwan under Grant No's. NSC 100-2221-E-002-088-MY3, NSC 99-2221-E-002-207-MY3, NSC 99-2221-E-002-210-MY3, and NSC 98-2221-E-002-119-MY3.

1. INTRODUCTION

As integrated circuit (IC) process nodes continue to shrink to 22nm and below, the IC industry will face severe manufacturing challenges with conventional optical lithography technologies. Electron beam lithography (EBL) is one of the most anticipated next-generation lithography (NGL) technologies, as the electron beam (e-beam) can be easily focused into nanometer diameters by using electromagnetic or electrostatic lenses [5,6]. In addition, in comparison with optical lithography, EBL is not limited by light diffraction [10]. As a result, the e-beam can define very fine, high-resolution patterns in a resist.

The dramatically decreasing minimum feature size of semiconductor devices has made mask making one of the most challenging tasks. The advantages of e-beam lithography have made high-voltage electron beams popular in photomask fabrication. However, these high-voltage beams deposit a considerable amount of heat in a small area and result in critical dimension (CD) distortion [8, 11].

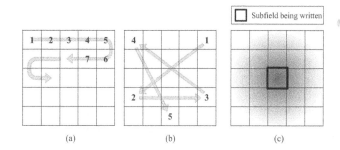

(a)   (b)   (c)

Figure 1: (a) The sequential e-beam writing process. (b) Writing process reordering with subfield scheduling to avoid successive writing of neighboring subfields. (c) The writing process of a subfield raises the temperature of neighboring regions.

To maximize throughput, the writing process is usually performed in a sequential manner, as shown in Figure 1(a). In this paper, we refer to this as contiguously sequential writing: where the writing proceeds in order from one subfield to the next adjacent subfield. However, the contiguously sequential writing process generates heat centralized in a region, which can aggravate the CD distortion problem. To solve this, Babin et al. proposed subfield scheduling, which reorders the sequence of the writing process to avoid the successive writing of subfields which are close to each other [2,3], as shown in Figure 1(b).

Figure 2 illustrates the effectiveness of subfield scheduling in alleviating the successive heating problem. Figures 2(a) and (b) show two scanning electron micrograph (SEM) images of 1-D lines with contiguously sequential and non-contiguously-sequential writing processes, respectively. From the highlighted SEM images shown in Figures 2(c) and (d), it can be seen that the resulting line end roughness is improved (much smaller) with the non-contiguously-sequential writing process. To further quantify the improvement in pattern quality with subfield scheduling, Table 1

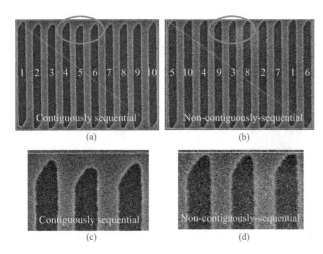

Figure 2: SEM images of 1-D lines with different writing orders. SEM images with (a) contiguously sequential and (b) non-contiguously-sequential writing processes. (c)(d) The highlighted SEM images of the two corresponding writing processes. The resulting line end roughness is improved (much smaller) with a non-continuously-sequential writing process.

lists the numerical data of line edge roughness obtained from the SEM images. The results show that the non-contiguously-sequential writing process significantly reduces the line edge roughness by about 63% and 68% for 1-D line patterns with 100 and 120 nm spacings, respectively. Subfield scheduling is hence an effective technique to mitigate the successive heating problem in EBL and thus reduce CD distortion.

Table 1: Impact of the heating effect on line edge roughness. Line edge roughness can be much improved with a non-contiguously-sequential writing process.

Line Spacing (nm)	Line edge roughness (nm)		Improv.
	Contiguously Sequential	Non-Contiguously-Sequential	
120	11.28	3.51	68.88%
100	13.73	4.99	63.66%

The objective of the subfield scheduling problem is to find a labelling (ordering) of subfields where the minimum distance between two subfields with successive labels is maximized. Babin et al. first applied Lagarias scheduling to derive a well-spaced labelling [2], in which the resulting minimum Manhattan distance between subfields with successive labels is at most one unit length (edge length of a subfield) away from the optimal (maximum possible distance) [9]. However, using the Manhattan measure in this problem might not be appropriate, as the e-beam writing head moves freely (the Euclidean distance is a more appropriate metric in this case). Later, they proposed a greedy local improvement method for this problem [3], where a random subfield scheduling is generated at the beginning and the scheduling is iteratively improved by swapping the orders of pairs of subfields. As the problem size grows, nevertheless, its running time grows prohibitively.

Another issue addressed in the previous work is the raised temperature of neighboring regions due to the writing process of a subfield [2]. As shown in Figure 1(c), the writing process of the middle subfield significantly raises the temperature of neighboring subfields. Those subfields with temperatures higher than a threshold value are called blocked subfields for a certain processing duration: they should not be written in successive writing processes before the temperature drops below the threshold. However, no existing work has optimized the subfield scheduling problem while simultaneously considering blocked subfields during the writing process.

This paper presents the first work to solve the subfield schedul-

ing problem taking into account blocked regions by formulating the problem as a constrained maximum scatter travelling salesman problem (constrained MSTSP for short). To tackle the constrained MSTSP which can be shown to be NP-complete in general, we identify a special case of the constrained MSTSP with points on two parallel lines and solve it optimally in linear time. We then decompose the constrained MSTSP into subproblems conforming to the special case, solve each subproblem optimally and efficiently by a graph-based algorithm, and then merge the sub-solutions into a complete scheduling solution. Experimental results show that our algorithm is effective and efficient in finding good subfield scheduling solutions that significantly alleviate the successive heating problem (and thus reduce CD distortion) for e-beam photomask fabrication.

The rest of this paper is organized as follows: Section 2 outlines the preliminaries and the problem formulation of this paper. In Section 3 is presented an exact algorithm that optimally solves the MSTSP for a special case; Section 4 details the graph-based subfield scheduling algorithm. Experimental results are reported in Section 5. Section 6 concludes our work.

2. PRELIMINARIES

In this section, the preliminaries of the subfield scheduling problem are given. First, a proven thermal model is introduced in Section 2.1. Then, the blocked box of a subfield is defined in Section 2.2. Section 2.3 introduces the traditional and constrained MSTSPs, and the problem formulation is presented in Section 2.4.

2.1 Thermal Model

Babin et al. [3] developed a greedy local improvement algorithm which yields better subfield scheduling results than any other methods. In the algorithm is embedded a model of temperature computation in which two basic principles are considered: (1) The writing process of a subfield causes temperature increases for all other subfields; these increases depend mainly on the amount of energy deposited through the writing process and the distance from the subfield being written. (2) The temperature of each subfield decays exponentially during the travelling time between two successive writing processes.

In the thermal model, $\pi = (\pi_1, \pi_2, \cdots, \pi_n)$ denotes a scheduling order of subfields, and $T_{i,j}$ represents the temperature of a subfield π_i before the writing process of a subfield π_j. According to the first principle of the model, the amount of the raised temperature of a subfield π_i due to the writing process of a subfield π_j, denoted by $T_{i,j}^{rise}$, is proportional to the temperature difference and is inversely proportional to the squared Euclidean distance between the two subfields. We have

$$T_{i,j}^{rise} \propto \frac{T_{j,j} + T_{j,j}^{rise} - T_{i,j}}{dist(\pi_i, \pi_j)^2}, \qquad (1)$$

where $T_{j,j}^{rise}$ denotes the raised temperature of a subfield π_j due to its own writing process, and $dist(\pi_i, \pi_j)$ is the Euclidean distance between subfields π_i and π_j. As the temperature of a subfield π_i decays exponentially during the travelling time between two successive writing processes, the temperature of a subfield π_i right before the writing process of a subfield π_j can be formulated as

$$T_{i,j} = (T_{i,j-1} + T_{i,j-1}^{rise})f, \qquad (2)$$

where f is the decay factor which depending on the length of the travelling time between two successive writing processes; that is, f depends on the distance between two subfields with successive labels.

Equations (1) and (2) show that increasing the distance between each pair of subfields with successive labels may not only prevent the writing of a subfield with excessively high temperature but also allow for a longer time to cool down all subfields. In this paper, therefore, we formulate this problem as a constrained maximum scatter travelling salesman problem and propose an algorithm to maximize the distance between each pair of successive subfields.

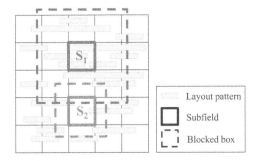

Figure 3: Blocked boxes for two subfields. The size of a blocked box varies with the pattern density of the corresponding subfield.

2.2 Blocked Boxes of Subfields

First, we let S be the set of given subfields. As mentioned in Section 1, the writing process of a subfield raises the temperature of neighboring subfields. Successive writing processes should not be performed on unwritten subfields whose temperatures are higher than a threshold value. As a result, we define a *blocked box* b_i for each subfield $S_i \in S$ which indicates the area that is blocked because of the writing process of the subfield S_i, as illustrated in Figure 3. Since the temperature increase depends on the energy deposited through the writing process, and the deposited energy is proportional to the volume of the patterns, the area of a blocked box varies with the pattern density of the corresponding subfield. In addition, each $S_i \in S$ is associated with a *blocking duration* d_i. The subfields included in b_i cannot be written within the time from t_i to $t_i + w_i + d_i$, referred to as the *blocked box constraint*, where t_i is the writing start time of S_i and w_i is the time required for the writing process of S_i. In addition to satisfying the blocked box constraint, the time difference between the writing start times of two subfields with overlapped blocked boxes should be as large as possible to further control the temperature during photomask fabrication.

2.3 Constrained Maximum Scatter Travelling Salesman Problem

The maximum scatter travelling salesman problem (MSTSP), a variation of the classical travelling salesman problem (TSP), finds a Hamiltonian cycle or path that is most scattered. Its objective is to maximize the minimum edge weight in the cycle or the path. In an edge-weighted complete graph $G = (V, E)$, let $\mathbb{F}$ be the family of Hamiltonian cycles (or paths) in G. For each edge $e \in E$, the edge weight w_e is predefined. For a Hamiltonian cycle or path $\mathcal{H}$, we denote $c(\mathcal{H})$ as the cost of $\mathcal{H}$, which is the minimum edge weight in $\mathcal{H}$; i.e., $c(\mathcal{H}) = \min\{w_e : e \in \mathcal{H}\}$. Then, the MSTSP is to find a Hamiltonian cycle (or path) $\mathcal{H} \in \mathbb{F}$ such that $c(\mathcal{H})$ is maximized.

Observing that the objective of the subfield scheduling in e-beam photomask fabrication is similar to that of MSTSP, we transform the scheduling problem into an MSTSP. However, due to the blocked box consideration, an optimal solution derived from the corresponding MSTSP may not be feasible. Figure 4 shows an example. For a case with seven subfields on a line, we assume that $w_i = 1$, $d_i = 2$, and that b_i only covers the adjacent subfields for each subfield S_i; that is, if subfield S_i is written from time t to time $t + 1$, then subfields S_{i-1} and S_{i+1} are blocked from time $t + 1$ to time $t + 3$. If the blocked box constraint is not considered, an optimal scheduling solution for the corresponding MSTSP is given in Figure 4(a), which is $< S_1, S_5, S_2, S_6, S_3, S_7, S_4 >$. In this scheduling order, at time $t = 2$, the subfield S_2 is written while it is blocked; this solution is thus infeasible. Another scheduling solution $< S_1, S_3, S_5, S_7, S_2, S_4, S_6 >$ is given in Figure 4(b). Take subfield S_2 for example: it is blocked from $t = 2$ to $t = 4$ due to the writing process of subfield S_3 which finishes at $t = 2$. Thus, the writing process of S_2 starting at $t = 4$ is valid since S_2 is unblocked after $t = 4$. As shown in Figure 4(b), the scheduling solution is feasible since no subfield is written while it is blocked. Although the shortest distance between any two successive subfields of the scheduling solution in

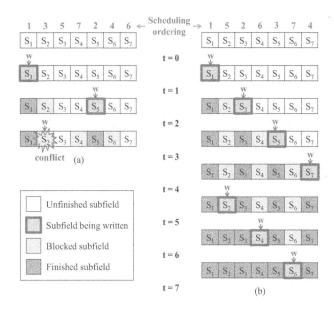

Figure 4: (a) An infeasible subfield scheduling with the shortest distance between successive subfields being 3. (b) A feasible subfield scheduling with the shortest distance between successive subfields being 2.

Figure 4(a) is longer than that in Figure 4(b), the former solution does not satisfy the blocked box constraint and is thus not feasible. Therefore, we must consider the blocked box constraint while finding a subfield scheduling solution; we define the problem as the *constrained MSTSP*.

It has been proven that there exists no polynomial-time, constant-performance-bound approximation algorithm for the general MSTSP [1]. Even given a graph satisfying the triangle inequality, there exists no polynomial-time algorithm with a performance bound smaller than two [1]. Since the MSTSP can be reduced to the constrained MSTSP with blocked box constraints by restricting d_i to be zero for each S_i, the constrained MSTSP can be shown to be NP-complete.

THEOREM 1. *The constrained MSTSP is NP-complete.*

2.4 Problem Formulation

We use a point to represent a subfield and the Euclidean measure to define the distances between subfields. We transform the subfield scheduling problem into a constrained MSTSP problem as follows:

PROBLEM 1. *Given a photomask layout and the predefined size of a subfield, find a Hamiltonian path containing all points such that the length of the shortest edge in the Hamiltonian path is maximized, and the blocked box constraint is satisfied.*

By solving this constrained MSTSP problem, we can alleviate the problem with successive heating and thus also control CD distortion.

Note that the moving time of a writer is negligible compared to the writing time of a subfield; therefore, the moving of an e-beam writer will not harm the throughput. As a result, we prefer to maximize the minimum distance between two subfields with successive labels (orders) of subfields (i.e., the length of the shortest edge in the Hamiltonian path) to alleviate the problems with successive heating and CD distortions.

3. EXACT MSTSP ALGORITHM FOR VERTICES ON TWO PARALLEL LINES

In this section, we develop an exact algorithm that optimally solves the MSTSP for a special case. In the case where points are on two parallel lines with an odd number of points on each line, and the points on different lines are aligned, the algorithm can find an optimal solution of the MSTSP under the Euclidean measure in linear time. Figure 5(a) shows an example. There are an odd number of points on Lines a and b, and the two lines are parallel. We denote as a_i and b_i the i-th points on Line a and Line b respectively. For each point pair (a_i, b_i), a_i and b_i are aligned; that is, the two points have the same x-coordinate.

In this special case, let V be the input points of size $|V| = 2n$, where $n = 2k + 1$. Then, an optimal Hamiltonian path can be derived by using the following algorithm:

Step 1. For each point a_h, connect a_h to points b_i and b_j, where

$$i = \begin{cases} h + (1 + k), & \text{if } h + (1 + k) \leq n \\ h + (1 + k) - n, & \text{if } h + (1 + k) > n, \end{cases} \quad (3)$$

and

$$j = \begin{cases} h + k, & \text{if } h + k \leq n \\ h + k - n, & \text{if } h + k > n. \end{cases} \quad (4)$$

Step 2. After connecting the edges, delete the shortest edge in the generated Hamiltonian cycle. The derived path is an optimal Hamiltonian path.

As illustrated in Figure 5(b), a Hamiltonian cycle is first constructed by applying Equations (3) and (4). After deleting the shortest edge, (a_5, b_3), on the path, an optimal Hamiltonian path is then generated.

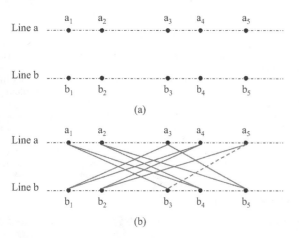

(a)

(b)

Figure 5: (a) A special case in which points are on two parallel lines and points on different lines are aligned. (b) An optimal Hamiltonian path generated by using the proposed algorithm.

To prove the optimality of the algorithm, we first denote d^* as the length of the shortest edge of an optimal Hamiltonian path, and denote $diam(R)$ as the diameter of the set $R \subseteq V$. In addition, we say that an edge (u, v) *spans* n points if there are n points on the boundary of the bounding box of the point u and the point v. In Figure 5(b), for example, the edge (a_1, b_3) spans six points and the edge (a_1, b_4) spans eight points. We utilize the following lemma and corollary stated in [1]:

LEMMA 1. *Let $R \subset V$ be a subset of points with $|R| > \lceil \frac{|V|}{2} \rceil$. Then, in any Hamiltonian path on V, there must exist an edge joining two points of R.*

COROLLARY 1. *Let $R \subset V$ be a subset of points with $|R| = \lceil \frac{|V|}{2} \rceil + 1$. Then, $d^* \leq diam(R)$.*

These yield the following theorem:

THEOREM 2. *For a set of points V with $|V| = 2n$ ($n = 2k + 1$ is an odd integer), if the points are on two parallel lines and points on the two lines are aligned, the algorithm finds an optimal Hamiltonian path for the MSTSP under the Euclidean measure in linear time.*

PROOF 1. *It is easy to verify that the edges generated by using Equations (3) and (4) consist of a Hamiltonian path. Thus, we only need to show that the upper bound of d^* stated in the corollary is achievable by applying the algorithm.*

Let the subset of points spanned by an edge e_i be R_i. Each edge e_i constructed by using Equation (3) or (4) spans at least $\lceil \frac{|V|}{2} \rceil + 1 = 2k + 2$ points, and the length of e_i is equal to $diam(R_i)$. With the corollary, we have

$$d^* \leq diam(R_i), \forall i. \quad (5)$$

Since each edge is an upper bound of d^*, the generated Hamiltonian path is optimal. Also, since an optimal Hamiltonian path can be specified in $O(1)$ time without knowing the positions of the points, and the path can be constructed in $O(n)$ time, an optimal Hamiltonian path can be found in linear time by using the proposed algorithm.

4. GRAPH-BASED SUBFIELD SCHEDULING FOR E-BEAM PHOTOMASK FABRICATION

In this section, we introduce our graph-based algorithm for the subfield scheduling for e-beam photomask fabrication. In Section 4.1 we describe the two-stage algorithm, and in Sections 4.2 and 4.3 we detail the sub-MSTSP and the post-processing stages, respectively.

4.1 Algorithm Flow

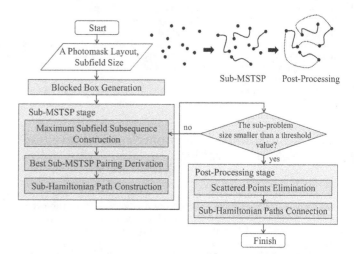

Figure 6: The subfield scheduling flow.

In this section, we present the overview of our subfield scheduling algorithm. Figure 6 shows our subfield scheduling flow.

Given a photomask layout and the predefined size of a subfield, we first construct the blocked boxes for all subfields according to their pattern densities. Our subfield scheduling algorithm consists of two major stages: the sub-MSTSP stage followed by the post-processing stage.

In the sub-MSTSP stage, by utilizing the exact algorithm of MSTSP for points on two parallel lines presented in Section 3, an optimal sub-Hamiltonian path for each subproblem can be constructed efficiently. First, for each row pair (r_i, r_j) of subfields, we apply a dynamic programming-based algorithm to find a set of subfields $\mathcal{S}_{ij}$ such that the arrangement of $\mathcal{S}_{ij}$ conforms to the special case mentioned in Section 3. Then we compute in linear time the length of the shortest edge of an optimal sub-Hamiltonian path of the set $\mathcal{S}_{ij}$. To simultaneously optimize all sub-Hamiltonian paths in an iteration, we propose a matching-based algorithm to find a best sub-MSTSP pairing result $\mathcal{M}$. Finally, for each $(r_i, r_j) \in \mathcal{M}$, an optimal Hamiltonian path of $\mathcal{S}_{ij}$ can be constructed by applying Equations (3) and (4).

If the subproblem sizes are small and the subfields are close to each other, the solution quality of subproblems may not be sufficient; thus, we add a post-processing stage. In this stage, we first connect each isolated subfield to one of the existing Hamiltonian paths. While scattered subfields are eliminated, all sub-Hamiltonian paths are merged into one Hamiltonian path by iteratively using the matching-based method.

In the following, we detail the two stages in Sections 4.2 and 4.3 respectively.

4.2 Sub-MSTSP Stage

Since the MSTSP is conjectured to be NP-hard for points in the plane with the Euclidean measure [1], an algorithm that can efficiently solve the MSTSP for all subfields of a photomask layout might not exist. In addition, since the problem we deal with is the constrained MSTSP, the blocked box constraint must also be taken into account. Thus, we propose an algorithm which can solve the constrained MSTSP efficiently by decomposing the problem into a set of sub-problems.

In the first stage of the algorithm, we first find the maximum subfield subsequence for each pair of subfield rows by using a dynamic programming-based algorithm. Then, we find the best MSTSP pairing by applying a matching-based method. The two processes are detailed in the following two sections.

4.2.1 Maximum Subfield Subsequence Construction

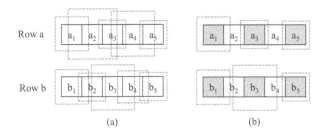

(a)   (b)

Figure 7: Subproblem extraction. (a) The blocked boxes of a pair of chosen rows do not overlap vertically. (b) The blocked boxes of a maximum subfield subsequence do not overlap, and the total area of the boxes is maximized.

Due to the inefficiency of solving the MSTSP for all subfields in a photomask layout, we extract subproblems and find the corresponding optimal solutions by using the algorithm presented in Section 3. To extract a subproblem conforming to the special case, two rows of subfields are selected first. In addition, since the problem we deal with is the constrained MSTSP, we expect that the blocked boxes of the subfields in an extracted subproblem are not overlapped, and thus the subproblem can be solved optimally with Equations (3) and (4). Thus, the blocked boxes of the first subfield row cannot overlap with those of the second subfield row. As shown in Figure 7(a), the blocked boxes of the chosen rows do not overlap vertically. The blocked boxes should not overlap horizontally either. We propose a dynamic programming-based algorithm to find a set of non-overlapping subfields in which the subfields in the first row are aligned with those in the second row. Furthermore, to maximize the subproblem size and to achieve the maximum blocked box uniformity in a subproblem, we choose the subfields whose total area of blocked boxes is maximized: in Figure 7(b), these subfields are rendered in red. We refer to the set of obtained subfields as the *maximum subfield subsequence*.

Let $A(a_i)$ denote the area of the blocked box of the i-th subfield in row a, and $A(b_i)$ denote the area of the blocked box of the i-th subfield in row b. In the dynamic programming formulation, we let $Area[i] = A(a_i) + A(b_i)$. The problem can thus be formulated as

$$C_0[i] = \begin{cases} 0, & \text{if } i = 1, \\ \max(C_1[i-1], C_0[i-1]), & \text{otherwise.} \end{cases} \quad (6)$$

$$C_1[i] = \begin{cases} Area[i], & \text{if } i = 1, \\ Area[i] + C_1[t], & \text{otherwise.} \end{cases} \quad (7)$$

$$\pi[i] = \begin{cases} Nil, & \text{if } i = 1, \\ t, & \text{otherwise.} \end{cases} \quad (8)$$

If the blocked boxes of the i-th subfields of the two rows are not chosen, $C_0[i]$ records the maximum total area of non-overlapping blocked boxes (likewise, $C_1[i]$ records the maximum total area for those of the two rows that *are* chosen). $t < i$ is an index such that $C_1[t]$ is maximized and the blocked boxes of a_t and b_t do not overlap with those of a_i and b_i. The maximum subfield subsequence can be derived by finding the largest i such that $C_1[i] \geq C_0[i]$ and tracing back the indexes with Equation (8). Note that if the number of selected subfields in a row is even, we omit subfields a_i and b_i with the minimum $Area[i]$.

4.2.2 Best Sub-MSTSP Pairing Derivation

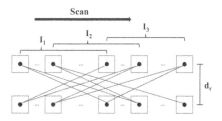

Figure 8: The length of the shortest edge in an optimal sub-Hamiltonian path can be computed by scanning $\lceil \frac{n}{2} \rceil$-intervals in linear time.

After generating the maximum subfield subsequence of an extracted subproblem, we can compute the length of the shortest edge in an optimal sub-Hamiltonian path of the subproblem. Suppose that the size of a subproblem is $|V| = 2n$. Let the l-interval be an interval containing l subfields in a row. By scanning $\lceil \frac{n}{2} \rceil$-intervals from left to right of a row, the length of the shortest $\lceil \frac{n}{2} \rceil$-interval d_s can be found. Then, the length of the shortest edge in an optimal sub-Hamiltonian path is $\sqrt{d_s^2 + d_r^2}$, where d_r is the distance between the two subfield rows. As shown in Figure 8, for a subproblem of size $|V| = 10$, the length of the shortest edge can be computed by scanning the three 3-intervals.

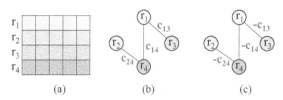

(a)   (b)   (c)

Figure 9: Best sub-MSTSP pairing deviation. (a) Four rows of subfields. (b) Compatible graph construction. (c) Modified compatible graph for use with maximum cardinality bottleneck matching algorithm.

For each pair of subfield rows, we find the maximum subfield subsequence and compute the length of the shortest edge of an optimal sub-Hamiltonian path. To simultaneously optimize each subproblem, we need a pairing for all subfield rows that maximizes the length of the shortest edge among all sub-Hamiltonian paths. To solve this *best sub-MSTSP pairing problem*, we propose a matching-based algorithm.

First, we construct a compatible graph in which a point represents a subfield row, and two points are connected through an edge if the blocked boxes of the two rows are non-overlapped vertically. Each edge is associated with an edge weight, which is the length of the shortest edge in the sub-Hamiltonian path of the corresponding maximum subfield subsequence. Figure 9 illustrates an example. Four rows of subfields are shown in Figure 9(a): suppose that the blocked boxes of each row only overlap vertically with those of adjacent rows. Thus, the corresponding compatible graph is shown in Figure 9(b). By utilizing the compatible graph, the best sub-MSTSP pairing problem can be solved by finding a maximum matching $\mathcal{M}$ in which the smallest edge weight is maximized. We apply the algorithm proposed in [7] to solve the maximum cardinality bottleneck matching problem to find a matching solution by simply multiplying the edge weights by -1 and slightly modifying the algorithm. Hence, the best pairing of the example in Figure 9 can be found by using the algorithm in [7] and the compatible graph in Figure 9(c).

For each pair of subfield rows in $\mathcal{M}$, an optimal sub-Hamiltonian path is then constructed with Equations (3) and (4) in linear time. We iteratively apply the above process for those subfields not contained in any sub-Hamiltonian paths constructed in previous iterations.

4.3 Post-Processing Stage

If the subproblem sizes are small and the subfields are close to each other such that the length of the shortest $\lceil \frac{n}{2} \rceil$-interval d_s is smaller than a threshold value, the solution quality of subproblems may not be sufficient; thus, the subfield scheduling algorithm flow enters the second stage, the post-processing stage. In this stage, the main objective is to maintain the good solution quality derived from the sub-MSTSP stage while constructing a complete subfield scheduling solution. We first eliminate scattered subfields by connecting each of them to an endpoint of existing sub-Hamiltonian paths. Then, we merge all sub-Hamiltonian paths into one Hamiltonian path, which is the final subfield scheduling solution. We detail the scattered subfield elimination and the sub-Hamiltonian paths merging in the following subsections.

4.3.1 Isolated Subfield Elimination

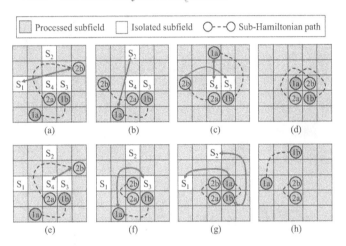

Figure 10: Scattered point elimination. (a)–(d) Eliminating the isolated points in the order $< S_1, S_2, S_3, S_4 >$ (from the margins to the center). (e)–(h) Eliminating the isolated points in the order $< S_4, S_3, S_2, S_1 >$ (from the center to the margins).

At the beginning of this stage, each subfield is either contained in a sub-Hamiltonian path, or left as an isolated subfield. We sequentially connect each isolated subfield to an endpoint of existing sub-Hamiltonian paths such that the distance between the subfield and the endpoint is not shorter than the shortest edge in existing sub-Hamiltonian paths. Since the elimination sequence determines the final positions of sub-Hamiltonian paths, the sequence may greatly affect the solution quality in the post-processing stage. Figure 10 gives an example. There are two sub-Hamiltonian paths, path 1 and path 2, with their endpoints, $1a, 1b, 2a,$ and $2b$, and four isolated subfields are left, $S_1, S_2, S_3,$ and S_4. Figures 10(a)–(d) show the process in which the isolated subfields are eliminated from the margins to the center of the chip; that is, the isolated subfields are connected to endpoints of sub-Hamiltonian paths with the sequence $< S_1, S_2, S_3, S_4 >$. In the example, we simply connect each isolated subfield to the farthest endpoint of path 1 and path 2. After connecting all isolated subfields, the four endpoints are close to each other, as shown in Figure 10(d). Close endpoints may cause very short edges as we merge sub-Hamiltonian paths. It can easily be seen in Figure 10(d) that the length of the edge connecting paths 1 and 2 must be one-unit long. On the other hand, if we eliminate the isolated subfields with the sequence $< S_4, S_3, S_2, S_1 >$, as illustrated in Figures 10(e)–(h), the endpoints of the final sub-Hamiltonian

paths are much more scattered, as shown in Figure 10(h). To merge paths 1 and 2, we can connect the endpoints $1b$ and $2a$ such that the length of the connecting edge is three units, which is much better than the previous one.

To maintain the solution quality of the sub-Hamiltonian paths merging, therefore, we eliminate isolated subfields from the center to the margins of a chip such that the final endpoints of sub-Hamiltonian paths tend to lie on the margins of the chip. This heuristic maintains the solution quality derived in the sub-MSTSP stage in most cases.

4.3.2 Sub-Hamiltonian Paths Merging

After eliminating all isolated points, the last step is to merge sub-Hamiltonian paths into one Hamiltonian path which contains all subfields in the given photomask layout. By observing that the operation of merging sub-Hamiltonian paths corresponds to connecting the endpoints of those paths, we transform the problem into a matching problem again. Figure 11 illustrates an example. There are four sub-Hamiltonian paths in Figure 11(a), and the endpoints of sub-Hamiltonian path i are denoted as i_1 and i_2. A compatible graph is then constructed, in which a node denotes an endpoint of a sub-Hamiltonian path and an edge between two nodes means that the two corresponding endpoints can be connected together, as shown in Figure 11(b). The weight of edge (i, j) is set to be the distance between endpoints i and j. To maximize the length of the shortest edge connecting sub-Hamiltonian paths, the maximum bottleneck maximum matching algorithm is then applied again to obtain a matching solution $\mathcal{M}$. Connecting the endpoints of sub-Hamiltonian paths according to $\mathcal{M}$, the resulting graph is composed of paths and cycles. As shown in Figures 11(c) and (d), the graph corresponding to the matching result $\mathcal{M} = \{(a_1, b_1), (a_2, b_2), (c_1, d_2), (c_2, d_1)\}$ consists of two cycles. We break a cycle by deleting the shortest edge in the cycle, and thus the sub-Hamiltonian paths in the cycle are merged into one longer sub-Hamiltonian path. After deriving a set of longer sub-Hamiltonian paths, an updated compatible graph is constructed as illustrated in Figure 11(e). The above merging process is performed iteratively until all sub-Hamiltonian paths are merged into one Hamiltonian path. Since during each iteration's merging process the number of sub-Hamiltonian paths is cut at least in half, the number of iterations required for the merging process is less than or equal to $\lceil \lg n \rceil$ times, where n is the number of sub-Hamiltonian paths constructed in the sub-MSTSP stage.

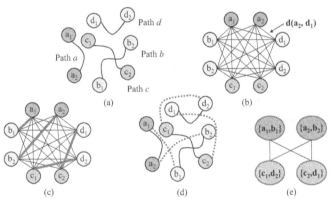

Figure 11: Sub-Hamiltonian paths merging. (a) Four sub-Hamiltonian paths. (b) The compatible graph corresponding to the four sub-Hamiltonian paths. (c) A matching solution. (d) The resulting graph with two cycles. (e) Updated compatible graph after breaking the cycles.

5. EXPERIMENTAL RESULTS

Our algorithm was implemented in the C++ programming language on a 2.13 GHz Linux workstation with 16 GB memory. The experiment was based on the 11 benchmark circuits used in [4].

Table 3: The subfield scheduling results of our graph-based subfield scheduling algorithm flow. Min.: shortest length between two successive subfields in a subfield scheduling solution. Avg.: average length among all successive subfields.

Circuit	# subfields											
	32×32			64×64			128×128			256×256		
	Min.	Avg.	CPU (s)	Min.	Avg.	CPU (s)	Min.	Avg.	CPU (s)	Min.	Avg.	CPU (s)
Mcc1	18.868	21.845	0.020	32.016	42.679	0.460	64.070	83.236	15.990	128.004	165.815	33.900
Mcc2	12.166	22.072	0.010	23.770	43.722	0.680	51.865	86.652	24.080	128.082	168.988	414.730
Struct	19.209	22.373	0.020	32.203	43.648	0.290	64.885	84.741	4.350	128.144	167.653	8.580
Primary1	16.156	21.621	0.030	32.016	39.965	0.330	64.008	77.578	1.460	128.160	154.234	3.320
Primary2	19.417	22.459	0.030	35.903	43.627	0.240	64.078	83.525	7.360	128.156	164.165	50.460
S5378	19.209	22.454	0.030	33.051	43.432	0.510	64.031	83.050	2.930	128.141	162.937	12.580
S9234	18.439	21.952	0.020	32.203	42.843	0.380	64.031	82.430	2.620	128.818	162.078	6.320
S13207	21.260	22.666	0.030	33.838	45.028	0.380	64.073	88.778	4.270	128.550	169.550	27.370
S15850	19.209	22.519	0.020	33.838	44.807	0.490	64.031	88.053	5.560	128.144	170.343	30.920
S38417	17.889	22.701	0.030	32.016	45.022	0.430	64.141	89.445	7.050	134.380	175.806	43.070
S38584	21.260	22.656	0.030	33.242	45.151	0.420	64.070	89.375	4.990	128.160	174.855	104.010
Avg.	18.462	22.302	0.025	32.190	43.629	0.419	63.026	85.169	7.333	128.794	166.948	66.842

Table 2: Statistics of the benchmark circuits.

Circuit	Size (μm²)	# Layers	# Nets	# Pins
Mcc1	45000×39000	4	1694	3101
Mcc2	152400×152400	4	7541	25024
Struct	4903×4904	3	3551	5471
Primary1	7522×4988	3	2037	2941
Primary2	10438×6488	3	8197	11226
S5378	4330×2370	3	3124	4734
S9234	4020×2230	3	2774	4185
S13207	6590×3640	3	6995	10562
S15850	7040×3880	3	8321	12566
S38417	11430×6180	3	21035	32210
S38584	12940×6710	3	28177	42589

Table 2 lists the set of benchmark circuits. We routed the circuits by using a two-pass bottom-up gridless router, and we used the metal-1 of each circuit as the input photomask layout. A layout is first divided into 32×32, 64×64, 128×128, and 256×256 subfields. The pattern density was analyzed, and the blocked box of each subfield was constructed. We let the blocked box of a subfield being written cover 15% of the total number of subfields in its neighboring region if the pattern density of the subfield was one, and let the area of a blocked box equal the area of the corresponding subfield if the pattern density was zero.

The subfield scheduling results of the benchmark circuits with different numbers of subfields are listed in Table 3. In the table, "Min." gives the shortest length between two successive subfields in a subfield scheduling solution, and "Avg." gives the average length among all successive subfields. Note that the lengths are measured in unit subfield edge length.

We first compared our graph-based algorithm with the modified greedy algorithm for subfield scheduling (denoted as GSS) proposed in [3]. This algorithm first randomly generates a subfield scheduling order and then iteratively improves the solution by swapping pairs of subfields. We modified the cost function in [3] from temperature measurement to distance measurement. The cost function is given by

$$\alpha \cdot D_{min} + \beta \cdot D_{avg}, \qquad (9)$$

where D_{min} is the minimum distance and D_{avg} is the average distance between two successive subfields. We set $\alpha \gg \beta$ to normalize the two values. Since GSS does not take into account the blocked box constraint, we simply ignore whether the scheduling order generated from GSS is feasible or not. In addition, since GSS does not consider the pattern density in each subfield, the scheduling results are independent of different circuit layouts. Thus, we compared their scheduling results with our average results derived from Table 3.

Table 4 shows the comparison results. Although the average lengths derived from our algorithm are slightly shorter than those derived from GSS, the minimum lengths obtained from our algorithms are much longer than those of GSS. This is because the objective of our algorithm is to maximize the minimum distance

between any two successive subfields such that the problem with successive heating and thus CD distortion can be alleviated. Furthermore, the CPU time required by our algorithm is much less than that required by GSS. These results show that our algorithm is effective and efficient in finding a good subfield scheduling solution for e-beam photomask fabrication.

Table 4: Comparison of the subfield scheduling results between GSS [3] and our graph-based subfield scheduling algorithm flow.

# subfields	GSS [3]			Ours		
	Min.	Avg.	CPU (s)	Min.	Avg.	CPU (s)
32×32	3.162	24.355	33.561	18.462	22.302	0.025
64×64	3.605	48.835	>1hr	32.190	43.629	0.419
128×128	2.000	96.431	>1hr	63.026	85.169	7.333
256×256	1.000	164.922	>1hr	128.794	166.948	66.842

To verify that the subfield scheduling solution obtained from our algorithm flow indeed mitigates the heating problem effectively, we used the thermal model introduced in Section 2.1 to calculate the maximum temperature of each subfield in a scheduling order, which is defined as the critical temperature of a subfield. Then, we computed the average critical temperature and the maximum critical temperature for each scheduling solution. Also, we implemented the original greedy subfield scheduling (GSS) algorithm developed in [3] to make a comparison. The cost function is

$$\alpha \cdot T_{max} + (1 - \alpha) \cdot T_{avg}, \qquad (10)$$

where T_{max} is the maximum critical temperature and T_{avg} is the average critical temperature of all subfields ($\alpha = 0.5$, the same setting as in [3]). In each iteration, the greedy algorithm generates the next scheduling order by swapping the order of the subfield with the maximum critical temperature and some other subfield such that the cost in Equation (10) is minimized. For this experiment, we set the initial temperature of each subfield to zero, and set the maximum running time to one hour. In addition, the maximum raised temperature of a subfield due to its own writing process was normalized with the average subfield density for each testcase.

The experimental results are shown in Tables 5 and 6. Column "Cost" lists the cost evaluated by Equation (10), column "Max. Temp." gives the maximum critical temperature in a scheduling order, and column "Imp. Rate" shows the improvement ratio of our results over those derived from the greedy algorithm. From the two tables, most of the scheduling solutions generated by our algorithm flow yield better solutions than those obtained from GSS for cases of size 32×32. For other cases, moreover, our algorithm approximately achieves 30% temperature reductions on average compared with GSS. This substantial improvement may result from the expensive computation of the cost function which requires $O(|\mathcal{S}|^2)$ time to evaluate, and thus it requires $O(|\mathcal{S}|^3)$

Table 5: Comparison of thermal results between GSS [3] and our algorithm flow for cases of sizes 32x32 and 64x64. Cost: the cost evaluated by Equation (10). Max. Temp.: maximum critical temperature in a scheduling order. Imp. Rate.: improvement ratio.

Circuit	# subfields											
	32×32						64×64					
	Cost			Max. Temp.			Cost			Max. Temp.		
	GSS [3]	Ours	Imp. Rate	GSS [3]	Ours	Imp. Rate	GSS [3]	Ours	Imp. Rate	GSS [3]	Ours	Imp. Rate
Mcc1	103.004	89.195	13.41%	104.662	97.110	7.22%	124.024	70.3574	43.27%	132.813	73.158	44.92%
Mcc2	119.431	94.747	20.67%	123.959	100.088	19.26%	145.258	83.048	42.83%	158.376	88.461	44.14%
Struct	98.341	95.066	3.33%	99.756	99.477	0.28%	125.969	91.818	27.11%	133.113	95.984	27.89%
Primary1	151.756	148.816	1.94%	163.374	165.521	-1.31%	160.745	132.145	17.79%	177.386	150.080	15.39%
Primary2	127.054	113.769	10.46%	138.489	128.274	7.38%	153.090	118.777	22.41%	165.705	131.182	20.83%
S5378	107.218	101.657	5.19%	109.458	106.753	2.47%	122.889	87.053	29.16%	130.706	94.855	27.43%
S9234	108.099	100.420	7.10%	111.583	107.087	4.03%	122.271	78.688	35.64%	131.529	79.893	39.26%
S13207	103.915	99.520	4.23%	106.197	111.106	-4.62%	125.131	99.086	20.81%	133.746	112.966	15.54%
S15850	115.847	109.990	5.06%	123.310	124.889	-1.28%	124.853	97.327	22.05%	134.952	107.494	20.35%
S38417	89.743	74.001	17.54%	91.611	78.898	13.88%	114.424	75.855	33.71%	119.891	78.602	34.44%
S38584	90.920	75.024	17.48%	93.657	82.014	12.43%	126.868	98.821	22.11%	142.166	118.502	16.65%
Avg.			9.67%			5.43%			28.809%			27.894%

Table 6: Comparison of thermal results between GSS [3] and our algorithm flow for cases of sizes 128x128 and 256x256.

Circuit	# subfields											
	128×128						256×256					
	Cost			Max. Temp.			Cost			Max. Temp.		
	GSS [3]	Ours	Imp. Rate	GSS [3]	Ours	Imp. Rate	GSS [3]	Ours	Imp. Rate	GSS [3]	Ours	Imp. Rate
Mcc1	101.466	60.214	40.66%	108.517	64.061	40.97%	83.018	50.369	39.33%	88.082	55.171	37.36%
Mcc2	126.901	73.714	41.91%	138.788	80.109	42.28%	97.631	57.985	40.61%	104.068	61.470	40.93%
Struct	102.845	81.905	20.36%	109.924	88.781	19.23%	80.093	58.185	27.35%	85.763	60.184	29.83%
Primary1	135.645	98.875	27.11%	150.154	108.426	27.79%	105.671	76.899	27.23%	117.174	84.446	27.93%
Primary2	140.234	116.800	16.71%	153.815	130.237	15.33%	116.804	96.719	17.20%	126.607	104.731	17.28%
S5378	102.551	65.731	35.90%	109.285	67.851	37.91%	71.185	42.820	39.85%	74.851	44.064	41.13%
S9234	99.212	70.448	28.99%	106.551	79.655	25.24%	71.446	43.546	39.05%	75.694	45.142	40.36%
S13207	107.167	69.180	35.45%	115.782	71.480	38.26%	73.159	43.899	40.00%	77.546	45.192	41.72%
S15850	96.729	68.973	28.69%	101.331	73.162	27.80%	79.858	55.656	30.31%	85.880	62.456	27.28%
S38417	103.480	69.109	33.22%	109.180	72.101	33.96%	81.691	54.485	33.30%	82.293	56.137	31.78%
S38584	75.447	55.656	26.23%	80.010	62.456	21.94%	85.457	65.816	22.98%	92.770	74.621	19.56%
Avg.			30.48%			30.07%			32.47%			32.29%

time to perform each swap operation in GSS, where $|\mathcal{S}|$ is the number of subfields. As a result, GSS can only perform a small number of iterations to improve the solution quality for the cases of sizes larger than 32×32 within an acceptable running time. These results show that our algorithm can effectively mitigate the heating problem and that it scales well with the number of subfields.

6. CONCLUSIONS

This paper has presented a graph-based algorithm that deals with the subfield scheduling problem by solving the constrained MSTSP. Compared with the prior work, our algorithm elegantly solves the problem by decomposing it into subproblems and solving the subproblems optimally. Experimental results show that our algorithm is effective and efficient in finding good subfield scheduling solutions that alleviate the successive heating problem (and thus reduce CD distortion) for e-beam photomask fabrication.

7. REFERENCES

[1] E. M. Arkin, Y.-J. Chiang, J. S. B. Mitchell, S.S. Skiena and T.-C. Yang, "On the maximum scatter travelling salesperson problem," *SIAM Journal on Computing*, Vol. 29, No. 2, pp. 515–544, 1999.

[2] S. Babin, A. B. Kahng, I. I. Mandoiu and S. Muddu, "Subfield scheduling for throughput maximization in electron-beam photomask fabrication," in *Proceedings of SPIE Conference on Emerging Lithography Technologies*, Vol. 5037, pp. 934–942, 2003.

[3] S. Babin, A. B. Kahng, I. I. Mandoiu and S. Muddu, "Resist heating dependence on subfield scheduling in 50kV electron beam maskmaking," in *Proceedings of SPIE Conference on Photomask and Next-Generation Lithography Mask Technology*, Vol. 5130, pp. 718–726, 2003.

[4] J. Cong, J. Fang, M. Xie, and Y Zhang, "MARS–A multilevel full-chip gridless routing system," *IEEE Transactions on Computer-Aided Design of Integrated Circuits and Systems*, Vol. 24, No. 3, pp. 382–394, March 2005.

[5] A. Fujimura, "Beyond light: the growing improtance of e-beam," in *Proceedings of International Conference on Computer-Aided Design*, November 2009.

[6] A. Fujimura, "Design for e-beam: getting the best wafers without the exploding mask costs," in *Proceedings of International Symposium on Quality Electronic Design*, March 2010.

[7] H. N. Gabow and R. E. Tarjan, "Algorithms for two bottleneck optimization problems," *Journal of Algorithms*, Vol. 9, No. 3, pp. 411–417, 1988.

[8] N. Kuwahara, H. Nakagawa, M. Kurihara, N. Hayashi, H. Sano, E. Murata, T. Takikawa and S. Nohuchi, "Preliminary evaluation of proximity and resist heating effects observed in high acceleration voltage e-beam writing for 180nm-and-beyond rule reticle fabrication," in *SPIE Symposium on Photomask and X-Ray Mask Technology VI*, Vol. 3784, pp. 115–125, 1999.

[9] J. C. Lagarias, "Well-spaced labelings of points in rectangular grids," *SIAM Journal of Discrete Mathematics*, Vol. 13, No. 4, pp. 521–534, 2000.

[10] H. C. Pfeiffer, "New prospects for election beams as tools for semiconductor lithography," in *Proceedings of the SPIE*, Vol. 7378, pp. 737802-737802-12, 2009.

[11] H. Sakurai, T. Abe, M. Itoh, A. Kumagae, H. Anze and I. Higashikawa, "Resist heating effect on 50kV EB mask writing," in *SPIE Symposium on Photomask and X-Ray Mask Technology VI*, Vol. 3748, pp. 126–136, 1999.

A Polynomial Time Exact Algorithm for Self-Aligned Double Patterning Layout Decomposition

Zigang Xiao, Yuelin Du, Hongbo Zhang, Martin D. F. Wong
Department of Electrical and Computer Engineering
University of Illinois at Urbana-Champaign
{zxiao2, du6, hzhang27, mdfwong}@illinois.edu

ABSTRACT

Double patterning lithography (DPL) technologies have become a must for today's sub-32nm technology nodes. There are two leading DPL technologies: self-aligned double patterning (SADP) and litho-etch-litho-etch (LELE). Among these two DPL technologies, SADP has the significant advantage over LELE in its ability to avoid overlay, making it the likely DPL candidate for the next technology node of 14nm. In any DPL technology, layout decomposition is the key problem. While the layout decomposition problem for LELE has been well-studied in the literature, only few attempts have been made to address the SADP layout decomposition problem. In this paper, we present the first polynomial time exact (optimal) algorithm to determine if a given layout has an overlay-free SADP decomposition. All previous exact algorithms were computationally expensive exponential time algorithms based on SAT or ILP. Other previous algorithms for the problem were heuristics without having any guarantee that an overlay-free solution can be found even if one exists.

Categories and Subject Descriptors

B.7.2 [**Design Aids**]: Layout; J.6 [**Computer-Aided Engineering**]: Computer-aided Design

General Terms

Algorithms, Design

Keywords

Self-aligned Double Patterning, Layout Decomposition, Polynomial time Algorithm

1. INTRODUCTION

Because of the increasing printing difficulties in the current IC industry, double patterning lithography (DPL) has become more and more important for the current sub-$32nm$

nodes in the $193nm$ micro-lithography process [13]. The conventional DPL, such as LELE (litho-etch-litho-etch), which splits the intended patterns into two exposures, has drawn lots of attention [15, 8, 14]. Nevertheless, conventional DPL technology suffers from the inevitable overlay problem during the process. Due to issues such as mask placement and wafer thickness variations, the features printed between two masks may be misaligned. This may result in short circuits or connection failures, which will in turn lead to yield degradation.

SADP (Self-aligned double patterning) is a promising alternative double patterning lithography that can significantly avoid overlay. As a DPL, it also contains two mask steps, namely *core* mask and *trim* mask. Figure 1 shows an example of the SADP process.[1] To generate the target features in Figure 1(a), core mask is first used to generate the core patterns. *Sidewall* patterns are then deposited around the sides of the core patterns. These steps are shown in Figure 1(b). The blue rectangular shapes are core patterns

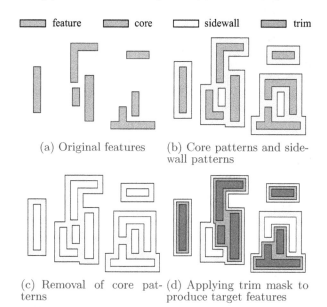

(a) Original features

(b) Core patterns and sidewall patterns

(c) Removal of core patterns

(d) Applying trim mask to produce target features

Figure 1: Example of SADP process.

---

[1]There are two types of SADP process: positive tone and negative tone. Since the positive tone process is more flexible, it is thus considered as a more promising technique. We assume positive tone process is used in our paper. Refer to [10] for more detail.

generated by core masks, while the yellow shapes are sidewall patterns deposited adjacent to the corresponding core patterns. For instance, a rectangle core pattern is generated for the leftmost feature, and stripes of sidewalls are generated around the sides of this core pattern. Similarly, sidewall patterns are generated and may be merged for the rest of the core patterns. Note that these sidewall patterns protect the underlying layer from being etched in the next phase. Next, the core patterns will be removed as shown in Figure 1(c). Finally, trim mask is used to trim out the desired region as shown in Figure 1(d). Note that *only the area that is covered by trim patterns but not covered by sidewalls will be etched*. As a result, the features can be printed exactly even if overlay happens, as long as the amount of overlay is small enough that the trim mask boundaries are contained in the sidewall. In other words, the sidewall in SADP provides tolerance to the unexpected misalignment of two masks. Therefore, by carefully placing core and trim patterns, overlay can be avoided in SADP. This example illustrates the SADP process and demonstrates its mechanism of avoiding overlay. Various works [6, 11, 12, 7] have demonstrated the advantages of SADP in the sub-$30nm$ process. Note that without further notice, we will continue to use the same color coding from Figure 1 in the subsequent figures.

Although SADP has been successfully deployed in 1D patterns and has several applications, applying it to 2D patterns is relatively new and needs research attention. Unlike traditional DPL, the core patterns and trim patterns in SADP may not obtained directly from the original features due to the process rules. As illustrated in Figure 1, some features have corresponding core patterns that match their shapes, while others do not. One reason is that two features cannot be printed simultaneously in the same mask step if they are too close. However, notice that some extra core patterns that do not correspond to any features are used to generate sidewalls for the nearby features. As a result, all of the target features can still be protected by sidewalls and printed correctly without overlay. Indeed, these extra cores play a vital part in SADP layout decomposition, as we will see later. We refer this type of core patterns as *auxiliary cores*, as they are essentially helping to generate sidewalls to avoid overlay.

Given a layout, it is of our interest to find a set of core patterns and trim patterns that can produce the layout *exactly without overlay* after SADP process. We refer such this set of core (trim) patterns as an *overlay-free decomposition* of the layout. Note that there may exist many different overlay-free decompositions for a given layout, but sometimes there may not. In the following, we will refer the problem of finding an overlay-free decomposition of a given layout for SADP process as *SADP decomposition problem*. In our paper, we are interested in finding overlay-free decompositions. The motivation is two-fold: (1) Overlay elimination is the major advantage of SADP, we want to maximize the effect of this advantage; (2) Designers may want to check whether a given design is SADP-compliant, especially in an interactive (real-time) fashion. Without further notice, in the following we focus on finding decompositions that do not have overlay.

The SADP decomposition problem is difficult because of the following reasons. Due to the technology constraints, two core or trim patterns cannot be printed simultaneously if their distance is smaller than a certain value. In other words, the placement of the core patterns and trim patterns

are subject to stringent process rules. As a result, to obtain a decomposition, we need to find auxiliary cores that help to generate the sidewalls with respect to the process rules. One possible approach to the problem is to use a graph formulation similar to the one used in LELE [9]. A pair of features that cannot be printed simultaneously in the core mask can be viewed as a conflict. Hence, it is tempting to construct a *constraint graph* according to the conflicts, and apply two-coloring algorithm to find out the set of features that can be printed simultaneously in the core mask. However, as we will discuss in detail later, the above graph formulation is actually incorrect and may produce patterns that differ from the target features. In our paper, we propose a correct graph formulation for the problem, and show that the two-colorability of the graph is a *necessary condition* for SADP decomposition problem. Note that the problem remains unsolved till this point, since we still do not have a decomposition. In fact, it is non-trivial to find a decomposition from the two-coloring solutions. A coloring solution only tells us what features can be assigned core patterns for simultaneous printing. For the other features, we still need to find auxiliary cores to generate the sidewalls. Furthermore, there may be many connected components in a graph, and the number of two-coloring solutions will be *exponential*. Meanwhile, not every one of them correspond to a feasible decomposition. We will further discuss the details in Section 3, where we present an efficient algorithm that addresses these issues.

Currently, all of the previous works addressed the problem indirectly by *minimizing* overlay. Zhang *et al.* proposed a Satisfiability (SAT) and Integer Linear Programming (ILP) formulation for the problem in [16, 17]. However, SAT and ILP are NP-hard problems in general and thus their methods cannot scale to large size problems. Ban *et al.* presented a graph two-coloring approach in [5, 4]. However, their algorithm is based on the above-mentioned incorrect graph formulation. The merging technique is problematic since it is only applicable when the technology used to print the trim mask has a smaller pitch than that is used for the core mask. Moreover, the two-coloring algorithm used in their approach is heuristic-based, which means it cannot guarantee to find an overlay-free decomposition even if one exists.

In our paper, we present a correct graph formulation of the problem, and propose an algorithm that solves the problem exactly in polynomial-time. To the best of our knowledge, this is the first work that addresses this problem directly. The major contributions of this paper include the following:

- We present a careful study of the SADP decomposition problem, and propose a correct graph formulation for the problem. We show that the two-colorability of a graph called SW-graph is a necessary condition for the SADP decomposition problem.

- We propose a polynomial-time exact algorithm that solves the SADP decomposition in general 2D layout. Our algorithm computes the decompositions for each graph components, and finds a final decomposition in polynomial-time using a 2SAT approach.

The rest of the paper is organized as follows. In Section 2, we give a detailed introduction to overlay issues and design rules in SADP. The details of our algorithm for general 2D layout will be covered in Section 3. Section 4 shows

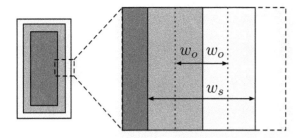

Figure 2: Trim mask overlay and sidewall thickness.

the experimental results of the proposed algorithm. Finally, conclusion will be drawn in Section 5.

2. PRELIMINARIES

2.1 Overlay in SADP Process

Overlay may happen in the second phase of SADP. We refer the maximum possible overlay of the trim mask as *trim mask overlay*, and quantify it as w_o. Overlay control in SADP is achieved by providing sidewall protection to feature boundaries. We denote the thickness of the sidewall as w_s. Figure 2 provides a detailed scenario of printing the left-most feature in Figure 1. The feature boundaries are protected by the sidewall, and the boundaries of trim pattern locate inside the sidewall. The trim mask may shift left or right within the distance of w_o during the SADP process. However, it is still inside the sidewall and will not generate incorrect feature. Ensuring feature boundaries to be protected by the sidewall is the key to avoid overlay when generating the final feature patterns. In particular, we observe that the sidewall must have thickness $w_s \geq 2 \cdot w_o$ to fully tolerate the overlay. Furthermore, the trim pattern must be put in such a location that even if overlay happens, it is still contained in the sidewall. In other words, the trim pattern must be at least w_o apart from the two boundaries of the sidewall to avoid overlay.

2.2 Process Rules

Design rules are always necessary for manufacturing. The SADP process should be performed with respect to these constraints. This is one of the major problems we face in the problem. We will adopt the following process rules in this paper:

- The width of a core pattern is at least w_c.
- The width of a trim pattern is at least w_t.
- The distance between core patterns is at least d_c.
- The distance between trim patterns is at least d_t.
- The width of sidewall is w_s.
- The trim overlay is w_o.

In practice, w_c and w_t are usually the same, without further notice we refer them as $w_{\min}$. d_c and d_t should also be the same, and we refer them as $d_{\min}$. Follows from the discussion in the previous subsection, we know that $w_s \geq 2 \cdot w_o$ must be satisfied in order to guarantee the existence of

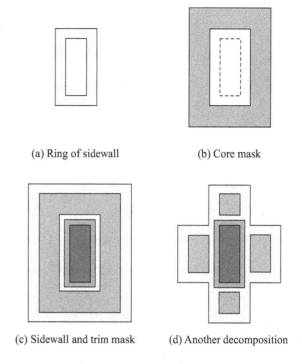

(a) Ring of sidewall   (b) Core mask

(c) Sidewall and trim mask   (d) Another decomposition

Figure 3: Alternative layout decompositions.

a decomposition. In this paper, we assume that the original features are all rectilinear polygons and satisfy the minimum mask width constraint.

3. A POLYNOMIAL-TIME EXACT ALGORITHM FOR SADP DECOMPOSITION

3.1 Overview

In this section, we first revisit and illustrate the concept and the usefulness of auxiliary cores. Next, we carefully examine a set of distance requirements according to the design rules, and their impact to the core and trim patterns. Based on the above observations, we propose a concept *SW-graph* that captures the distance requirements. We further show that the two-colorability of the SW-graph is a necessary condition for SADP layout decomposability. We then present an exact polynomial-time algorithm that finds a decomposition for a given layout using the SW-graph. The proposed algorithm contains two stages. In the first stage, the algorithm finds decompositions for each of the connected components in the SW-graph separately. Despite the fact that there may be many decompositions from the coloring solutions of a component, we show that there are actually at most two special decompositions with the property that all the other decompositions are a subset of either one of them. In the second stage, a set of decompositions are selected from the component decompositions and combined as a final complete decomposition. In particular, exactly one decomposition will be selected for a component in such a way that their combination is still an overlay-free decomposition. In other words, the core and trim patterns in the resulting decomposition will not violate process rules or cause overlay. Although there may be exponentially number of combinations, we show that the problem of choosing compatible decompositions can be reduced to 2SAT (2-Satisfiability) problem and thus can be solved efficiently in polynomial-time. Finally, we provide a brief analysis of the correctness and complexity of our algorithm.

3.2 Auxiliary Cores

To avoid overlay, we rely on using the *auxiliary cores*. An auxiliary core is a core pattern that is placed outside of a feature to generate parts of the sidewall required for the feature. To distinguish, we call a core pattern as a *main core* when it is placed at the same location as a feature and generate the required sidewall pattern for that feature directly. We have seen that one way to produce a feature is to use a main core as shown in Figure 2. However, sometimes we cannot use this approach due to the process rules. An alternative way is to use auxiliary cores. Observe that to print a feature exactly without overlay, there must be a *ring of sidewall* surrounding the target feature as shown in Figure 3(a). Instead of using a main core, we use an auxiliary core outside of the feature to generate the ring of sidewall as shown in Figure 3(b). The sidewall generated and trim mask are shown in Figure 3(c). Note that from now on we include the core mask, sidewalls and trim mask into one figure for compactness. This example demonstrates how to generate a feature using an auxiliary core instead of a main core. We will the approaches used in Figure 2 and Figure 3 as core-method and aux-method in the following. Notice that using the aux-method, we can obtain different decompositions. For instance, Figure 3(d) illustrates another possible decomposition that can generate the ring of sidewall. This time, we are using four auxiliary cores to generate the ring of sidewall. Clearly, this is also a valid decomposition that will not cause overlay.

The above example shows the basic usage of auxiliary cores. They become extremely important when there are multiple features. An example is illustrated in Figure 4. In Figure 4(a), the two features are too close to be printed simultaneously in core mask. Figure 4(b) shows how we can obtain an overlay-free solution by using auxiliary cores. If we assign a core pattern to feature A, the sidewall generated can then protect the left edge of feature B. Another way to interpret this is that the left portion of the ring of sidewall required by feature B will be provided by this core pattern. Then, to avoid overlay for the other three edges of feature B, we introduce an auxiliary core that can generate the remaining part of the ring of sidewall. The final result is shown in Figure 4(c). We use a single trim pattern to trim out both features. Note that there is a dual of this decomposition, *i.e.*, we place a core pattern for feature B, and an auxiliary core to generate a sidewall for feature A. This again shows the flexibility of SADP decomposition. Nevertheless, the distance between an auxiliary core and the feature that is being assisted must satisfy some requirements. This is discussed in the following section.

3.3 A Graph Formulation for SADP Decomposition

Now we carefully examine the impact of the distance between features. The distance d between two features is defined as the distance of two closest edges between two features. We will first show that the constraint graph used in LELE cannot be used directly in SADP decomposition problem. Instead, constraints of feature distance must be

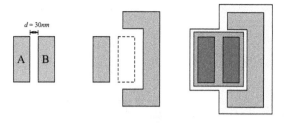

(a) Target feature   (b) Core mask   (c) Sidewall and trim mask

Figure 4: Using auxiliary cores to produce target features.

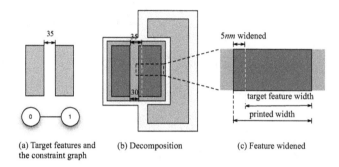

(a) Target features and the constraint graph   (b) Decomposition   (c) Feature widened

Figure 5: Example of widened feature.

included. For demonstration purpose, in the following we assume the $d_{\min} = w_{\min} = 40nm$, and $w_s = 30nm$. Consider the two target features in Figure 5(a), where the distance between them is $35nm$. According to the constraint graph formulation, a graph with two vertices will be constructed. Without loss of generality, assume that the left feature is assigned a core pattern in the coloring solution. Note that we cannot use two separate trim patterns to trim out the features, for otherwise they will be too close and violate the process rules. Thus, we can only use a single trim pattern to cut out both features as shown in Figure 5(b). However, the sidewall generated by the left feature cannot touch the right feature since $w_s = 30nm$. As a result, after trimming the right feature generated will be widened $5nm$ as shown in Figure 5(c). This incorrect result severely damages the original design. We can further conclude that there is no way to decompose this layout and print the features exactly. Hence, we cannot adopt the previous constraint graph formulation as used in LELE. The distance between the features must be carefully considered and reflected in the constraint graph.

Suppose we are now given two features in Figure 4(a). To decompose the layout, we have the following cases depending on the value of d:

1. If $d \geq 2w_s + w_{\min}$, the features are far away from each other. We can use either method to produce the features separately.

2. If $w_{\min} \leq d < 2w_s + w_{\min}$, we cannot use method aux-method for both features. But we can still use method core-method for each feature.

3. If $w_s < d < w_{\min}$, this layout cannot be decomposed, unless the feature can be widened. This is the case we discussed in Figure 5.

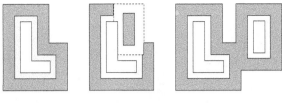

(a) Ring of core   (b) Removing conflicting part   (c) Merging with other auxiliary core

Figure 6: Illustration of ring of core.

4. If $d = w_s$, we can use core-method for one feature while use aux-method for another.

5. If $d < w_s$, this layout cannot be decomposed because neither core-method nor aux-method will work.

Clearly, the distance between features are critically related with w_s, *i.e.*, the width of the sidewall. Given a layout that consists of a set of target features, we construct $G_s = (V, E)$ as follows:

- For each feature in the layout, include a vertex in V.

- If the distance d between two features u and v is exactly w_s, insert an edge $e = (u, v)$ in E. If $d < w_s$ or $w_s < d < w_{\min}$, the layout is not decomposable.

We refer the above graph as *SW-graph*. The following theorem shows the relationship between G_s and SADP decomposition problem:

THEOREM 1. *Two-colorability of G_s is a necessary condition for SADP layout decomposability.*

PROOF. If the feature distances does not satisfy the condition we discussed above, we know the layout cannot be decomposed. Now, we assume the feature distances satisfy the condition. Given a decomposition of the layout, we simply color a vertex as 'core' when a core is assigned to its corresponding feature, and the remaining vertices as 'space'. For each $e = (u, v)$ in G_s, u and v must be colored differently for otherwise the decomposition is invalid. Hence, we obtain a two-coloring solution from the decomposition. As result, two-colorability of G_s is a necessary condition of whether the layout is SADP-decomposable. □

Clearly, building the graph and finding two-coloring solutions for a graph can be done efficiently in polynomial-time. Note that a graph is two-colorable if and only if there is no odd cycle in the graph. In the following, we will refer the two colors of the graph as 'core' and 'space', where the 'core' color denotes a feature that is assigned a main core.

3.4 Our Algorithm

In this section, we detail our algorithm for the SADP decomposition problem. The two-colorability we discussed so far is only a necessary condition for the problem. From a two-coloring solution, we can only obtain a set of non-conflicting core patterns and trim patterns that produce the corresponding features exactly. Since we are interested in overlay-free decompositions, we need to find auxiliary cores for those features that are not assigned core patterns to avoid overlay. Based on the SW-graph formulation, our algorithm

---

Algorithm 1: DecomposeLayout

Input: A set of target features
Output: A set of core patterns and trim patterns

Construct G_s;
Report undecomposable if feature distances are invalid;
foreach *Component c in G_s* **do**
    TwoColor (*c*);
    Report undecomposable if *c* is not two-colorable;
    foreach *Two-coloring solution of c* **do**
        foreach *Feature colored as 'space'* **do**
            Generate a ring of auxiliary core;
            Process the interaction of this core pattern with other core patterns and features;
        end
    end
end
Post process auxiliary cores and trim patterns;
Report failure if cannot meet the design rules;
Combine a set of compatible decompositions as final decomposition and return it;

---

handles the problem in two stages. In the first stage, the decompositions of each connected component in the SW-graph are computed separately. In the second stage, one decomposition from each component will be selected to combine as a complete decomposition of the whole layout. The outline of our algorithm is given in Algorithm 1.

We now introduce how we find decompositions for components in the first stage. Consider a two-coloring solution of a connected component in G_s, while temporarily ignore features and patterns in other components. We assign a main core to a feature that is colored as 'core'. A trim pattern that has the same shape as the feature is also assigned and scaled such that it has a distance w_o away from the feature edges. For each feature that is assigned a 'space', we generate *a ring of auxiliary core*, which is defined as the auxiliary core that has a width w_c and generates the ring of sidewall. Figure 6(a) illustrates the ring of auxiliary core of an 'L' shape feature. Note that the ring of core may interact with other cores and features. We handle the interaction according to the following two cases:

1. The ring of auxiliary core conflicts with some main cores or other features in the same component. Since the positions of the main cores and features are fixed, the *only* workaround is to remove the conflicting parts from this ring. An example is shown in Figure 6(b), where a rectangle core pattern locates near the feature.

2. The ring of auxiliary core overlaps with some other auxiliary cores. In this case, we can safely *merge* them together Figure 6(c), where a auxiliary core pattern locates near the feature.

Note that due to Case 1, the shape of the auxiliary core patterns may be trimmed and no longer satisfy the process rules. If the width of the remaining ring of auxiliary core violates the minimum width rule, we conclude that there is no overlay-free decomposition for this component. This is because removing the auxiliary cores will result in failure to generate the required sidewall it provides. As a result, the whole layout will not have a decomposition. Nevertheless, if the lengths of the remaining auxiliary cores do not satisfy the minimum width rule, we should not remove it. The reason is that we can expand their lengths to meet the constraint later on, or they may merge with other auxiliary cores in other components and become valid. This will be discussed in detail in the second part of the algorithm. Finally, we will join the trim patterns between a pair of conflicting features. As we have seen before, there may be various sets of auxiliary cores that can generate the ring of sidewall for a given feature, and thus there may be many different decompositions. However, we argue that each of these sets must be a subset of the ring of auxiliary core. The reason is that a sidewall pattern can only be generated by an adjacent core pattern, and the ring of auxiliary core is defined using the ring of sidewall. In other words, the ring of auxiliary core is actually the union of all possible decompositions. Note that in the above discussion, we have assumed that the width of the auxiliary cores are w_c. Using the minimum width is already enough to satisfy the process rules. Any set of auxiliary cores using a larger width is just a superset of a set of auxiliary cores that has a minimum width. If the auxiliary core with minimum width does not exist, so will the one that has a larger width. In summary, the ring of auxiliary cores contains all the possible decompositions that correspond to a two-coloring solution. A complete example of two such decompositions for a component is shown in Figure 7.

Note that for each component, we will have exactly two two-coloring solutions. It follows from the above discussion that there are at most two corresponding decomposition. In the second stage of our algorithm, we want to combine the decompositions and form a complete final decomposition. Observe that the interaction between two decompositions are only caused by the auxiliary cores, *i.e.*, the auxiliary cores from one decomposition may (1) conflict with the main cores or features in another decomposition, or (2) overlap with some auxiliary cores in another decomposition. Hence, the 'interaction' between two decompositions is the same as what we have discussed in the first stage. We can thus apply the same process to obtain a combined decomposition. Similar arguments suggest that if some auxiliary cores in the combined decomposition do not satisfy the minimum width constraint, a combined overlay-free decomposition cannot be found from the two decompositions. We now define *compatibility* between two decompositions from two different components. We say two decompositions are *compatible* if they can be merged as a consistent overlay-free decomposition. Otherwise, they are *incompatible*. Our objective is to include exactly one decomposition from each component to form a final decomposition, and all the decompositions selected are mutually compatible. Recall that in Section 1, we have already realized that the number of final decompositions may be huge. Because there are at most two decompositions in a component, and the number of components may be as large as the number of features n, the number of final decompositions will be $O(2^n)$. However, this problem can be solved efficiently, as the following theorem indicates.

THEOREM 2. *Given decompositions of all components of G_s, the final decomposition can be found in polynomial-time.*

PROOF. We show this by reducing the problem to 2SAT (2-Satisfiability), which is a special case of the general Boolean Satisfiability problem. 2SAT asks to determine whether a

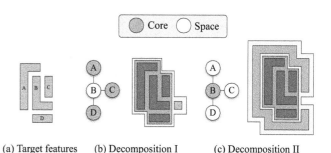

(a) Target features   (b) Decomposition I   (c) Decomposition II

Figure 7: Illustration of the proposed algorithm, where there is a single component in the layout. Note that the cores that are not covered by the core patterns are *auxiliary cores*.

collection of binary variables with paired constraints can be assigned values satisfying all the constraints. Assume that there are k components in G_s, each with at most two decompositions. We construct a boolean formula f as follows. For component i, we denote its two decompositions as x_i and $\bar{x}_i$. If x_i ($\bar{x}_i$) does not exist, include it in f as a disjunction. For each decomposition, we check whether it is compatible with all the decompositions except the one in the same component. Hence, there will be at most $O(n^2)$ incompatible pairs. Disjunct a clause $(x_i \wedge x_j)$ in f if x_i is incompatible with x_j where $i \neq j$. Let $g = \bar{f}$. Clearly, g is in *2-conjunctive normal form* (2CNF). We claim that g is satisfiable if and only if there exists a final decomposition; the satisfiable assignment for g is the final decomposition we want. The argument is straightforward: if there is a satisfiable assignment for g, it is also an unsatisfiable assignment for f, which means all the clause in f must be false. In other words, the decomposition that corresponds to this assignment does not violate any constraint and is thus a compatible final decomposition. Conversely, if we have a final decomposition, where all sub-decompositions in it are compatible, no incompatible pair will be included in f. Hence, this final decomposition corresponds to a satisfiable assignment for g. Finally, there exist efficient algorithms for 2SAT problem such as [3]. This concludes the proof. □

Now we have introduced the two stages of the proposed algorithm. The correctness and complexity of our algorithm is concluded in the following theorem.

THEOREM 3. *The proposed algorithm solves the SADP layout decomposition problem exactly in polynomial time.*

PROOF. In the first stage of the algorithm, we are generating ring of auxiliary cores in the decomposition, which are essentially the combination of all possible decompositions. We only remove the minimal parts that violate the process rules. When the remaining auxiliary cores do not satisfy the process rules, it means all of the possible decompositions are removed due to conflicts. Hence, if there exists a feasible decomposition, it must be contained in the decompositions found so far. In the second stage, the way we define compatible decompositions is the same as in the previous stage. Hence, the consequence is the same, *i.e.*, we will not wrongly eliminate feasible decomposition during the procedure. Finally, it is straightforward to understand that the set of compatible decompositions is a feasible final decomposition of the layout since it does not violate process rule any-

Table 1: Comparison between our method and ILP [17]

Name	# Fs.	# Cs.	Runtime (s) Ours	Runtime (s) ILP
INV_X1	4	2	0.05	7.18
BUF_X1	5	1	0.03	3.45
BUF_X16	5	3	0.04	38.26
NAND2_X1	5	3	0.16	36.87
AND2_X1	6	2	0.07	25.33
AND3_X1	7	3	0.15	48.90
AND4_X1	8	4	0.35	55.90
OR2_X1	6	3	0.20	4.65
OR4_X4	8	2	0.18	5.22
XOR2_X1	7	2	0.13	4.56

where. Hence, the proposed algorithm finds a decomposition if one exists and is thus exact. The complexity analysis of our algorithm is analyzed as follows. Checking the distance of features is essentially collision detection problem, which can be done in $O(n^2)$ using existing computational geometry techniques. Constructing the graph and performing two-coloring can be done in $O(n^2)$ and $O(n)$, respectively. In the first stage, we mainly deal with boolean polygon operations, *e.g.*, intersecting, removing and merging the polygons. All of these can be done in polynomial time. Similar process exists in the second stage. Furthermore, the second stage requires constructing a 2SAT formula from the component. Since the number of components is at most $O(n)$, the pairwise compatibility check requires $O(n^2)$. Finally, the 2SAT problem can be solved using Tarjan's strongly connected component based algorithm [3], which runs in linear time. In conclusion, the proposed algorithm runs in polynomial time. □

4. EXPERIMENTAL RESULTS

To demonstrate the efficiency of our algorithm, we implement our algorithm in C++, and conduct experiments on a 3.20 GHz Intel Xeon CPU with 32GB memory. We test our program with the following process rules. The minimum space between the core (trim) patterns and the minimum width of the core (trim) patterns are all set to be $40nm$, while the sidewall width and trim mask overlay are set to be $30nm$ and $10nm$. We use the $45nm$ Nangate Open Cell Library [2] as our test data. Since the library is not designed for SADP, most of the layouts do not have an overlay-free decomposition. As a result, we scaled and adjusted the cells for our test. In particular, the features are scaled to meet the minimum width requirement $40nm$. We implemented the ILP method in [17] for comparison purpose, and used Gurobi Optimizer [1] as the ILP solver. Table 1 shows the performance comparison between our method and the ILP method [17], where '#Fs.' refers to number of features and '#Cs.' refers to the number of components in the SW-graph. From the table, we can see that our algorithm achieves a 167X speed up on average comparing to the ILP method.

Since the cells in the library only contain a trivial number of features, we also tested our program on larger 2D data sets. The number of features ranges from a hundred to a thousand. Note that ILP cannot scale to solve problems of this size. Figure 8 shows a part of the decomposition generated by our program. In this experiment, we observed that for the undecomposable layouts, the running time of our pro-

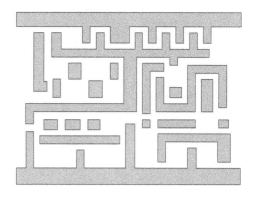

(a) Features

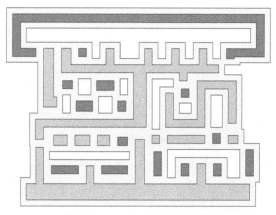

(b) Core mask and sidewalls

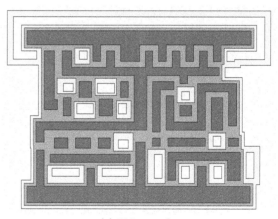

(c) Trim mask

Figure 8: Running our algorithm on general 2D layout. Note that a darker color is used for auxiliary cores to better distinguish from main cores.

gram is within seconds. Notice that there are two possible reasons that a layout cannot be decomposed. The first case is that the feature distance does not satisfy the constraint required for SW-graph. For this type of layout, we can check the decomposability when constructing the graph, and thus can be done very efficiently. Another case is that the feature distances are valid and the SW-graph is two-colorable, however, some of the components in the graph may not have a decomposition, and thus a final decomposition does not exist. This can also be detected efficiently when finding the decompositions for components. For the layouts that are decomposable, the average runtime is 7.25 minutes for layouts that contains one thousand features. Due to time limit, we did not have a very efficient implementation of the algorithm. In particular, we did not implement the computational geometry functions such as intersecting, merging and subtraction by ourselves. Instead, we called a library to perform these operations. However, the overhead in the library significantly slows down the program, since the library is designed for general 2D or 3D geometry but not optimized for rectilinear shapes. The program is expected to run much faster if we implement our own geometry operations and efficient geometry queries. There exists potential performance improvement over the current implementation. We will optimize the program and provide improved results in the journal version of the paper.

5. CONCLUSION

In this paper, we studied the SADP decomposition problem. We proposed a graph formulation that correctly models the problem, and showed that the two-colorability of the SW-graph is a necessary condition for SADP layout decomposability. We proposed the first polynomial-time exact algorithm that solves the SADP decomposition problem. Experimental results show the efficiency of our algorithm. Our algorithm is expected to aid the designers to efficiently check whether their designs are SADP-friendly. Our future research will focus on extension of our algorithm such as hot spot detection and allowing overlay for non-critical edges.

6. ACKNOWLEDGEMENT

This work was partially supported by the National Science Foundation under grant CCF-1017516 and a grant from the semiconductor Research Corporation (SRC).

7. REFERENCES

[1] Gurobi optimizer. http://www.gurobi.com.
[2] Nangate open cell library. http://www.si2.org/openeda.si2.org/projects/nangatelib.
[3] B. Aspvall. A linear-time algorithm for testing the truth of certain quantified boolean formulas. *Inform. Process. Lett.*, 8:121–123, 1979.
[4] Y. Ban, K. Lucas, and D. Z. Pan. Flexible 2d layout decomposition framework for spacer-type double patterning lithography. In *Proc. DAC*, pages 789–794, Jun 2011.
[5] Y. Ban, A. Miloslavsky, K. Lucas, S. Choi, C.-H. Park, and D. Z. Pan. Layout decomposition of self-aligned double patterning for 2d random logic patterning. In *Proc. SPIE*, Jan 2011.
[6] C. Bencher. Sadp: The best option. *Nanochip Technology Journal*, 5(2):1–6, Oct 2007.
[7] Y.-S. Chang, J.-C. Lai, C.-C. Lin, J. Sweis, and J. Yu. Full-area pattern decomposition of self-aligned double patterning for 30nm node nand flash process. *Proc. SPIE*, Jan 2010.
[8] C.-H. Hsu, Y.-W. Chang, and S. R. Nassif. Simultaneous layout migration and decomposition for double patterning technology. In *Proceedings of the 2009 International Conference on Computer-Aided Design*, ICCAD '09, pages 595–600, New York, NY, USA, 2009. ACM.
[9] A. B. Kahng, C.-H. Park, X. Xu, and H. Yao. Layout decomposition for double patterning lithography. In *Proc. ICCAD*, ICCAD '08, pages 465–472, Piscataway, NJ, USA, 2008. IEEE Press.
[10] Y. Ma, J. Sweis, C. Bencher, H. Dai, Y. Chen, J. P. Cain, Y. Deng, J. Kye, and H. J. Levinson. Decomposition strategies for self-aligned double patterning. In *Proc. SPIE*, volume 7641, 76410T, Jan 2010.
[11] M. C. Smayling, C. Bencher, H. D. Chen, H. Dai, and M. P. Duane. Apf pitch-halving for 22nm logic cells using gridded design rules. *Proc. SPIE*, 6925(1):69251E–69251E–8, 2008.
[12] S. Sun, C. Bencher, Y. Chen, H. Dai, M.-P. Cai, J. Jin, P. Blanco, L. Miao, P. Xu, X. Xu, J. Yu, R. Hung, S. Oemardani, O. Chan, C.-P. Chang, , and C. Ngai. Demonstration of 32nm half-pitch electrical testable nand flash patterns using self-aligned double patterning. *Proc. SPIE*, 7274(1):72740D–72740D–7, 2009.
[13] Y. Wei and R. L. Brainard. *Advanced Processes for 193-nm Immersion Lithography*, chapter 9, pages 215–225. SPIE Press Book, 2009.
[14] Y. Xu and C. Chu. Grema: graph reduction based efficient mask assignment for double patterning technology. In *Proceedings of the 2009 International Conference on Computer-Aided Design*, ICCAD '09, pages 601–606, New York, NY, USA, 2009. ACM.
[15] K. Yuan, J.-S. Yang, and D. Pan. Double patterning layout decomposition for simultaneous conflict and stitch minimization. *Computer-Aided Design of Integrated Circuits and Systems, IEEE Transactions on*, 29(2):185 –196, feb. 2010.
[16] H. Zhang, Y. Du, M. Wong, R. Topaloglu, and W. Conley. Effective decomposition algorithm for self-aligned double patterning lithography. In *Proceedings of SPIE*, volume 7973, page 79730J, 2011.
[17] H. Zhang, Y. Du, M. D. F. Wong, and R. O. Topaloglu. Self-aligned double-patterning decomposition for overlay minimization and hot spot detection. In *Proc. DAC*, 2011.

Flexible Self-aligned Double Patterning Aware Detailed Routing with Prescribed Layout Planning

Jhih-Rong Gao and David Z. Pan
ECE Dept. Univ. of Texas at Austin, Austin, TX 78712
{ jrgao, dpan}@cerc.utexas.edu

ABSTRACT

Self-aligned double patterning (SADP) is a promising manufacturing option for sub-22nm technology nodes. Studies have shown that SADP provides better overlay control than traditional litho-etch-litho-etch double patterning. However, the use of stitch is not allowed, which makes layout decomposition for SADP more difficult. It is necessary to find a new solution to handle pattern conflicts and consider SADP in earlier stages. In this paper, we propose a novel multi-layer SADP-aware detailed routing with prescribed layout planning. Our method is based on a correct-by-construction approach to take SADP compliancy into account during routing, and to achieve layout decomposition simultaneously. The experimental result shows that the proposed approach consistently achieves SADP-compliant solutions on both single-layer and multi-layer designs.

Categories and Subject Descriptors

B.7.2 [**Hardware, Integrated Circuit**]: Design Aids - Placement and Routing

General Terms

Algorithms, Design, Performance

Keywords

Detailed Routing, Double Patterning, SADP, Design for Manufacturability

1. INTRODUCTION

Due to the delay in the next generation lithography technology such as Extreme Ultra Violate (EUV) [1], the manufacturing industry still relies on a 193nm (ArF) wavelength light source. As technology continues to scale to 22nm and 14nm, semiconductor manufacturing with ArF is greatly challenging because the required half pitch size is beyond the resolution limit of ArF. Double Patterning Lithography (DPL) has been a promising solution for 22nm/14nm node volume production.

The working principal of DPL is to decompose dense layout patterns into two masks. The decomposition process is referred to as coloring. Since each mask contains sparse patterns with doubled spacing, the lithography resolution can be improved. There are two main DPL schemes in current IC manufacturing: litho-etch-litho-etch (LELE) double patterning, and self-aligned double patterning (SADP). LELE consists of two exposure and two etch processes [2–4], and it allows stitch insertion to resolve the conflict after pattern decomposition. Stitch is used to split a pattern into two masks, which makes patterns even sensitive to process variation. Fig. 1 shows how stitch insertion resolves conflicts. Possible variations may occur at the stitch inserting point, such as line-end shortening, CD shrinking [4], and overlay error due to two consecutive mask exposure processes. Several layout decomposition methods [3–6] have been proposed for LELE DPL. However, the alignment and magnification errors on the second mask exposure cause LELE to induce significant pattern overlay error [7] and thus degrade yield rate.

SADP is similar to LELE, which also requires layout decomposition into two masks, core mask and trim mask. For the mandrel pattern that is lithographically defined on the core mask, sidewall spacers are applied on each side to effectively double the pattern density. The trim mask is then used to remove unnecessary patterns. Because the most critical patterning control in SADP is not governed by lithography, but by the deposition of the sidewall spacer, it has less overlay error and excellent variability control compared to LELE [8]. However there is no stitch allowed in SADP to resolve conflicts, making its layout decomposition difficult. Moreover, SADP layout decomposition is not intuitive in the sense that the decomposition result does not have a direct relation to the original layout. SADP requires assist mandrels [9] during its patterning process and these unwanted mandrels need to be trimmed out by the trim mask. There-

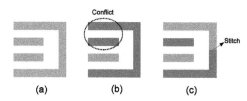

Figure 1: Conflict and stitch in LELE. (a) Target layout. (b) A conflict occurs after layout decomposition. (c) The conflict is resolved by splitting a pattern with stitch insertion.

fore, the core mask and the trim mask cannot be obtained simply from the target layout. For 2D patterns, this mask assignment process would be more complicated.

Recently, SADP has attracted more and more interest because of its good overlay controllability. Starting from 1D pattern decomposition [10], the flexibility of SADP is further extended to 2D random logic patterns [11, 12]. The layout decomposition problem is solved with an ILP formulation [13] to minimize overlay. A lithography friendly algorithm to automatically generate core pattern, trim patterns, and assist core patterns for better manufacturability is proposed in [9].

Most existing SADP-related works focus on post layout optimization. As described above, the prohibition against using stitch makes SADP decomposition extremely challenging. Therefore, it is necessary to consider SADP in earlier design stages, especially in detailed routing, to guarantee high layout decomposability. Double patterning friendly routing has been proposed in [14, 15], but their methods cannot be applied to solve SADP induced issues. A SADP-friendly detailed routing flow [16] is presented by performing detailed routing and layout decomposition concurrently. However, [16] simply works for single-layer designs and does not provide a solution for resolving layout decomposition conflicts.

In this paper we propose a robust multi-layer SADP-aware detailed routing algorithm which includes the following features:

- We propose a novel SADP-aware detailed routing approach that can handle 2D patterns on multi-layer designs in the presence of obstacles.

- We solve routing and layout decomposition simultaneously based on the correct-by-construction approach.

- We incorporate layer assignment to resolve potential pattern conflicts, which increases the flexibility of layout decomposition for SADP.

- We present a set of SADP-aware routing guidelines, which helps improve the pattern quality of SADP.

The rest of the paper is organized as follows. Section 2 gives preliminary information on SADP technology and our work. Our prescribed layout planning techniques are explained in Section 3. The details of the proposed SADP-aware routing framework are presented in Section 4. The experimental results are discussed in Section 5, followed by the conclusion in Section 6.

2. PRELIMINARIES

Two main process flows are available for SADP, positive tone process and negative tone process. These two processes are analyzed in [8] which suggested positive tone process is preferred due to its more cost effective and better controllability of overlay. Fig. 2 shows the process flow of the positive tone SADP. Assume (a) is the target layout, it can be generated by the process from (b) - (e). First of all, the core mask shown in blue in (b) is derived by selecting a subset of target layout patterns and assist patterns. The patterns on the core mask are called main mandrel and are generated through a lithography process. Then the sidewall spacers are deposited into both sides of the main mandrel as shown

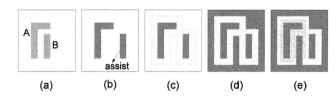

Figure 2: Flow of positive tone SADP. (a) Target layout. (b) Core mask. (c) Spacer deposition. (d) Substrate material filling. (e) Trim mask.

in (c). Next, main mandrels are removed and the non-spacer region is filled with substrate materials. In the final step (e), the trim mask shown in red is applied to remove unnecessary patterns and retain the desired patterns.

Similar to LELE, SADP requires layout decomposition to separate patterns into two subsets. The decomposition result is usually obtained by performing coloring on all patterns. One subset can be generated by directly assign the patterns into core mask. While the other has to be generated by forming a trim mask which can retain the patterns on the final layout. In this paper, we define patterns that are printed by core mask as *mandrel patterns*; and patterns that are printed by the assistant of trim mask as *trim patterns*. A route can be assigned as either mandrel pattern or trim pattern. For example, Fig. 2(e) shows the case when pattern A in (a) is assigned as mandrel pattern and B as trim pattern.

Our approach adopts the grid-based routing model. Because multi-layer designs are taken into consideration, a three-dimensional grid graph is constructed. Our routing not only searches solutions on single metal layer, but also allows solutions crossing multiple layers through vias. Each pin is mapped to one grid, and a routing solution of a multi-pin net is composed of grids connecting all of its pins. Here we define the minimum width of mandrel pattern and trim pattern as W_{min}. We also assume that the width of sidewall spacer equals to W_{min}. A minimum spacing S_{min} must be kept between any neighboring routing patterns. For a legal layout decomposition result, patterns within the minimum DPL spacing S_{dp} must be assigned to different types of pattern (mandrel pattern or trim pattern).

3. PRESCRIBED LAYOUT PLANNING FOR SADP COMPLIANCY

Our objective is to achieve better SADP compliancy by performing routing and SADP layout decomposition simultaneously. As a result, the routing solutions are able to take advantage of SADP's good overlay control. In this section, we present SADP-friendly routing guidelines to improve pattern quality and reduce decomposition conflicts.

3.1 SADP-aware routing guidelines

Mandrel patterns and trim patterns are fabricated by different manufacturing processes. The interaction between these two types of patterns may affect the printing images. Therefore, simply determining whether a layout is decomposable is not adequate for SADP-friendly routing. We analyze the impact of different pattern assignments on the pattern quality. The following three layout planning guidelines provide a systematic procedure to construct a SADP-friendly routing. Incorporating these guideline into our rout-

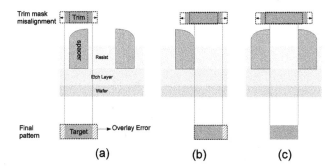

Figure 3: Overlay error due to trim mask misalignment. (a) No feature boundary aligned to spacer. (b) One feature boundary aligned to spacer. (c) Both feature boundaries aligned to spacer.

ing framework enables us to take advantage of SADP technology.

1. If both mandrel pattern and trim pattern are conflict-free when being assigned to a route, the mandrel pattern is preferred.

2. If the candidate routes have the same routing cost and can only be assigned as trim patterns, the route with more spacer protection is preferred.

3. The distance between a trim pattern and a mandrel pattern is suggested to be larger than the forbidden spacing S_{forb}; although a valid routing solution only requires the minimum spacing $S_{dp} < S_{forb}$ to be satisfied.

These guidelines are explained below. The simulation result in [17] observes the printability degradation for the second mask lithography due to the presence of topography generated from the first mask on the wafer. One degradation can be seen from the CD variation, where patterns from the second lithography tend to have wider width when there are underlying patterns from the first litho/etch step. As a result, SADP prefers mandrel patterns from the first lithography for better printability control, and is different from LELE which prefers two balanced subsets of patterns [4].

Another advantage of SADP is the use of spacer. Three decomposition cases showing how a trim mask can be formed [8] are presented in Fig. 3. The grey rectangle represents the target pattern that will be generated by the trim mask shown in red. In (a), a wide line is formed by the trim mask overlapping both sides of a spacer. Consequently, the trim mask misalignment may affect both the left and right boundary of the printing image. The possible overlay errors on the final pattern are shown in the slash area. In (b), a line is formed by the trim mask side-lapped with one spacer, causing overlay errors only on one side of the printing image. If both sides of the target pattern are aligned to spacers, as shown in (c), the overlay error can be totally avoided. Given this auto-alignment property of the spacer, a trim pattern protected by multiple spacers is preferred.

The minimal spacing S_{dp} in DPL constraints the minimum allowable distance between any two identical type patterns. A conflict occurs if two patterns within S_{dp} are assigned to the same mask. In addition, a forbidden spacing needs to be considered. The simulation results in [16] show that the printed image of a trim pattern would be affected by a close

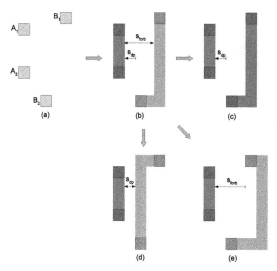

Figure 4: Prescribed layout planning. (a) Unrouted nets. (b) Legal patterns with bad quality. (c) - (e) Improved patterns by our prescribed layout planning.

mandrel pattern even if S_{dp} is satisfied. In contrast, the quality of a trim pattern can be improved if its neighboring mandrel patterns are kept at a sufficient distance. Therefore, we define a forbidden spacing $S_{forb} > S_{dp}$ such that any distance $d_{mt} < S_{forb}$ is discouraged, where d_{mt} denotes the distance between a neighboring trim and mandrel pattern.

These layout planning techniques work as prescriptions for our routing engine to generate SADP-compliant layout patterns and to prevent patterns with bad quality. The example in Fig. 4 shows how routing patterns can be improved by our approach. The pin locations are given in (a) for two unrouted nets, and (b) is one routing and layout decomposition solution without considering SADP. Mandrel pattern is shown in blue and trim pattern is shown in red in our following explanation. Although (b) is a legal solution by satisfying S_{dp} constraint, the mandrel pattern and the trim pattern may affect each other because their distance are within S_{forb}. Three alternative solutions with better pattern quality are shown in (c) - (e); where (c) adopts more mandrel patterns; (d) acquires more spacer protection; and (e) enlarges the distance between neighboring mandrel and trim patterns.

3.2 Simultaneous layer assignment for conflict prevention

The biggest challenge of SADP is the prohibition against using stitches. For a route $path_1$ on a single layer, either all grids in $path_1$ are assigned as mandrel patterns or all are assigned as trim patterns. This limitation dramatically decreases the possibility of generating a decomposable layout for SADP. In order to increase the flexibility of SADP layout decomposition, we perform simultaneous layer assignment during routing. In contrast to single-layer layout decomposition, multi-layer layout decomposition allows patterns to be assigned independently if they are on different layers. For example, a route $path_2$ is composed of seg_1- via_{12}-seg_2, where seg_1 is on metal 1, seg_2 is on metal 2, and via_{12} is used to connect seg_1 and seg_2. Since seg_1 and seg_2 are on different metal layers, they can be decomposed independently without introducing any conflict. Via can be viewed

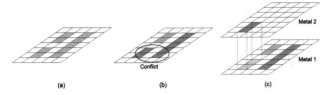

Figure 5: Prevent conflicts by simultaneous layer assignment. (a) Target layout. (b) Conflict occurs in single-layer layout decomposition. (c) Conflict removed by proper layer assignment.

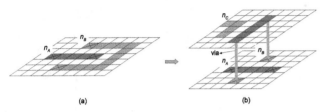

Figure 6: Increase flexibility of layout decomposition by simultaneous layer assignment. (a) Single layer. (b) Multiple layers.

as a splitting point similar to the function of stitch in LELE layout decomposition.

Performing layer assignment during layout decomposition on multi-layer designs has two advantages for SADP compliancy. First, a conflict can be easily resolved by assigning conflicting patterns into different metal layers. Fig. 5 shows a conflict that is solvable by our simultaneous layer assignment. Fig. 5(a) is the target layout that needs to be printed by DPL. A conflict occurs after single-layer layout decomposition in (b). By properly assign the patterns to different layers as shown in (c), the conflict can be prevented. The second advantage of considering layer assignment is that it increases the flexibility of layout decomposition. Fig. 6(a) shows an example after routing and layout decomposition on a single layer. Net n_B needs to detour to prevent intersecting net n_A. By assigning a section of patterns on n_B to an upper layer as shown in (b), wirelength is reduced. Besides, the patterns on different layers are not restricted to a single color. In Fig. 6(b), patterns of n_B on metal 1 is assigned as trim patterns (shown in red); while the pattern on metal 2 is assigned as mandrel pattern (shown in blue) to provide spacer protection for the routed net n_C and to prevent conflicts.

The simultaneous layer assignment technique increases the solution space of both routing and layout decomposition, and thus helps prevent conflicts. This layer assignment is integrated into our three-dimensional path finding process, which will be explained in the next section.

4. MULTI-LAYER SADP-AWARE DETAILED ROUTING

This section gives the detail of our proposed routing framework. We first introduce the overall flow, and then present the techniques incorporated in the flow.

4.1 Overall flow

We adopt a correct-by-construction approach to build our routing flow. When a net is routed, its layout decomposition

---

Algorithm 1 SADP-aware detailed routing

Input: A set of blockages B, and a set of nets N
1: Layout decomposition for B
2: $Q \leftarrow$ An arbitrary net $n_{begin} \in N$
3: **while** !$Q.empty()$ **do**
4:    $n \leftarrow Q.pop()$
5:    **for** each 2-pin net $k \in n$ **do**
6:       Three-dimensional A* search for k
7:    **end for**
8:    **for** each $n_{neighbor} \in N$ where bbox of $n_{neighbor}$ overlaps bbox of n **do**
9:       $Q \leftarrow Q + n_{neighbor}$
10:    **end for**
11: **end while**

Figure 7: SADP-aware detailed routing.

is done simultaneously. During path finding, a rule checking procedure ensure not only a route is legal but also its patterns are decomposable. Consequently, once the routing is done, its layout decomposition result is also obtained.

Algorithm 1 in Fig. 7 describes the overall flow of our approach. First, we perform initial layout decomposition for the blockages composed of pre-routed nets. Since pre-routed nets in this stage are usually sparse, most would be assigned as mandrel patterns according to Guideline 1. Next, we process the input nets sequentially according to the routing order determined in line 8-9 (Section 4.3). Each multiple-pin net is decomposed into 2-pin nets and then routed using our three-dimensional A* search in line 5-7 (Section 4.4). The routing cost in A* search is a combination of wirelength and SADP cost, which will be illustrated in Section 4.2. After the A* search, the pattern assignment with lowest cost will be chosen.

4.2 SADP-aware weighted cost

When performing A* search, the cost of routing on a grid is calculated. Suppose an edge connecting grid g_i to g_j is considered, the cost of routing g_j as a mandrel and as a trim pattern is defined as follows:

$$\begin{cases} cost_j(m) &= cost_i(m) + \alpha \cdot WL_{ij} + \beta \cdot SADPC_j(m) \\ cost_j(t) &= cost_i(t) + \alpha \cdot WL_{ij} + \beta \cdot SADPC_j(t) \end{cases}$$
(1)

if g_i and g_j are on the same layer.

$$\begin{cases} cost_j(m) &= \min\{cost_i(m), cost_i(t)\} + \\ & \alpha \cdot WL_{ij} + \gamma \cdot VIA + \beta \cdot SADPC_j(m) \\ cost_j(t) &= \min\{cost_i(m), cost_i(t)\} + \\ & \alpha \cdot WL_{ij} + \gamma \cdot VIA + \beta \cdot SADPC_j(t) \end{cases}$$
(2)

if g_i and g_j are on different layers.

The pre-calculated cost $cost_i(m)$ and $cost_i(t)$ represent the cost when g_i is assigned as a mandrel pattern and a trim pattern, respectively; WL_{ij} is the wirelength between neighboring grids g_i and g_j; VIA is the via cost and $SADPC$ can be either positive or negative to represent a bad or good impact on pattern quality, respectively. User-defined parameters α, β and γ adjust the weight between wirelength and SADP awareness. As mentioned previously, stitch is not allowed in SADP. Therefore, g_j must be assigned as the same pattern of g_i if they are on the same layer, just as defined in Equation 1. When multi-layer designs are involved, more optimization options are available. Therefore Equation 2 provides more solution space when searching on multiple layers.

$SADPC$ is the double patterning cost when a grid is assigned as a mandrel/trim pattern and is determined by the guidelines provided in Section 3.1,which is defined as follows:

$$SADPC = \begin{cases} C_{mandrel} \\ C_{trim} \end{cases} - m \cdot C_{spr} + n \cdot C_{forb} \quad (3)$$

$C_{mandrel}$ and C_{trim} are the unit cost of assigning a grid as a mandrel or a trim pattern, respectively. The weight of $C_{mandrel}$ is set to be less than the weight of C_{trim} according to Guideline 1 such that more mandrel patterns will be used. C_{spr} represents the benefit of a self-aligned spacer and thus it reduces the total $SADPC$ according to Guideline 2. The number of newly generated spacer-protected grids m can be optimized by routing more mandrel patterns next to existing trim patterns, or routing more trim patterns next to existing mandrel patterns. C_{forb} represents the penalty for patterns violating W_{forb} according to Guideline 3. Similar to m, n is the total number of newly generated forbidden grids by the current routing path. Note that violating W_{forb} is not encouraged, but it is valid for double patterning.

In general, the weight of these SADP costs differs depending on the technology. However, we may adjust the weight according to the routing density. For example, a larger C_{spr} encourages the binding of mandrel and trim patterns, and thus helps generate a tighter layout. In contrast, larger C_{forb} encourages a detour to prevent violating forbidden spacing, and thus consumes more routing resources. In our experiments, we set $C_{mandrel}=C_{spr}=C_{forb}$ and $C_{trim}=2C_{mandrel}$.

4.3 Neighborhood-based net ordering

How a routing algorithm explores its solution defines how important net ordering is. For an ILP-based algorithm, solutions are calculated currently, thus net ordering is unnecessary. However, ILP-based algorithms usually have high runtime overhead. On the other hand, a sequential routing algorithm that processes nets one by one relies on a good net ordering method. The better the net ordering is, the less rip-up and reroute are required and the less the runtime is needed. According to the cost function defined in Section 4.2, a preferred routing path should keep a low wirelength and has more spacer-protected grids. Fig. 8 shows the comparison of a bad and a good net ordering. In (a), net n_A is routed first and then n_B is routed. The bold line in the grid boundary shows where the grid boundary is protected. The net order of Fig. 8(b) is contrary to (a). We can see that with the same wirelength, the solution in (b) obtains much more spacer protection.

To achieve SADP-friendly net ordering, we propose an ordering method based on the geographic relation among nets. First, an arbitrary net n_i is selected to be routed. After n_i is routed, we obtain the next net to be routed n_j by finding every $bbox_{n_j}$ overlapping $bbox_{n_i}$. Here $bbox_n$ is determined by enlarging the net bounding box by a specific width w_{enl}. This ordering method encourages nets within a certain distance to be routed in a sequence, so that the probability to provide spacer protection for these neighboring nets can be increased. In our implementation, we set w_{enl} slightly larger than S_{forb} so that the enlarged area is sufficient but not causes too much computational burden.

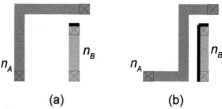

Figure 8: Net ordering impact on pattern quality. Bolder lines show grid boundaries that are protected by spacers. (a) Net n_a is routed first. (b) Net n_b is routed first.

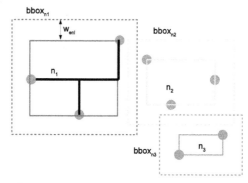

Figure 9: Neighborhood-based net ordering. n_2 allows more spacer protection to be provided for n_1.

Fig. 9 shows an example of neighborhood-based net ordering. In the beginning, net n_1 is routed and the next routing net will be determined. It can be seen that $bbox_{n_2}$ overlaps $bbox_{n_1}$ and thus n_2 will be routed next. Finally, n_3 will be routed because its $bbox_{n_3}$ overlaps $bbox_{n_2}$.

Because the searching for overlapping $bbox$ needs to be done whenever a net is routed, it is important to reduce the overhead of this search. We adopt R-tree [18] for fast indexing $bbox$ information.

4.4 Efficient three-dimensional path finding by dynamic programming

During path finding, when a routing grid g is considered, the validity of assigning g as a mandrel pattern (blue) and a trim pattern (red) is checked simultaneously. The combined routing and layout decomposition result is denoted as $R(path, LD(path))$, where $path$ is the routing path composed of grids, and $LD(path)$ is the coloring result for $path$. If a solution candidate $R(path_1, LD(path_1))$ generates any conflict, a high routing cost defined in Section 4.2 would be applied to prevent this candidate being selected.

The solution space for $R(path_i, LD(path_i))$ $\forall i$ in single-layer SADP is limited because all grids $g_j \in path_i$ must be assigned as the same color. However, the solution space on multi-layer designs would be much larger. As discussed in Section 3.2, simultaneous layer assignment with routing enables more flexible layout decomposition. Therefore, we adopt a three-dimensional path finding so that layer assignment can be integrated into the routing process. Fig. 10 shows a routing path connecting pins p_1 and p_2. Because the path is composed of three independent segments, seg_1, seg_2, seg_3, which are connected by vias, each segment is flexible to be assigned as either a mandrel or a trim pattern. It can be seen that in total 8 candidate solutions are available for the case in Fig. 10.

The time and space complexity would be an issue if we

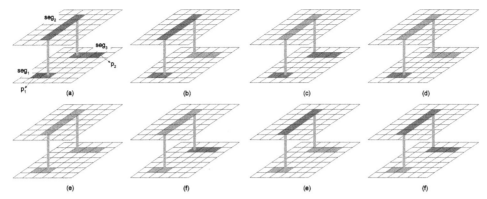

Figure 10: Solution candidates for multi-layer SADP.

simply explore all possible solutions during three-dimensional path finding. We find that, in fact, it is not necessary to maintain all combination of $R(path_i, LD(path_i))$ during simultaneous routing and coloring. Given this observation, we develop an efficient three-dimensional path finding based on dynamic programming.

Assume a grid g_i is considered to be routed by a 2-pin net $n(g_s, g_t)$ where g_s and g_t are the source and sink pins, respectively. We first evaluate the costs of assigning g_i as a mandrel pattern and as a trim pattern. According to the definition in Equation 1 and 2, we then obtain the accumulated cost along the path from g_s to g_i. Although there are many solution candidates for the routing path through g_i, we only need to maintain two solutions, $cost_i(m)$ and $cost_i(t)$, where $cost_i(m)$ and $cost_i(t)$ are the accumulated routing costs when g_i is assigned as a mandrel pattern and a trim pattern, respectively. By keeping the minimum $cost_i(m)$ and $cost_i(t)$ in $path_{s,i}$ for each traversed grid g_i, we are guaranteed to obtain the minimum cost solution for $path_{s,t}$. The solution for the routing path of $n(g_s, g_t)$ can be expressed as the following recursive form of dynamic programming:

$$R(path_{s,t}, LD(path_{s,t})) =$$
$$R(path_{s,i}, LD(path_{s,i})) + R(path_{i,t}, LD(path_{i,t}))$$
$$(4)$$

, for any g_i in the routing grid

According to Equation 4, we only need to maintained two minimum cost solutions $cost_i(m)$ and $cost_i(t)$ for any grid g_i traversed during A* search. This makes our three-dimensional path finding more efficient on both time and space.

5. EXPERIMENTAL RESULTS

We implemented the proposed algorithm in C++ and tested it on the machine with 2.66GHz CPU and 4G memory. The parameters in Equation 1 and Equation 2 are set as follows: $\alpha = \beta = 1$ and $\gamma = 0.3$. Two experiments test the performance and robustness of our approach. The first experiments contains only single-layer and obstacle-free designs, while the second experiment includes multi-layer designs in the presence of obstacles. For single-layer design, the method in [16] is implemented and compared with our approach. For multi-layer designs, our results are compared with a wirelength-driven routing method.

First we compare our result with [16] which simply works for single-layer designs. Because [16] also adopts A* search technique, we are able to incorporate its cost function into our routing flow. However, due to the unavailability of the

Table 2: Benchmark statistics for multi-layer designs.

Circuit	Size(um^2)	#Nets	#Blockages		
			M1	M2	Tot
CK1	20x20	29	279	26	305
CK2	48x48	306	3528	210	3699
CK3	100x100	872	13207	766	13813
CK4	160x160	1937	38792	2029	40370

benchmark in [16], we randomly generate test cases to perform the comparison. Four cases are generated with different number of nets as shown in Table 1. Note that the layout size of these cases is the same; in which Case1 has the lowest routing density while Case4 has the highest routing density. We compare the result in terms of wirelength (WL) and double patterning performance including (1) the number of spacer-protected trim patterns (#SP-trim), (2) the number of non-spacer-protected trim patterns (#NSP-trim), (3) the number of forbidden grids (#FORB grid), and (4) the number of conflicts (#conflict). The result shows our approach consistently generates better pattern quality with only a 3% wirelength increase. On average, our result generates 51% more spacer-protected trim patterns than [16], in which spacer protection implies better pattern quality. In addition, we reduce the number of non-spacer-protected trim patterns and forbidden grids by 39% and 55%, respectively.

We then test the performance of our approach on multi-layer designs in the presence of blockages. Since there is no previous routing work taking double patterning into consideration on multi-layer designs, we implement a multi-layer wirelength-driven routing method followed by SADP layout decomposition as our comparison baseline. A set of two-layer industrial designs are scaled down to 22nm technology for the experiment. Table 2 gives the statistics of these designs. Each design contains two metal layers, M1 and M2, and blockages appear on both layers. Table 3 shows the comparison between our approach and the wirelength-driven routing in terms of wirelength, the number of vias (#Via), double patterning performance and runtime. Our approach achieves a great improvement in the results of double patterning. On average, the number of spacer-protected trim is increased by 2.87X; and the number of non-spacer-protected trim patterns and forbidden grids are reduced by 31% and 49%, respectively. The runtime of WL-driven is less than our approach because it does not perform any decomposability

Table 1: Result comparison with [16] on single-layer designs.

Testcase	#Nets	Router	WL	Double Patterning			
				#SP-trim	# NSP-trim	# FORB grid	#conflict
Case1	300	[16]	3770	28	63	3	0
		Ours	3820	40	33	0	0
Case2	600	[16]	7258	209	346	26	0
		Ours	7330	250	216	12	0
Case3	800	[16]	9704	427	727	48	0
		Ours	10130	725	464	25	0
Case4	1000	[16]	12171	750	1107	122	0
		Ours	12929	1291	702	101	0
Avg Ratio			1.03	1.51	0.61	0.45	1

Table 3: Result comparison of routing and layout decomposition on multi-layer designs.

Circuit	Router	WL	#Via	Double Patterning				Runtime(s)
				#SP-trim	# NSP-trim	# FORB grid	#conflict	
CK1	WL-driven	22911	48	320	262	13	22	2.3
	Ours	23045	60	480	179	3	15	6.6
CK2	WL-driven	126215	616	2397	5248	794	251	37.8
	Ours	133893	906	9397	3539	518	136	208
CK3	WL-driven	530555	1788	6222	10588	1772	757	190.8
	Ours	536215	2292	18162	7491	923	290	1021.6
CK4	WL-driven	1269046	4484	13740	25005	4375	1682	556.2
	Ours	1297775	5708	43238	17587	2787	670	2802.5
Avg Ratio		1.02	1.32	2.87	0.69	0.51	0.50	4.69

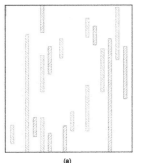 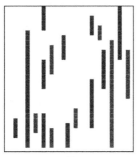

Figure 11: Sample Layout decomposition result by (a) [9] and (b) our approach.

checking. It is worth mentioning that the benchmarks are quite dense and some areas contain congested pins which are difficult for double patterning technology. Table 3 also shows that unresolvable conflicts exist in both of our result and wirelength-driven result, which may be fixed by post-routing techniques. Our approach outperforms wirelength-driven routing with fewer conflicts. The number of vias is increased by 32% because we utilize layer assignment to prevent conflicts and to improve the pattern quality.

Fig. 11 shows a 1D layout generated by SADP-friendly layout decomposition [9] and our approach. Our result tends to generate more mandrel patterns and reduces the number of non-spacer-protected trim patterns, which implies our result obtains better pattern quality according to the proposed routing guidelines.

Overall, our approach consistently achieves SADP-compliant results with negligible wirelength overhead. We provide more flexibility on layout decomposition by taking layer assignment into consideration. In addition, our prescribed lay-out planning techniques greatly improve the pattern quality and thus can benefit lithography manufacturing for SADP.

6. CONCLUSION

In this paper, we propose a novel multi-layer SADP-aware detailed routing approach. A set of SADP-aware routing guidelines are presented, which improves SADP compliancy. We adopt a multi-layer routing model and present simultaneous layer assignment to increase the flexibility of SADP layout decomposition. Our work simultaneously solves routing and layout decomposition problems using a correct-by-construction methodology. The experimental results show that the proposed approach achieves promising results on both single-layer and multi-layer designs.

7. ACKNOWLEDGEMENTS

This work is supported in part by NSF, Oracle, and NSFC.

8. REFERENCES

[1] Jo Finders, Micrea Dusa, and Stephen Hsu. Double patterning lithography: The bridge between low k1 ArF and EUV. *Microlithography World*, February 2008.

[2] George E. Bailey, Alexander Tritchkov, Jea-Woo Park, Le Hong, Vincent Wiaux, Eric Hendrickx, Staf Verhaegen, Peng Xie, and Janko Versluijs. Double pattern EDA solutions for 32nm HP and beyond. In *Proc. of SPIE*, volume 6521, 2007.

[3] A. B Kahng, C.-H. Park, X. Xu, and Hailong Yao. Layout decomposition for double patterning lithography. In *Proc. Int. Conf. on Computer Aided Design*, November 2008.

[4] Jae-Seok Yang, K. Lu, Minsik Cho, Kun Yuan, and D. Z. Pan. A new graph-theoretic, multi-objective layout decomposition framework for double patterning lithography. In *Proc. Asia and South Pacific Design Automation Conf.*, January 2010.

[5] Szu-Yu Chen and Yao-Wen Chang. Native-conflict-aware wire perturbation for double patterning technology. In *Proc. Int. Conf. on Computer Aided Design*, November 2010.

[6] Kun Yuan and D. Z. Pan. WISDOM: wire spreading enhanced decomposition of masks in double patterning lithography. In *Proc. Int. Conf. on Computer Aided Design*, November 2010.

[7] Martin Drapeau, Vincent Wiaux, Eric Hendrickx, Staf Verhaegen, and Takahiro Machida. Double patterning design split implementation and validation for the 32nm node. In *Proc. of SPIE*, volume 6521, 2007.

[8] Yuansheng Ma, Jason Sweis, Chris Bencher, Huixiong Dai, Yongmei Chen, Jason P. Cain, Yunfei Deng, Jongwook Kye, and Harry J. Levinson. Decomposition strategies for self-aligned double patterning. In *Proc. of SPIE*, 2010.

[9] Yongchan Ban, K. Lucas, and D. Z. Pan. Flexible 2D layout decomposition framework for spacer-type double patterning lithography. In *Proc. Design Automation Conf.*, June 2011.

[10] Michael C. Smayling, Christopher Bencher, Hao D. Chen, Huixiong Dai, and Michael P. Duane. APF pitch-halving for 22nm logic cells using gridded design rules. In *Proc. of SPIE*, volume 6925, 2008.

[11] Yayi Wei and Robert L. Brainard. Advanced processes for 193-nm immersion lithography. In *Proc. of SPIE*, February 2009.

[12] Yue Xu and C. Chu. GREMA: graph reduction based efficient mask assignment for double patterning technology. In *Proc. Int. Conf. on Computer Aided Design*, November 2009.

[13] Hongbo Zhang, Yuelin Du, Martin D. F. Wong, and Rasit Topaloglu. Self-aligned double patterning decomposition for overlay minimization and hot spot detection. In *Proc. Design Automation Conf.*, 2011.

[14] Minsik Cho, Yongchan Ban, and D. Z. Pan. Double patterning technology friendly detailed routing. In *Proc. Int. Conf. on Computer Aided Design*, November 2008.

[15] Kun Yuan, K. Lu, and D. Z. Pan. Double patterning lithography friendly detailed routing with redundant via consideration. In *Proc. Design Automation Conf.*, July 2009.

[16] Minoo Mirsaeedi, J. Andres Torres, and Mohab Anis. Self-aligned double-patterning (SADP) friendly detailed routing. In *Proc. of SPIE*, 2011.

[17] Kevin Lucas, Christopher Cork, Alexander Miloslavsky, Gerard Luk-Pat, Levi Barnes, John Hapli, John Lewellen, Greg Rollins, Vincent Wiaux, and Staf Verhaegen. Double-patterning interactions with wafer processing, optical proximity correction, and physical design flows. *Journal of Micro/Nanolithography, MEMS and MOEMS*, 8, 2009.

[18] A. Guttma. R-Trees: a dynamic index structure for spatial searching. In *Proc. SIGMOD Int. Conf. on Management of data*, 1984.

3D CMOS-Memristor Hybrid Circuits:
Devices, Integration, Architecture, and Applications

Kwang-Ting Cheng and Dmitri B. Strukov
Department of Electrical and Computer Engineering
University of California, Santa Barbara, CA 93106
{timcheng, strukov}@ece.ucsb.edu

ABSTRACT

In this paper, we give an overview of our recent research efforts on monolithic 3D integration of CMOS and memristive nanodevices. These hybrid circuits combine a CMOS subsystem with several layers of nanowire crossbars, consisting of arrays of two-terminal memristors, all connected by an *area-distributed interface* between the CMOS subsystem and the crossbars. This approach combines the advantages of CMOS technology, including its high flexibility, functionality and yield, with the extremely high density of nanowires, nanodevices and interface vias. As a result, the 3D hybrids can overcome limitations pertinent to other 3D integration techniques (such as through-silicon vias) and enable 3D circuits with unprecedented memory density (up to 10^{14} bits on a single 1-cm^2 chip) and aggregate interlayer communication bandwidth (up to 10^{18} bits per second per cm^2) at manageable power dissipation. Such performance represents a significant step towards addressing the most pressing needs of modern compact electronic systems.

Categories and Subject Descriptors

B.3.m [**Memory Structures**]: Miscellaneous. B.7.1 [**Integrated Circuits**]: Types and Design Styles.

General Terms

Design, Reliability.

Keywords

3D Integration; Memristors.

1. Introduction

Three-dimensional (3D) circuits are a natural way of increasing integration density to overcome the inevitable limitations in the lateral scaling of electron devices [1] . In comparison with 2D planar circuits, 3D integrated circuits (ICs) offer the potential benefits of better performance, higher connectivity, reduced interconnect delays, lower power consumption, better space utilization, and more flexible heterogeneous integration [2] [3] [4] . Applications

that can directly benefit from 3D integration [5] include those demanding significant amounts of memory access (such as imaging, networking, and computing) as well as those whose performance is dominated by interconnects (such as switches and FPGAs [6]). As both military and industry are moving toward multi-core, multi-threaded, and multi-media applications, the demand for lower latency and higher bandwidth between computing elements and memory is growing fast.

Beyond these immediate needs, there is tremendous potential in the field of neuromorphic circuits. Indeed, in many cases, conventional signal processing using digital architectures are inadequate for real-time applications such as pattern classification (including threat detection and face recognition). Neuromorphic architectures may be a viable solution to address challenges associated with inherent requirements for massive, parallel information processing, provided that dense integration between memory and processing elements can be obtained.

In order to realize the true potential of 3D integration, vertical stacking must maintain a sufficiently high density of vertical interconnects to provide high-bandwidth and low-latency communication to and from each layer ("tier") in the stack, without sacrificing too much area for vias. Presently, vertically stacking of multiple tiers of 3D ICs is implemented using through-silicon vias (TSVs) (e.g. [7]), micro-bumps, or capacitive/inductive couplings (e.g. [8]). Such vertical integration solutions suffer from power dissipation problems, low interconnect density (because of a poor accuracy of wafer alignment, as compared with that of photolithographic masks defining features on a single wafer for multiple metallization layers), low yields (due to the lack of the known-good-die assembly), and poor cost efficiency [9] . As a result, such solutions cannot provide adequate density and throughput that will be required, for example, for real time signal processing from focal plane arrays [10] .

3D integration could fully benefit from monolithic ICs, but previous approaches to 3D monolithic integration have been limited by the requirement to integrate active components in a vertical stack (e.g. [11] [12] [13]). Such vertical integration of active devices faces severe thermal management, design complexity, and testability challenges. Moreover, multilayer CMOS circuits with thin-film transistors have inadequate characteristics for high performance memory and logic applications.

These challenges can be overcome with 3D hybrid circuits that monolithically integrate conventional CMOS circuits with quasi-passive, two-terminal, nanoscale devices distributed in additional layers. To pursue this idea, we formed the Center of 3D Hybrid CMOS-Nano Circuits (*HyNano*), funded by the AFOSR with an award from the 2011 Multidisciplinary Research Program of the University Research Initiative (MURI). *HyNano* consists of 9 faculty members from multiple disciplines, including computer engineering, electrical engineering, materials science, and physics, from four universities: University of California, Santa Barbara (the lead institution), Stony Brook University (SUNY), University of Michigan at Ann Arbor, and University of Massachusetts at Amherst. In this paper we give a brief overview of the main ideas underpinning this approach and discuss preliminary results, including the conceptual design, simulations and key experimental milestones.

2. Overview of 3D Hybrid Circuits

Figure 1 demonstrates the main idea of CMOL circuits [14] [15] [16] - the acronym stands for Cmos+MOLecular scale devices, which were recently extended to 3D circuits [27] . Such circuits are based on the combination of a single conventional CMOS chip and several layers of quasi-passive nanoscale crosspoint devices. In the simplest case these are two-terminal resistive switching ("memristive") devices (Fig. 1c), which can be reversibly switched between high and low

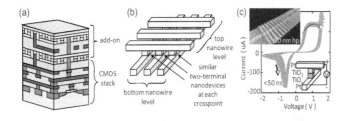

Fig. 1. The basic idea of 3D CMOL circuits [15] [16] [26] : (a) hybrid circuit cartoon, (b) crossbar topology, and (c) micrograph of array of metal oxide memristive devices, and typical switching *I-V* curves [21] .

resistive states [16] [17] [18] [21] [23] .

The main advantage of memristive devices is their very high density. This high density may be sustained with their interconnection by crossbar circuits, in which nanodevices are sandwiched between two layers of parallel nanowires (Fig. 1b). Because of crossbar regularity and lack of overlay (alignment) requirements, they allow for simple and cost-efficient fabrication.

Moreover, in CMOL circuits, high density and low cost nanodevices are combined with the most attractive properties of CMOS technology, including high flexibility, functionality and yield. Such synergy can be effectively exploited to build terabit scale digital memories [24] [30]

[31] , high performance reconfigurable logic circuits [25] [33] [34] [35] [38] [39] and to achieve high bandwidth between monolithically integrated memory and logic in a System-On-Chip fashion. In addition, due to its unique properties, CMOL may be the first technology to enable bio-inspired massively parallel advanced information processing [14] [37] [40] .

The main objectives of our recent efforts are extending the CMOL concept to the third dimension, and to simultaneously make this approach practical by improving memristive crosspoint devices, understanding the solid state physics of their operation, and developing those computer architectures with the greatest benefit from 3D CMOL technology.

3. Technical Challenges and Recent Progress

3.1 Resistive Switching "Memristive" Devices

In the simplest form the memristive device consists of three layers, top and bottom (metallic) electrodes and a thin film of some material sandwiched in between them (Fig. 2). By applying voltage bias across electrodes of such device, the electrical conductivity of a thin film can be changed reversibly and retained for sufficiently long time between high conductive (ON) state and the highly resistive (OFF) state. Figure 2b shows schematically an I-V for a bipolar switching device. In bipolar devices

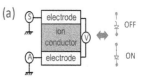

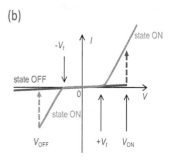

Fig. 2. (a) Device structure and (b) typical hysteretic I–V behavior for bipolar switching, shown schematically [15] [16] [19] [26] .

electrical stress (voltage or current bias) of opposite polarity is required to switch the device between ON and OFF states and a particular shape of I-V in the ON/OFF states might vary and will depend on the particular device and/or additional layers integrated in the device stack. Note that in addition to the digital mode operation most of the memristive devices can be switched continuously between ON and OFF states (Fig. 2b) by applying gradually increasing electrical stress. Also, Fig. 2 does not show the so-called "forming step," which might be required before the devices can be switched reversibly. Such a forming step is essentially a one-time application of relatively large voltage bias and might be eliminated in properly engineered devices.

Resistive switching has been observed experimentally for at least fifty years [20] . The interest in devices exhibiting resistive bistability was greatly revived as further scaling of CMOS technology has become increasingly challenging. Although, many material systems exhibit resistive switching and several are compatible with the CMOS processes, these devices face their own challenges. In particular, they suffer from low yield and high device-to-device variability. Moreover, the device operation physics, which is likely different for different material systems, is still not well understood.

There are three main classes of the materials systems for memristive devices. The first one is based on *chalcogenide materials*, with resistance modulation in this materials is induced by transition from the crystalline state to a disordered amorphous state [22] . While this concept is the most mature and best understood, rather slow set process (i.e. annealing time to get to crystalline state) and limited endurance may be a problem in the context of the proposed hybrid circuits.

Resistive switching in *metal-oxide* devices has been observed in a wide range of material systems including non-stoichiometric binary/ternary oxides and perovskites [21] . This group includes the most promising candidates for hybrid circuits, due to CMOS compatibility. The switching mechanisms in these devices are, however, the least understood, in part due to very rich experimental phenomena observed.

In *solid state electrolytes* and some *a*-Si and SiO_2 devices, the most likely mechanism is metallic filament formation via redox reaction at the interface. Such reaction results in the penetration of electrode material into the insulating film, with subsequent diffusion upon bias reversing [18] . While these devices exhibit some of the best endurance and low reset currents, the main concern here is volatility due to high mobility cations (which may also create problems for CMOS integration). A similar mechanism is likely behind switching in some organic films. Such material systems may be very attractive because of potentially low fabrication costs, but their temperature stability is a concern.

While there is an abundance of literature on bistable *I-V* curves, statistics and yield data for memristive devices are rarely reported, most likely due to poor results [16] . Still, there are some very encouraging data, e.g. as obtained very recently for *p*-Si/*a*-Si/Ag junctions [23] , the sample-to-sample distribution of the threshold voltage in these devices is very narrow (r.m.s. deviation ~10%). In addition, these devices combine long retention time, high ON/OFF ratio (~10^4), and large endurance (> 10^8 cycles). Also, excellent reproducibility has been reported in [24] which demonstrates integration of such devices in a 30 by 30 crossbar structure.

Though reproducibility of crosspoint memristive devices is perhaps still the most critical challenge at the moment, in some cases even relative large amount of device-to-device variations might be mitigated by using approach proposed in [41] . In this work a simple feedback algorithm, which is based on memristive properties common for Pt/TiO_{2-x}/Pt, was design to tune device conductance at a specific bias point to 1% relative accuracy (which is roughly equivalent to seven-bit precision) within its dynamic range even in the presence of large variations in switching behavior.

3.2 Crossbar Circuits

The footprint of the memristive devices could be very small and essentially defined by the overlap area of the two electrodes (wires), i.e. close to $4F_{nano}^2$, where F_{nano} is a wire half pitch. The most natural way of sustaining the density of single devices is to integrate them into a passive crossbar structures, which are implemented with mutually perpendicular layers of parallel wires (electrodes) with thin film sandwiched between two layers (Fig. 3a). Unlike active devices of CMOS circuits (MOSFETs), two-terminal memristive devices could have only one critical dimension, i.e. film thickness [15] , which can be controlled without the use of expensive fabrication techniques. This is why crossbars can be built with advanced patterning techniques such as nanoimprint lithography, so that F_{nano} may be potentially scaled down to just a few nanometers (essentially limited by quantum mechanical tunneling), which far exceeds the limits of conventional optical lithography. Finally, fabrication of many thin-film, memristive devices does not require high temperatures, thus enabling back-end monolithic integration of multiple layers on a CMOS base (Fig. 1a), so that the effective device footprint may be reduced even further to $4F_{nano}^2/K$, where K is the number of vertically integrated layers.

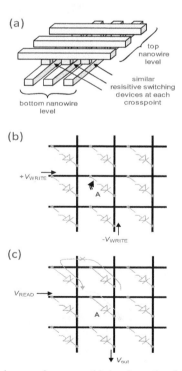

Fig. 3. Passive crossbar array: (a) A schematic of the structure and the idea of (b) writing and (c) reading a particular bit [26] [31] .

The basic operation (i.e. reading/writing the state of the crosspoint devices) of passive crossbar circuits can be explained using simplified equivalent circuits shown on Fig. 3 [26] [31] . Assuming digital mode operation, in the ON state (representing logic 1) memristive device is essentially a diode, so that the application of a voltage V_{READ}, with $V_t <$ $V_{READ} < V_{ON}$, where V_t and V_{ON} are denoted in Figure 2, to one nanowire (say, the second horizontal wire in Fig. 3c) leading to the particular crosspoint memristive device gives a substantial current injection into the second wire (Fig. 3c). This current pulls up voltage V_{out} which, e.g., can now be read out by a sense amplifier. To have low current at voltages above $\sim V_t$, the diode property prevents parasitic currents which might be induced in other ON-state cells by the output voltage (see the red line in Fig. 3c). In OFF state (which represents logic zero) the crosspoint current is very small, giving a nominally negligible contribution to output signals at readout. In order to switch the cell into ON state, the two nanowires leading to the device are fed by voltages V_{WRITE}, with $V_{WRITE} < V_{ON} < 2V_{WRITE}$ (Fig. 3b). The left inequality ensures that this operation does not disturb the state of "semi-selected" devices contacting just one of the biased nanowires. The write 0 operation is performed similarly using the reciprocal switching with threshold V_{OFF}.

3.3 CMOS-Nano Crossbar Integration

One of the most important challenges of the hybrid circuits is the efficient implementation of the integration of CMOS and nano crossbars. First of all, there is a conceptual problem that arises from the mismatch of half-pitches F_{nano} and F_{CMOS} of the nano and CMOS subsystems. Even if $F_{nano} \approx F_{CMOS}$, and the high area density of memristive devices is achieved by stacking many layers, the overhead for programming circuitry, which is required to program all the devices, should be small so as not to comprise the density advantages. Secondly, actual integration of memristive devices, placed on top of the CMOS stack is a substantial challenge by itself.

Area-Distributed Interface. The most promising solution to the first problem is an area-distributed interface between CMOS and nano subsystems [14] [15] (Fig. 4a-c), which is an essential feature of the proposed 3D CMOL circuits [27] Such interface enables high vertical bandwidth, and

potentially very low cost and low overhead. The area interface is enabled by (i) the crossbar array which is rotated by an angle α with respect to the mesh of CMOS-controlled vias; and (ii) a double decoding scheme that provides unique access to each crosspoint device. More specifically, as Figs. 4a-c show, two types of vias, one connecting to the lower (shown with blue dots) and the other to the upper (red dots) wire level in the crossbar, are arranged into a square array with a side length of $2\beta F_{CMOS}$, which is also equal to the side length of the "cells" grouping two vias of each kind. At the crossbar rotation angle $\alpha = \arcsin(1/\beta)$, the vias naturally subdivide the wires into fragments of length $2(\beta F_{CMOS})/F_{nano}$. Here F_{CMOS} is the CMOS half pitch, while $\beta > 1$ is a dimensionless number that depends on the cell size (i.e. complexity) in the CMOS subsystem; it is not arbitrary, but is chosen from the spectrum of possible values $\beta = (r^2 + 1)^{1/2} \times F_{nano}/F_{CMOS}$, where r is an integer so that the precise number of devices on the wire fragment is $r^2 - 1 \approx \beta^2 (F_{CMOS}/F_{nano})^2$.

The decoding scheme in CMOL is based on two separate address arrays (one for each level of wire in the crossbar so that there are a total of $4N$ edge channels to provide access to two different via controllers (one 'blue' and one 'red') in each of N^2 addressing cells in the CMOS plane. In contrast to standard memory arrays, in CMOL each control and data line pair electrically connects the peripheral input/outputs to a via instead of a single memory element. In turn, each via is connected to a wire fragment in the crossbar. The two perpendicular sets of wire fragments provide unique access to any crosspoint device even for large values of β. For example, selecting pins δv and b4 (which are highlighted with blue and red circles, respectively) provides access to the leftmost of the two shown devices on Fig. 4c, while pins δv and c4 for the rightmost device.

The total number of crosspoint devices that can be accessed by the $N \times N$ array of CMOS addressing cells is $\sim N^2 \beta^2 (F_{CMOS}/F_{nano})^2$, which would be much larger than N^2, if $F_{nano} < F_{CMOS}$. Consequently, one can use complex CMOS circuitry built with a significantly larger feature size to address regular crossbars built on a finer lithographic scale.

Experimental Milestones. Recently, significant progress has been made in the experimental realization of CMOL-like

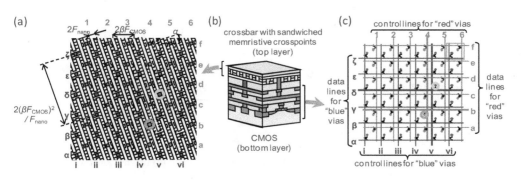

Fig. 4. Original CMOL circuits: (a) top view of the crossbar structure, (b) cut-away illustration, and (c) corresponding equivalent circuit diagram of the configuration logic in CMOS layer for the $N = 5$ primitive cell array.

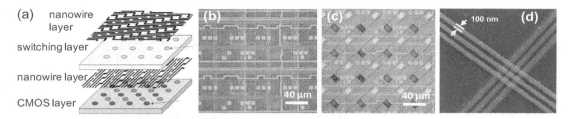

Fig. 5. Hybrid circuit demonstration [25] : (a) Conceptual illustration of the hybrid circuit, (b) optical micrograph of the as-received CMOS chip, (c) a hybrid chip with the memristor crossbars built on top; and (d) scanning electron microscope image of a fragment of the memristor crossbar array (where 3 nanowires cross 3 other nanowires, forming 9 memristors) with junction areas of 100×100 nm^2.

interfaces. For example, a 100-gate-scale hybrid CMOS/memristor circuit has been fabricated, in which memristive TiO$_{2-x}$ film is integrated onto a foundry-built CMOS platform using nanoimprint lithography, with materials and processes that are compatible with CMOS [32] [25] . This implementation features an area-distributed CMOL-like interface with tilted crossbar fabricated with nanoimprint technology (Figure 5). To make alignment between the CMOS and crossbar layers feasible, interconnects between the memristor layer and the CMOS layer have been implemented using larger contact pads connected the nanowires to the tungsten vias in the CMOS substrate.

3.4 Applications

Memory Arrays. The most natural applications of the 3D hybrid circuits are embedded memories and stand-alone memory chips, with their simple matrix structure. Such memories are an extension of the so-called resistive memories (RRAM) [28] . (Memories based on phase change devices are already in the development stage [29] .) For the memory application, each memristive device corresponds to a single-bit memory cell, while the CMOS subsystem may be used for coding, decoding, line driving, sensing, and input/output functions.

Having larger crossbar arrays helps to approach the ideal density $1/(2F_{nano})^2$ because for $N \times N$ crossbar array peripheral area overhead (i.e. sense amplifiers, decoders etc.) is proportional to $N^* \log N$, while useful area scales as N^2. On

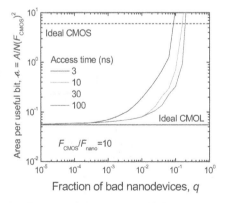

Fig. 6. Density (in terms of chip area per bit) for CMOL memory as a function of defective device fraction, for several memory access time values, and for a particular F_{CMOS}/F_{nano} ratio [31] .

the other hand, increasing N leads to larger readout delays, voltage drop across crossbar wires and, most importantly, leakage currents via semi-selected devices (Fig. 3). This is why implementing strong nonlinearity in the I-V is one of the most important goals for the resistive switching devices in the context of passive crossbar memories.

Another concern is the effect of the defective crosspoint devices on the memory performance. The defect density for memristive devices is likely to be much higher than that of conventional CMOS technology so that some novel defect and fault tolerance schemes must be considered. A detailed analysis of CMOL memories (which neglects an issue of leakage via semi-selected devices) with global and quasi-local ("dash") structure of matrix blocks was carried out [30] [31] . Both analyses rely on the combination of two major techniques for increasing their defect tolerance: the memory matrix reconfiguration (the replacement of some rows and columns with the largest number of bad memory cells for spare lines), and error correction codes. The best results were achieved in the later architecture using a synergy of bad bit exclusion with BCH error-correction codes: the defect tolerance is up to 8 to 12% depending on the required access speed (Fig. 6). As a result, CMOL memories may be the first technology to reach the terabit frontier.

Reconfigurable Circuits. To implement logic operations, each cell (or supercell) hosts some CMOS logic gate in addition to configuration (pass transistor) circuitry (Fig. 7). In the original CMOL field-programmable gate array (FPGA), each cell contains a CMOS inverter, and any multi-input NOR gate can be implemented with the inverter and a few memristive devices with diode-like functionality [33] [34] [35] . During the reconfiguration stage, all logic gates are disabled and any memristive device can be configured to the high or low resistive state, similar to the operation of the CMOL memory. Once configured, memristive devices do not change state and those set to the low resistive state represent electrical connection between logic gates.

For example, a particular input of the NOR gate can be connected to any of the outputs of other NOR gates that are within the "connectivity domain" by programming the corresponding memristive devices (Fig. 7d). The cells that are not within the connectivity domain of each other can be connected by dedicating some CMOS cells for routing purposes, e.g., to use two inverters in sequence as a buffer.

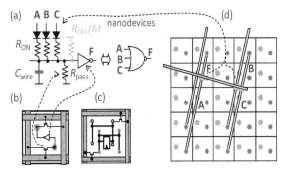

Fig. 7. CMOL FPGA circuits [35] : (a) The idea of diode NOR logic; (b) basic inverter and (c) latch CMOS cells; For clarity, panel (d) shows only nanodevices and nanowires participating in the NOR gate demonstrated on panel a.

Preliminary theoretical results show a very substantial density advantage (on the average, about two orders of magnitude) over the purely CMOS circuits, and a considerable leading edge over alternative hybrid circuit concept, so-called nanoPLA [36] .

On the other hand, the simulation results show that the speed of CMOL FPGA circuits is not much faster than that of CMOS. The situation may be rather different in custom logic circuits, were CMOL may lose a part of its density advantage, but become considerably faster than CMOS. An example of such a circuit, performing fast, parallel imaging processing, was presented in [42] . The simulated time of convolution of a large (1,024×1,024 pixel) image with a 32×32 window function (at 12-bit precision) is close to just 25 µs. This time has to be compared with estimated 3,500 us for a CMOS circuit based on the same design rules. This speed advantage is an explicit result of small CMOL footprint: the whole circuit processing one input pixel may be placed behind a 25×25 um^2 pixel sensor. As a result, the communication delays are cut to the bone.

Slightly modified original CMOL FPGA circuits have potentials to perform massively parallel high throughput pattern matching far exceeding the state-of-the-art conventional implementations [38] [39] . Pattern matching applications take the full advantage of the very wide fan-in

intrinsic to CMOL FPGA gates and thus enabling much more efficient use of nanosubsystem. Wide fan-in allows to compare an n-bit pattern (which is programmed as a binary weight in the memristive devices) with the streaming data in just one cycle and enables average nanodevice utilization close to ~15% for these circuits. Moreover, multiple patterns may be compared in parallel enabling a throughput of up to 3×10^{19} bits/second/cm^2 for pattern matching for an aggregate of 10^{10} bits for the 45-nm CMOS technology and practicable power density, even with conservative assumptions for nanodevice density and performance characteristics, and without any optimization.

Figure 8 shows the results of successful experimental demonstration of the CMOL FPGA-like concept [25] . In this slightly simplified version of the original CMOL FPGAs [43] complete logic gates are implemented in CMOS subsystem while memristive devices are used only to connect selectively CMOS gates. In particular, Figure 8 shows one of the many possible signal routings for NOT, AND, OR, NAND and NOR gates and a D-type flip-flop which were successfully programmed and tested. It is worth mentioning that this work is perhaps the first ever successful demonstration of using this novel technology at such scale.

Bio-inspired Information Processing. Perhaps the most exciting application of CMOL circuits is in bio-inspired information processing – for example, in the field of artificial neuromorphic networks (ANN). The motivation behind ANN comes from the fact that the mammalian brain still remains much more efficient (in power and processing speed) for a number of computational tasks, such as pattern recognition and classification, as compared to conventional computers, despite the exponential progress in the performance of the latter during the past several decades [14] [37] [40] . The main reason behind poor performance of the conventional CMOS-based circuits is that they cannot provide high complexity, connectivity and massive parallel information processing which is required typical for biological neural networks.

The very structure of CMOL circuits makes them uniquely suited for the implementation of ultrafast ANN [14] [37] and, in general, mixed-signal information processing systems [42] [41] . For example, the CMOS subsystem would implement less dense but also more complex somatic cells. The area-distributed interface and dense fragmented crossbar structure of CMOL ensure very rich interconnect among somas cells which are connected to ech other via nanowire-memristive device – nanowire links, i.e. with memristive devices acting as artificial synapses. Crude estimates have shown that artificial synapses smaller than 10 nanometres in length could be used to

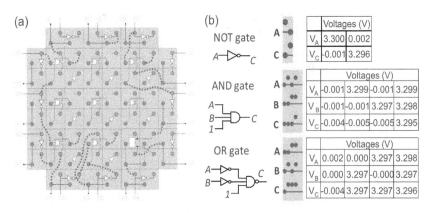

Fig. 8. Experimental demo of a CMOL FPGA-like circuit [25] : (a) CMOS layer fabric on a die; and (b) equivalent circuits and digital logic results from the chip tester.

make artificial neural networks of sufficient complexity and connectivity to challenge the computational performance of the human brain [37].

Very recently, a key operation for analog and mixed signal processing - multiply-and-add (i.e. dot-product) has been demonstrated with hybrid CMOS/memristor circuits (Fig. 9) [41]. To realize such circuitry the memristive devices implement density-critical configurable weights (e.g. artificial synapses), while CMOS is used for the summing amplifier, which provides gain and signal restoration (e.g. implementing simple soma function in the context of ANN). As a result, individual voltages applied to memristors can be multiplied by the unique weight (conductance) of memristor and summed up by CMOS amplifier - all in analog fashion.

4. Summary

The development of 3D CMOL circuits should greatly benefit digital memory and logic circuits and may, for the first time, enable large scale bio-inspired neuromorphic circuits. Practical introduction of such circuits will enable new compact, fast electronic systems which would open up a number of new applications. For example, the initial experimental demonstrations with just two crossbar layers, with F_{nano} = 50 nm, and F_{CMOS} = 130 nm, would enable a data density of up to 20 Gbit/cm^2 combined with a throughput, between CMOS and nano subsystems, of up to 5×10^{15} bits per second per cm^2, assuming a manageable power density [23] [24] [44]. With an assumption of more aggressive, though still realistic, technology parameters, i.e. 10 crossbar layers, F_{nano} = 10 nm, and F_{CMOS} = 45 nm, the maximum density will be close to 2.5 Tbit/cm^2 while ensuring similar or superior throughput. The high memory density and aggregate throughout of such circuits open unprecedented information processing capabilities, e.g., a pattern matching rate of about 10^{19} bits/second/cm^2 for 4×10^8 locally stored 500-bit-wide patterns.

While the initial simulation results and recent

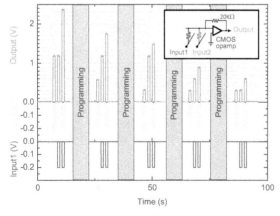

Fig. 9. Illustration of analog multiply and add circuitry operation with IC summing amplifier and memristive devices set with high precision to 15, 30 and 60μA states with adaptive variation tolerant algorithm [41].

experimental milestones are very encouraging, more work is certainly needed in order to fulfill the potentials of CMOL technology. The most critical issues for CMOL technology still remains yield, reproducibility of the memristive devices and their viable integration with conventional complementary metal oxide semiconductor technology.

5. Acknowledgments

The authors would like to acknowledge useful discussions with K. Likharev as well as other members of the HyNano team. This work is supported by the Air Force Office of Scientific Research (AFOSR) under the MURI grant FA9550-12-1-0038.

6. References

[1] D.J. Frank, R.H. Dennard, E. Nowak, P.M. Solomon, Y. Taur, and H.S.P. Wong, "Device scaling limits of Si MOSFETs and their application dependencies", *Proc. IEEE*, vol. 89, pp. 259–288, 2001.

[2] A.W. Topol et al., "Three-dimensional integrated circuits", *IBM Journal of Research and Development*, vol.50, pp.491-506, July 2006.

[3] V.F. Pavlidis and E.G. Friedman, "Interconnect-based design methodologies for three-dimensional integrated circuits", *Proc. IEEE*, vol.97, pp.123-140, Jan. 2009.

[4] S. Borkar, "3D integration for energy efficient system design", in: *Proc. 2009 VLSI Technology Symposium*, Kyoto, Japan, June 2009, pp. 58-59.

[5] R. S. Patti, "Three-dimensional integrated circuits and the future of system-on-chip designs", *Proceedings of the IEEE*, vol. 94, pp.1214-1224, June 2006.

[6] D.B. Strukov and A. Mishchenko, "Monolithically stackable hybrid FPGA", in: *Proc. Design Automation and Test in Europe*, Dresden, Germany, Mar. 2010, pp. 661-666.

[7] F. Liu et al., "A 300-mm wafer-level three-dimensional integration scheme using tungsten through-silicon via and hybrid Cu-adhesive bonding", in: *Proc. IEEE International Electron Devices Meeting*, Dec. 2008, pp. 1-4.

[8] W. R. Davis et al., "Demistifying 3D ICs: The pros and cons of going vertical", *IEEE Design and Test of Computers*, vol. 22, pp. 498-510, 2005.

[9] E.-K. Kim and J. Sung, "Yield challenges in wafer stacking technology", *Microelectron Reliab.*, Vol. 48, pp.1102–1105, 2008.

[10] S.M. Chai, A. Gentile, W.E. Lugo-Beauchamp, J. Fonseca, J.L. Cruz-Rivera, and D.S. Wills, "Focal-plane processing architectures for real-time hyperspectral image processing", *Appl. Optics*, vol. 39, pp. 835-849, 2000.

[11] K. Sakuma et al., "3D chip-stacking technology with through-silicon vias and low-volume leadfree interconnections", *IBM Journal of Research and Development*, vol. 52, pp. 611 – 622, 2008.

[12] H. Wei et al., "Monolithic three-dimensional integrated circuits using carbon nanotube FETs and interconnects", in: *Proc. IEEE International Electron Devices Meeting (IEDM)*, San Francisco, CA, Dec. 2009, pp.1-4.

[13] T. Naito *et al.*, "World's first monolithic 3D-FPGA with TFT SRAM over 90 nm 9 layer Cu CMOS", in: *Proc.*

Symposium on VLSI Technology (VLSIT), Honolulu, HI, June 2010, pp.219-220.

[14] K. Likharev, A. Mayr, I. Muckra, and O. Türel, "CrossNets: High-performance neuromorphic architectures for CMOL circuits", *Ann. NY Acad. Sci,* vol. 1006, pp. 146-163, 2003.

[15] K.K. Likharev and D.B. Strukov, "CMOL: Devices, circuits, and architectures", in: *Introducing Molecular Electronics,* G. Cuniberti, G. Fagas, and K. Richter, Eds., Berlin: Springer, pp. 447-478, 2005.

[16] K.K. Likharev, "Hybrid CMOS/nanoelectronic circuits: Opportunities and challenges", *J. Nanoelectronics and Optoelectronics,* vol. 3, pp. 203-230, 2008.

[17] D.B. Strukov, G. Snider, D. Stewart, R.S. Williams, "The missing memristor found", *Nature,* Vol. 453, pp.80-83, 2008.

[18] R. Waser, R. Dittman, G. Staikov, K. Szot, "Redox-based resistive switching memories – nanoionic mechanisms, prospects, and challenges", *Advanced Materials,* vol. 21, pp. 2632-2663, 2009.

[19] D.B. Strukov and H. Kohlstedt, "Resistive switching phenomena in thin films: Materials, devices, and applications", *Materials Research Society Bulletin,* vol. 37, Feb 2012.

[20] M.T. Hickmott, "Low-frequency negative resistance in thin anodic oxide films", *J. Appl. Phys.* vol. 33, pp. 2669-2682, 1962.

[21] J.J. Yang et al. "Memristive switching mechanism for metal/oxide/metal nanodevices", *Nature Nanotechnology,* vol. 3, pp. 429-433, 2009.

[22] Y.C. Chen et al., "Ultra-thin phase-change bridge memory device using GeSb", in: *Proc. Electron Device Meeting,* San Francisco, CA, Dec. 2006, pp. 1 – 4.

[23] S.H. Jo and W. Lu, "CMOS compatible nanoscale nonvolatile switching memory", *Nano Lett.,* vol. 8, pp.392-397, 2008.

[24] S.H. Jo, K.-H. Kim, and W. Lu, "High-density crossbar arrays based on a-Si memristive system", *Nano Lett.,* vol. 9, pp. 870-874, 2009.

[25] Q. Xia et al., "Memristor-CMOS hybrid integrated circuits for reconfigurable logic", *Nano Letters,* vol. 9, pp. 3640-3645, 2009.

[26] D.B. Strukov and K.K. Likharev, Reconfigurable nano-crossbar architectures, in: *Nanoelectronics,* R.Waser, Eds., 2012.

[27] D. B. Strukov and R. S. Williams, "Four-dimensional address topology for circuits with stacked multilayer crossbar arrays," *Proc. National Academy of Sciences,* vol. 106, pp. 20155-20158, Dec. 2009.

[28] K.K. Likharev, "Resistive and hybrid CMOS/nanodevice memories", in: J. E. Brewer and M. Gill (eds.), *Nonvolatile Memory Technologies with Emphasis on Flash, Wiley, Hoboken,* NJ, (2008), pp. 696-703.

[29] H.-B. Kang et al., "Core circuit technologies for PN-diode-cell PRAM", *J. Semiconductor Technology and Science,* vol. 8, pp. 128-133, 2008.

[30] D. Strukov and K. Likharev, "Prospects for terabit-scale nanoelectronic memories", *Nanotechnology,* vol. 16, pp. 137-148, Jan. 2005.

[31] D.B. Strukov and K.K. Likharev, "Defect-tolerant architectures for nanoelectronic crossbar memories", *J. Nanoscience Nanotechnology,* vol. 7, pp. 151-167, 2007.

[32] J. Borghetti et al., "A hybrid nanomemristor/transistor logic circuit capable of self-programming", *Proc. National Academy of Sciences,* vol. 106, pp. 1699-1703, 2009.

[33] D.B. Strukov and K.K. Likharev, "CMOL FPGA: A reconfigurable architecture for hybrid digital circuits with two-terminal nanodevices", *Nanotechnology,* vol. 16, pp. 888-900, 2005.

[34] D.B. Strukov and K.K. Likharev, "CMOL FPGA circuits", in: *Proc. Int. Conf. Computer Design,* Las Vegas, NE, Jun. 2006, pp. 213-219.

[35] D.B. Strukov and K.K. Likharev, "A reconfigurable architecture for hybrid CMOS/nanodevice circuits", in: *Proc. Field Programmable Gate Arrays,* Monterey, CA, Feb. 2006, pp. 131-140.

[36] A. DeHon, "Array-based architecture for FET-based nanoscale electronics", *IEEE Trans. Nanotechnology,* vol. 2, pp. 23-32, 2003.

[37] K. K. Likharev, "CrossNets: Neuromorphic hybrid CMOS/nanoelectronic networks", *Science of Advanced Materials,* vol. 3, pp. 322-331 (2011).

[38] F. Alibart, T. Sherwood, and D.B. Strukov, "Hybrid CMOS/nanodevice circuits for high throughput pattern matching applications", in: *Proc. AHS'11,* pp. 279-286, San Diego, CA, Jun. 2011.

[39] D.B. Strukov, "Hybrid CMOS/Nanodevice circuits with tightly integrated memory and logic functionality", in: *Proc. Nanotech'11,* vol. 2, pp. 9-12, Boston, MA, Jun. 2011.

[40] D.B. Strukov, "Smart connections", *Nature* 476, pp. 403-405, 2011.

[41] F. Alibart, L. Gao, B. Hoskins, and D.B. Strukov, "High-precision tuning of state for memristive devices by adaptable variation-tolerant algorithm", *Nanotechnology,* vol. 23, art. 075201, 2012.

[42] D.B. Strukov and K.K. Likharev, "Reconfigurable hybrid CMOS/nanodevice circuits for image processing", *IEEE Trans. Nanotechnology,* vol. 6, pp. 696-710, 2007.

[43] G.S. Snider, R.S. Williams, "Nano/CMOS architectures using a field-programmable nanowire interconnect", *Nanotechnology,* vol. 18, art. 035204, 2007.

[44] K.-H. Kim, S.H. Jo, S. Gaba and W. Lu, "Nanoscale resistive memory with intrinsic diode characteristics and long endurance", *Applied Physics Letters,* vol. 96, art. 053106, 2010.

A Fast Estimation of SRAM Failure Rate Using Probability Collectives

Fang Gong, Sina Basir-Kazeruni, Lara Dolecek, Lei He
Electrical Engineering
University of California, Los Angeles
Los Angeles, CA, 90095
{fang08, sinabk, dolecek, lhe}@ee.ucla.edu *

ABSTRACT

Importance sampling is a popular approach to estimate rare event failures of SRAM cells. We propose to improve importance sampling by probability collectives. First, we use "Kullback-Leibler (KL) distance" to measure the distance between the optimal sampling distribution and the original sampling distribution of variable process parameters. Further, the probability collectives (PC) technique using immediate sampling is adapted to analytically minimize the KL distance and to obtain a sampling distribution as close to the optimal as possible. The proposed algorithm significantly accelerates the convergence of importance sampling. Experiments demonstrate that proposed algorithm is $5200X$ faster than the Monte Carlo approach and achieves more than $40X$ speedup over other existing state-of-the-art techniques without compromising estimation accuracy.

Categories and Subject Descriptors: B.7.[Hardware]: - Integrated Circuits-Design Aids
General Terms: Algorithms, Verification
Keywords: SRAM, Failure probability, Importance Sampling, Kullback-Leibler distance

1. INTRODUCTION

It has become increasingly challenging to estimate the failure probability of SRAM cells under large-scale process variations, because SRAM bit-cell needs to be copied millions or billions of times as an array for higher integration density and the failure of a few cells could be catastrophic. Therefore, SRAM cell designs need to have extremely small failure probability [1, 2]. This failure is a rare event[3] that can only be captured with millions of samples through extremely long Monte Carlo (MC) simulations.

To avoid the expensive MC runs, importance sampling has been proposed based on the insight that only the "important samples" capturing relevant rare event (i.e., cell failure) can

*This work was partially supported by Cisco, ICscape Inc, and UC Discovery Program.

improve the estimation accuracy and further speed up the estimation convergence. This approach has been extensively used for rare event estimation problems [4, 5, 6, 7, 8, 9]. However, one critical issue that affects the efficiency of importance sampling is how to build an "optimal sampling distribution" so that more "important samples" of the relevant rare event can be chosen.

Many statistical methodologies have been developed to build the optimal sampling distribution for importance sampling and have been applied to failure rate estimation of SRAM cells [4, 5, 6, 7, 8, 9]. For example, [4] approximates the optimal sampling distribution by mixing a uniform distribution, the given sampling distribution and a "shifted" distribution centering around the failure region. Works in [5, 6] simply shift the mean values and keep the shape of original sampling distributions, and minimize the norm value of the shift vectors to find the optimal sampling distribution. The approach in [7] makes use of a "classifier" to block the Monte Carlo samples that are likely to satisfy the given performance constraints and runs simulations on remaining samples. In addition, "particle filtering"-based approach was proposed in [8] to tilt more samples towards the failure region. Moreover, it was recently proposed to adapt "Gibbs Sampling" in order to draw more failure region samples directly for improved performance [9]. While recent works made important advancements over the Monte Carlo approach, an efficient and low-complexity approach is still urgently needed to accurately estimate the failure rate of SRAM cells.

In this paper, we present a fast algorithm based on probability collectives (PC) method for the failure rate estimation of SRAM cells. First, "Kullback-Leibler (KL) distance" from probability theory [10] and information theory [11] is used to quantitatively measure the distance between the optimal sampling distribution and the given distribution of variable parameters. Then, a set of parameterized sampling distributions is analytically derived by minimizing the KL distance with a probability collective (PC) method using immediate sampling [12, 13], which is as close to the optimal sampling distribution as possible. Therefore, the convergence of the importance sampling approach can be significantly improved. The experimental results show that the proposed algorithm not only provides extremely high accuracy but also achieves $5200X$ speed-up over Monte Carlo. Moreover, the proposed method can be more than $40X$ faster than other state-of-the-art techniques (i.e., mixture importance sampling method [4] and spherical sampling method [6]).

Although the probability collective approach was initially

developed in the statistics field [12, 13], it was previously unknown how to interface it to the importance sampling method for the failure analysis of SRAM cells. In fact, there are three major issues that need to be resolved: first, one particular type of parameterized distribution should be chosen in order to approximate the optimal sampling distribution. Second, it is important but difficult to initialize the parameterized sampling distribution. Third, the minimization of the KL distance involves complicated optimization problems and usually requires expensive computational efforts. To resolve these issues, we select a set of Gaussian distributions parameterized by mean and sigma, and adapt the "norm minimization" from [5, 6] to initialize the distribution by shifting the given sampling distribution towards the failure region. Moreover, the immediate sampling-based probability collectives method [12, 13] is used to analytically solve for the optimal parameterized sampling distributions for importance sampling. To the best of our knowledge, this is the first work on successfully developing the probability collectives based importance sampling method for the failure probability estimation of SRAM cells.

The rest of this paper is organized as follows. In Section 2, we provide the necessary background on importance sampling, KL distance and probability collectives methods. Section 3 contains more details of the required techniques in the proposed method for SRAM failure analysis. The experiments and further discussion are provided in Section 4 to validate the accuracy and efficiency of the proposed method. The paper is concluded in Section 5.

2. BACKGROUND

2.1 Importance Sampling

Let ξ_i $(i = 1, \cdots, m)$ be independent random variables with probability density function (PDF) given by $p(\xi_i)$, characterizing circuit parameters under process variations, such as the threshold voltage and effective channel length of transistors. The joint PDF of ξ_i is denoted as $p(\boldsymbol{\xi})$ and can be expressed as follows due to the independence property:

$$p(\boldsymbol{\xi}) = \prod_{i=1}^{m} p(\xi_i). \tag{1}$$

The j-th Monte Carlo sample $\boldsymbol{\xi}^j = (\xi_1^j, \cdots, \xi_m^j)$ consists of one sample from each random variable distribution.

Let $f(\boldsymbol{\xi})$ be the performance merit of interest, such as static noise margin of SRAM cell. This quantity typically needs to be evaluated with expensive transistor-level circuit simulation.

Let f_0 be the performance constraint; the circuit failure $\{f(\boldsymbol{\xi}) < f_0\}$ event is designed to be "rare". Thereby, the indicator function $I(\boldsymbol{\xi})$ is defined to identify pass/fail of $f(\boldsymbol{\xi})$ as:

$$I(\boldsymbol{\xi}) = \begin{cases} 0 & \text{if } f(\boldsymbol{\xi}) \geq f_0 \text{ (pass)}, \\ 1 & \text{if } f(\boldsymbol{\xi}) < f_0 \text{ (fail)}. \end{cases} \tag{2}$$

Therefore, the probability of failure ($\mathcal{P}_r$) is estimated in (3):

$$\mathcal{P}_r = \int I(\boldsymbol{\xi}) \cdot p(\boldsymbol{\xi}) d\boldsymbol{\xi}. \tag{3}$$

In general, $p(\boldsymbol{\xi})$ is known but $I(\boldsymbol{\xi})$ is unknown since the indicator function $I(\boldsymbol{\xi})$ cannot be evaluated explicitly. When

$I(\boldsymbol{\xi})$ characterizes a failure region, extremely long Monte Carlo simulations on million samples of $\boldsymbol{\xi}$ are required.

To avoid massive Monte Carlo simulations, importance sampling has been proposed to sample from one "distorted" sampling distribution $g(\boldsymbol{\xi})$ that tilts towards the failure region where failures become more likely to happen. Then:

$$\mathcal{P}_r = \int I(\boldsymbol{\xi}) \cdot \frac{p(\boldsymbol{\xi})}{g(\boldsymbol{\xi})} \cdot g(\boldsymbol{\xi}) d\boldsymbol{\xi} = \int w(\boldsymbol{\xi}) \cdot I(\boldsymbol{\xi}) \cdot g(\boldsymbol{\xi}) d\boldsymbol{\xi}. \tag{4}$$

where $w(\boldsymbol{\xi})$ re-weights each sample of $\boldsymbol{\xi}$ to convert the sample into the original sampling distribution. Theoretically, the optimal sampling distribution $g^{opt}(\boldsymbol{\xi})$ [9], where only one sample is needed to provide the accurate estimation of failure probability, can be expressed as:

$$g^{opt}(\boldsymbol{\xi}) = \frac{I(\boldsymbol{\xi}) \cdot p(\boldsymbol{\xi})}{\mathcal{P}_r} \tag{5}$$

However, $g^{opt}(\boldsymbol{\xi})$ cannot be evaluated with (5) directly because $I(\boldsymbol{\xi})$ is unknown and $\mathcal{P}_r$ is indeed the desired failure rate. Instead, another sampling distribution $h(\boldsymbol{\xi})$ should be created to provide an approximation as close to $g^{opt}(\boldsymbol{\xi})$ as possible. For example, the Kullback-Leibler distance can be used to define the distance between $h(\boldsymbol{\xi})$ and $g^{opt}(\boldsymbol{\xi})$.

2.2 Kullback-Leibler Distance

The Kullback-Leibler (KL) distance was proposed in probability theory [10] and information theory communities [11] to measure the *directional* distance from one distribution to another.

The KL distance from distribution $g^{opt}(\boldsymbol{\xi})$ in (5) to $h(\boldsymbol{\xi})$ is expressed as:

$$\mathbb{D}_{KL}(g^{opt}(\boldsymbol{\xi}), h(\boldsymbol{\xi})) = \mathbb{E}_{g^{opt}} \left[log \left(\frac{g^{opt}(\boldsymbol{\xi})}{h(\boldsymbol{\xi})} \right) \right]. \tag{6}$$

Note that both distributions g^{opt} and h should be defined over the same random variable $\boldsymbol{\xi}$. In addition, $\mathbb{E}[\cdot]$ denotes the expectation operator and the subscript g^{opt} indicates that $\mathbb{E}[\cdot]$ is taken with respect to distribution g^{opt}.

Therefore, it is desired to minimize $\mathbb{D}_{KL}(g^{opt}(\boldsymbol{\xi}), h(\boldsymbol{\xi}))$ in order to achieve $\hat{h}(\boldsymbol{\xi})$ as the best approximation of $g^{opt}(\boldsymbol{\xi})$. To this end, the probability collective method can be adapted to solve the minimization problem efficiently.

2.3 Probability Collectives

In general, probability collectives (PC) method is an efficient optimization framework [12, 13], which can search for the optimal probability distributions of variable parameters in order to optimize the objective function.

As an illustration, we consider random variables $\boldsymbol{\xi} = (\xi_1, \cdots, \xi_m)$ and aim to minimize the KL distance as:

$$\arg \min \mathbb{E}_{g^{opt}} \left[log \left(\frac{g^{opt}(\boldsymbol{\xi})}{h(\boldsymbol{\xi})} \right) \right]. \tag{7}$$

By change of measure, the above minimization problem is equivalent to the statement in (8):

$$\arg \max \mathbb{E}_h \left[I(\boldsymbol{\xi}) \cdot log(h(\boldsymbol{\xi})) \right]. \tag{8}$$

It is highly prohibitive to perform exhaustive search for $h(\boldsymbol{\xi})$ since the search space is extremely large and contains arbitrary distributions. The PC method simplifies the search problem by utilizing a set of parameterized sampling distributions $h(\boldsymbol{\xi}, \theta)$ with additional parameters $\theta =$

$(\theta_1, \cdots, \theta_m)$. As such, the maximization problem in (8) becomes:

$$\hat{\theta} = \arg\max_{\theta} \mathbb{E}_h \left[I(\boldsymbol{\xi}) \cdot log(h(\boldsymbol{\xi}, \theta)) \right]. \qquad (9)$$

where $\hat{\theta}$ is the optimal parameter of the distribution $h(\boldsymbol{\xi}, \theta)$ that leads the minimum KL distance in (7).

Note that the expectation value $\mathbb{E}_h[\cdot]$ in (9) cannot be evaluated with analytical formula and thereby sampling techniques must be used. In fact, several sampling based PC methods such as delay sampling based PC, and immediate sampling based PC were proposed in [12, 13].

In this paper, we adapt the immediate sampling based PC method as summarized in Algorithm (1). Interested readers are referred to [12, 13] for other PC methods.

Algorithm 1 Immediate Sampling based PC Algorithm

1: Choose the initial parameter $\theta^{(1)}$ to build parameterized sampling distributions $h(\boldsymbol{\xi}, \theta^{(1)})$.
2: Draw random samples from $h(\boldsymbol{\xi}, \theta^{(1)})$ and set iteration index number $t = 2$.
3: **repeat**
4:     Evaluate values of indicator function $I(\boldsymbol{\xi})$ with chosen samples.
5:     Solve for $\theta^{(t)}$ by:

$$\theta^{(t)} = \arg\max_{\theta} \mathbb{E}_h \left[I(\boldsymbol{\xi}) \cdot log(h(\boldsymbol{\xi}, \theta^{(t-1)})) \right].$$

6:     Draw random samples from the parameterized distribution $h(\boldsymbol{\xi}, \theta^{(t)})$ and set $t = t + 1$.
7: **until** Converged (e.g., $\theta^{(t)}$ does not change for several subsequent iterations)
8: Obtain the optimum parameter $\hat{\theta}$ for the optimal sampling distribution.
9: Sample the final $h(\boldsymbol{\xi}, \hat{\theta})$ to get solution(s) in order to optimize the objective function.

Since the updated distribution $h(\boldsymbol{\xi}, \theta^{(t)})$ at the t-th iteration is sampled immediately, the procedure is called "immediate sampling" based PC method. However, there exist several issues that need to be resolved when immediate sampling PC method is used for failure analysis of SRAM cells:

- First, there exist many types of parameterized distributions (e.g., Gaussian distributions, Boltzmann distributions, etc.), and it remains unclear how to choose $h(\boldsymbol{\xi}, \theta)$ for the SRAM failure analysis.

- It is important and nontrivial to find $\theta^{(1)}$ which provides a "starting point" or a "heuristic initial solution" for the solution of (9). This quantity significantly affects the speed of convergence in Algorithm (1).

- The optimization problem in (9) is very difficult to solve and a closed-form solution is highly desired.

Therefore, it is of interest to develop an approach to use immediate sampling based PC method in a way that is suitable for SRAM failure analysis.

3. PROPOSED METHOD

In this section, we introduce several existing techniques and highlight our novel contributions that are needed to utilize the immediate sampling PC method for SRAM failure analysis.

3.1 Parameterized Distribution Selection

Before we move forward, let us first introduce the modeling of process variations in SRAM cells. In general, the variation sources of CMOS transistors can be threshold voltage V_{th}, effective channel length L_{eff} and other device parameters, but V_{th} variation is dominant so that the variability effects of other parameters are significantly dampened [2].

Moreover, V_{th} variations are typically modeled as independent random variables of Gaussian distributions [4, 5, 6, 7, 9]. As such, it is a natural choice to deploy a family of Gaussian distributions parameterized by mean (μ) and standard deviation (σ). In fact, parameterized Gaussian distributions can lead to a closed-form solution to the optimization problem in (9) as shown in following sections.

As an illustration, let ξ_i be the independent Gaussian random variable for i-th V_{th} variation source, which has the mean $\mu_i^{(0)}$ and the standard deviation $\sigma_i^{(0)}$. To build the parameterized Gaussian distribution for ξ_i, we shift the mean to $\hat{\mu}_i$ and reducing the standard deviation to $\hat{\sigma}_i$. This approach is motivated by the following insights:

- **Mean-shift** can tilt the sampling distribution towards the failure region where the rare failures are more likely to happen. This approach is similar to the finding in [5] and has been extensively used in previous works [5, 6, 8, 14, 15, 16].

- **σ-reduction** can concentrate the samples around a much smaller region where rare failures can happen with higher probability.

Therefore, the samples drawn from the parameterized Gaussian distribution $h(\xi_i, \hat{\mu}_i, \hat{\sigma}_i)$ are more likely to fail, and can thereby expedite the convergence of the failure probability estimation in the importance sampling. However, it is still unknown how to find the optimal parameters $\hat{\mu}_i$ and $\hat{\sigma}_i$ efficiently; this question will be investigated in following sections.

3.2 Parameterized Distribution Initialization

As discussed in Section 2.3, the first step is to initialize the parameters (μ_i, σ_i), which, in fact, provides a "starting point" or "heuristic initial solution" to search for the optimal parameters $(\hat{\mu}_i, \hat{\sigma}_i)$. As such, the initial parameters $\mu_i^{(1)}$ and $\sigma_i^{(1)}$ can significantly affect the efficiency of the iterative search in PC method or even lead to completely misleading results.

To this end, we propose an efficient initial parameter selection method inspired by the insights of "norm minimization" in [5], which can rapidly shift the given sampling distribution towards the failure region and make rare failures most likely to happen.

Assume random variables ξ_i follow Gaussian distributions $N(\mu_i^{(0)}, \sigma_i^{(0)})$. The proposed initial parameter selection can be summarized as following: first, a few hundred *uniformly-distributed* samples of ξ_i are generated using Quasi Monte Carlo method [17] in order to evenly cover the entire parameter range, such as the eight-sigma range from $(\mu_i^{(0)} - 4\sigma_i^{(0)})$ to $(\mu_i^{(0)} + 4\sigma_i^{(0)})$. Then, transistor level simulations are run on these samples and the failed samples are identified with given performance constraints. We can further choose one failed sample with the *minimum L_2-norm* and use its value as the initial parameter for $\mu_i^{(1)}$. In addition, the initial sigma parameter $\sigma_i^{(1)}$ can be the same as $\sigma_i^{(0)}$.

It is worthwhile to point out that the above "norm minimization" based method is a heuristic for obtaining an *initial* parameterized Gaussian distribution but cannot provide the *optimal* sampling distribution $h(\boldsymbol{\xi}, \hat{\boldsymbol{\mu}}, \hat{\boldsymbol{\sigma}})$ in (9) by any means. The optimization problem in (9) should be solved for $h(\boldsymbol{\xi}, \hat{\boldsymbol{\mu}}, \hat{\boldsymbol{\sigma}})$ and an efficient closed-form approach is needed.

3.3 Closed-Form Optimization Solution

Before we present the closed-form solution, it should be noted that the optimization in (9) must be revised as (10) because samples are generated from the parameterized distributions $h(\boldsymbol{\xi}, \boldsymbol{\mu}, \boldsymbol{\sigma})$ rather than from distributions $h(\boldsymbol{\xi})$:

$$\hat{\boldsymbol{\mu}} = \arg\max_{\boldsymbol{\mu}} \mathbb{E}_h[I(\boldsymbol{\xi}) \cdot w(\boldsymbol{\xi}, \boldsymbol{\mu}, \boldsymbol{\sigma}) \cdot log(h(\boldsymbol{\xi}, \boldsymbol{\mu}, \boldsymbol{\sigma}))],$$
$$\hat{\boldsymbol{\sigma}} = \arg\max_{\boldsymbol{\sigma}} \mathbb{E}_h[I(\boldsymbol{\xi}) \cdot w(\boldsymbol{\xi}, \hat{\boldsymbol{\mu}}, \boldsymbol{\sigma}) \cdot log(h(\boldsymbol{\xi}, \hat{\boldsymbol{\mu}}, \boldsymbol{\sigma}))] \quad (10)$$

where $w(\boldsymbol{\xi}, \boldsymbol{\mu}, \boldsymbol{\sigma})$ denotes the weights to unbias the samples from the parameterized distribution $h(\boldsymbol{\xi}, \boldsymbol{\mu}, \boldsymbol{\sigma})$ and can be expressed as:

$$w(\boldsymbol{\xi}, \boldsymbol{\mu}, \boldsymbol{\sigma}) = \frac{h(\boldsymbol{\xi})}{h(\boldsymbol{\xi}, \boldsymbol{\mu}, \boldsymbol{\sigma})}. \quad (11)$$

For the illustration purpose, let us consider following example:

- $\boldsymbol{\xi} = (\xi_1, \cdots, \xi_m)$: independent random Gaussian variables.

- $h(\boldsymbol{\xi}) = (h(\xi_1), \cdots, h(\xi_m))$: the given Gaussian sampling distributions of $\boldsymbol{\xi}$.

- $h(\boldsymbol{\xi}, \boldsymbol{\mu}, \boldsymbol{\sigma}) = (h(\xi_1, \mu_1, \sigma_1), \cdots, h(\xi_m, \mu_m, \sigma_m))$: the chosen parameterized Gaussian distributions for $\boldsymbol{\xi}$.

- $\xi_i^1, \cdots, \xi_i^j, \cdots, \xi_i^N$: the samples of ξ_i drawn from the parameterized Gaussian distribution $h(\xi_i, \mu_i, \sigma_i)$.

As such, the weights of j-th sample $\boldsymbol{\xi}^j = (\xi_1^j, \cdots, \xi_m^j)$ can be expressed as:

$$w(\boldsymbol{\xi}^j, \boldsymbol{\mu}, \boldsymbol{\sigma}) = \frac{h(\xi_1^j) \times \cdots \times h(\xi_m^j)}{h(\xi_1^j, \mu_1, \sigma_1) \times \cdots \times h(\xi_m^j, \mu_m, \sigma_m)}. \quad (12)$$

Moreover, the expectation value $\mathbb{E}_h[\cdot]$ in (10) cannot be evaluated directly in general, because there is no analytical formula for the integral operation, and sampling methods must be used. For instance, with the samples $\xi_i^j, (j = 1, \cdots, N)$, the optimization problem for μ_i becomes the sampled form as (13). Similar expression can be derived for σ_i.

$$\hat{\mu}_i = \arg\max_{\mu} \frac{1}{N} \sum_{j=1}^{N} \left(I(\boldsymbol{\xi}^j) w(\boldsymbol{\xi}^j, \boldsymbol{\mu}, \boldsymbol{\sigma}) \log(h(\xi_i^j, \mu_i, \sigma_i)) \right). \quad (13)$$

As proposed in [12], the above optimization problem is a convex optimization problem that can be solved with closed-form formula, because the parameterized distribution $h(\boldsymbol{\xi}, \boldsymbol{\mu}, \boldsymbol{\sigma})$, following Gaussian distribution, is a log-concave distribution.

Specifically, the optimal parameters $\hat{\mu}_i$ and $\hat{\sigma}_i$ can be analytically solved with closed-form formulae as [12, 13]:

$$\hat{\mu}_i = \frac{\sum_{i=1}^{N} I(\xi^j) \times w(\xi^j, \boldsymbol{\mu}, \boldsymbol{\sigma}) \times \xi_i^j}{\sum_{i=1}^{N} I(\xi^j) \times w(\xi^j, \boldsymbol{\mu}, \boldsymbol{\sigma})}. \quad (14)$$

where $\hat{\mu}_i$ can be asymptotically approached by iteratively updating the parameter $\boldsymbol{\mu}$ and evaluating the above formula. In practice, the iterative process can converge very fast within only a few iterations. Note that [11, 14, 15, 16] use the identical analytical formula to find the optimal parameter for mean shift.

Similarly, the closed-form formula can be derived to analytically compute $\hat{\sigma}_i$ as:

$$\hat{\sigma}_i = \sqrt{\frac{\sum_{i=1}^{N} I(\xi^j) \times w(\xi^j, \hat{\boldsymbol{\mu}}, \boldsymbol{\sigma}) \times (\xi_i^j - \hat{\mu}_i)^2}{\sum_{i=1}^{N} I(\xi^j) \times w(\xi^j, \hat{\boldsymbol{\mu}}, \boldsymbol{\sigma})}}. \quad (15)$$

It is obvious that the calculation of $\hat{\sigma}_i$ depends on the optimization result $\hat{\boldsymbol{\mu}}$ from (14). In other words, the potential error from the optimization of $\hat{\boldsymbol{\mu}}$ can propagate into the computation of $\hat{\sigma}_i$ and lead to completely misleading results, which is especially undesired because the performance of importance sampling is highly sensitive to the sampling distribution. This observation can further validate the necessity of the initial parameter selection presented in previous section.

Therefore, the optimal sampling distribution is obtained as $h(\boldsymbol{\xi}, \hat{\boldsymbol{\mu}}, \hat{\boldsymbol{\sigma}})$, which can be finally sampled to estimate the probability of SRAM rare event failures in the importance sampling to provide significant improvement on both accuracy and efficiency.

3.4 Overall Algorithm Flow

The proposed algorithm for the SRAM failure analysis is based on the above techniques. The overall algorithm flow is described in Algorithm (2), which consists of three stages:

(1) **Parameterized distribution initialization**: The first stage initializes the parameterized sampling distribution $h(\boldsymbol{\xi}, \boldsymbol{\mu}, \boldsymbol{\sigma})$ as a "heuristic initial solution" to search for the optimal parameterized sampling distribution $h(\boldsymbol{\xi}, \hat{\boldsymbol{\mu}}, \hat{\boldsymbol{\sigma}})$. Initialization adopts the insight of "norm minimization" from [5] and shifts the given sampling distribution towards the failure region where SRAM failures are more likely to happen.

(2) **Optimal parameter evaluation**: This stage starts with the initial parameterized sampling distribution and analytically solves the optimization problems in (14) and (15) to achieve the optimal parameterized sampling distribution $h(\boldsymbol{\xi}, \hat{\boldsymbol{\mu}}, \hat{\boldsymbol{\sigma}})$.

(3) **Failure probability estimation**: The conventional importance sampling method is performed with the obtained optimal sampling distribution $h(\boldsymbol{\xi}, \hat{\boldsymbol{\mu}}, \hat{\boldsymbol{\sigma}})$ to estimate the failure rate of SRAM cells.

As shown in Section 4, the proposed approach in Algorithm (2) can provide more than $40X$ speedup over the existing state-of-the-art techniques and be up to $5200X$ faster than Monte Carlo method without compromising any accuracy.

4. EXPERIMENTAL RESULTS

We have implemented our proposed algorithm using MATLAB and Hspice with BSIM4 model. Also, Monte Carlo (MC), spherical sampling (SS) [6] and mixture importance sampling (MixIS) [4] are all implemented. As an illustration, the threshold voltages of all MOSFETs are considered as variation sources and static noise margin (SNM) failure

Algorithm 2 Overall Algorithm for SRAM Failure Analysis

Input: random variables $\boldsymbol{\xi} = (\xi_1, \cdots, \xi_M)$ with given Gaussian distributions $h(\boldsymbol{\xi}, \boldsymbol{\mu}^{(0)}, \boldsymbol{\sigma}^{(0)})$, and sample counts (N_1, N_2, N_3).
Output: the estimation of failure probability $\mathcal{P}_r$.

1: /* **Stage 1: Initial Parameter Selection** */
2: Draw uniformly-distributed samples $\xi^j (j = 1, \cdots, N_1)$ from the given distributions $h(\boldsymbol{\xi})$ and run simulations on these samples.
3: Identify samples that fail with given performance constraints and calculate their L_2-norm values.
4: Choose the failed sample with the minimum L_2 norm and use the value of this sample as the initial $\boldsymbol{\mu}^{(1)}$.
5: Set the initial sigma $\boldsymbol{\sigma}^{(1)}$ to be the same as given $\boldsymbol{\sigma}^{(0)}$.
6:
7: /* **Stage 2: Optimal Parameter Finding** */
8: Draw N_2 samples ξ^j from the initial parameterized distribution $h(\boldsymbol{\xi}, \boldsymbol{\mu}^{(1)}, \boldsymbol{\sigma}^{(1)})$ and set the iteration index number $t = 2$.
9: **repeat**
10:    Evaluate the indicator function $I(\xi^j)$ in (14) and (15) with these samples.
11:    **for** $i = 1 \rightarrow M$ **do**
12:       Solve for $\mu_i^{(t)}$ and $\sigma_i^{(t)}$ with

$$\mu_i^{(t)} = \frac{\sum_{i=1}^{N} I(\xi^j) \times w(\xi^j, \boldsymbol{\mu}^{(t-1)}, \boldsymbol{\sigma}^{(t-1)}) \times \xi_i^j}{\sum_{i=1}^{N} I(\xi^j) \times w(\xi^j, \boldsymbol{\mu}^{(t-1)}, \boldsymbol{\sigma}^{(t-1)})}.$$

$$\sigma_i^{(t)} = \sqrt{\frac{\sum_{i=1}^{N} I(\xi^j) \times w(\xi^j, \boldsymbol{\mu}^{(t-1)}, \boldsymbol{\sigma}^{(t-1)}) \times (\xi_i^j - \mu_i^{(t)})^2}{\sum_{i=1}^{N} I(\xi^j) \times w(\xi^j, \boldsymbol{\mu}^{(t-1)}, \boldsymbol{\sigma}^{(t-1)})}}.$$

13:    **end for**
14:    Draw N_2 samples from the updated parameterized distribution $h(\boldsymbol{\xi}, \boldsymbol{\mu}^{(t)}, \boldsymbol{\sigma}^{(t)})$ and set $t = t + 1$.
15: **until** Converged; when $\boldsymbol{\mu}^{(t)}$ and $\boldsymbol{\sigma}^{(t)}$ do not change for several subsequent iterations.
16: Obtain the optimal parameter $\hat{\boldsymbol{\mu}}$ and $\hat{\boldsymbol{\sigma}}$ for parameterized sampling distribution.
17:
18: /* **Stage 3: Failure Probability Estimation** */
19: Draw N_3 samples from the obtained optimal sampling distribution $h(\boldsymbol{\xi}, \hat{\boldsymbol{\mu}}, \hat{\boldsymbol{\sigma}})$.
20: Run simulations on these samples ξ_j and evaluate the indicator function $I(\xi^j), (j = 1, \cdots, N_3)$.
21: Solve for the failure probability, $\mathcal{P}_r$, with sampled form:

$$\mathcal{P}_r = \frac{1}{N_3} \sum_{i=1}^{N_3} I(\xi^j) \times w(\xi^j, \hat{\boldsymbol{\mu}}, \hat{\boldsymbol{\sigma}}).$$

where $w(\xi^j, \hat{\boldsymbol{\mu}}, \hat{\boldsymbol{\sigma}})$ is the weight for sample ξ^j and is defined as

$$w(\xi^j, \hat{\boldsymbol{\mu}}, \hat{\boldsymbol{\sigma}}) = \frac{\prod_{i=1}^{M} h(\xi_i^j)}{\prod_{i=1}^{M} h(\xi_i^j, \hat{\boldsymbol{\mu}}, \hat{\boldsymbol{\sigma}})}.$$

---

is studied. Note that the same algorithm can be applied to

other variation sources (i.e. L_{eff}, T_{ox}, etc.) and other rare failures (i.e. reading/writing failures) as well.

4.1 SRAM Cell and Static Noise Margin

The typical design of a 6-transistor SRAM cell is shown in Fig.1. We introduce process variations to threshold voltage V_{th} of all MOSFETs as *independent* random variables of Gaussian distributions. Specifically, the nominal mean values of the threshold voltages for NMOS and PMOS are $0.466V$ and $-0.4118V$, respectively. The standard deviations (σ) of threshold voltage variations are 10% of nominal threshold voltage values.

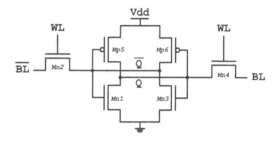

Figure 1: The schematic of the 6T SRAM cell.

The SRAM cell consists of six transistors: $Mn2$ and $Mn4$ control the access of the cell during reading, writing and standby operations; the remaining four transistors form two inverters and use two stable states (either '0' or '1') to store the data in this memory cell.

Static Noise Margin (SNM) is used to evaluate the stability of SRAM cell by describing the noise voltage that is needed to flip the stored data. More specifically, SNM can be measured by the length of maximum embedded square in the butterfly curves, which consist of the voltage transfer curve (VTC) of the two inverters in SRAM cell [18]. As such, when SNM is less than zero, the butterfly curve is collapsed and the data retention failure happens.

4.2 Accuracy Comparison

4.2.1 Comparison of Failure Rate Estimation

To validate the estimation accuracy of the proposed algorithm, we perform different methods, including Monte Carlo (MC), mixture importance sampling (MixIS) [4], spherical sampling (SS) [6], and the proposed algorithm on the same 6-T SRAM cell example in 45nm process to predict the probability of data retention failure due to SNM variation. Here, we choose $Vdd = 300mV$ as an example for comparison.

Evolutions of the probability estimation from different methods are plotted in Fig.2(a), the following observations can be made:

- First, the failure rate estimations from different methods closely match each other, which validates the estimation accuracy of our proposed method.

- Second, the proposed method in contrast to other methods starts with an estimation that is very close to the final accurate result, because it can find the *optimal* sampling distribution using probability collectives method for importance sampling.

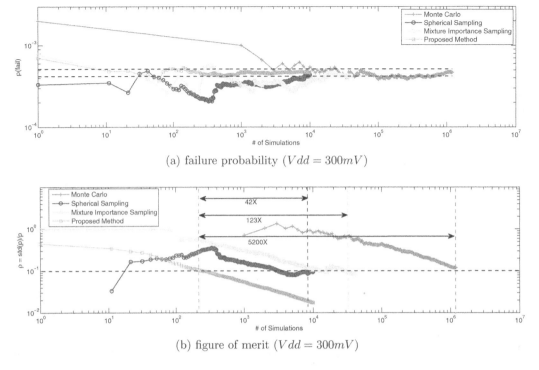

(a) failure probability ($Vdd = 300mV$)

(b) figure of merit ($Vdd = 300mV$)

Figure 2: Evolution comparison of the failure probability estimation and figure of merit for different methods.

- The comparisons among MixIS, SS and proposed method also reveal that the importance sampling is highly sensitive to the sampling distribution, which can affect both the accuracy and efficiency. This is the very motivation behind this paper to exploit the optimal sampling distribution.

4.2.2 Comparison of Figure-Of-Merit (FOM)

As stated in [5, 6], Figure-Of-Merit (FOM), ρ, has been extensively used to quantify the accuracy of probability estimation, which is defined as:

$$\rho = \frac{\sqrt{\sigma_{\mathcal{P}_r}^2}}{\mathcal{P}_r}. \tag{16}$$

where $\mathcal{P}_r$ is the estimation of failure probability and $\sigma_{\mathcal{P}_r}$ is the standard deviation of $\mathcal{P}_r$. In fact, the FOM can be treated as a *relative error* so that a smaller figure of merit means higher accuracy.

Similarly, we further calculate the evolution of FOM for different methods which are plotted in Fig.2(b). To clearly compare the accuracy of different methods, we plot a dashed line to indicate the 90% accuracy level with 90% confidence interval ($\rho = 0.1$). Two important observations can be made:

- MixIS, SS and proposed method quickly reach higher accuracy level ($> 90\%$) while Monte Carlo can only closely approach the 90% accuracy. This is because importance sampling based methods can choose more failed samples from the failure region to efficiently improve the accuracy, while Monte Carlo method wastes a large number of samples that are far from the failure region.

Table 1: Results of all methods with $10,000$ samples.

	MC	MixIS [4]	SS [6]	Proposed
prob. of failure	5.455E-4	3.681E-4	4.342E-4	4.699E-4
ρ	0.8129	0.1111	0.9831	0.021
accuracy	18.71%	88.53%	90.42%	98.2%
#runs	1.0e+4	1.0e+4	1.0e+4	1.0e+4

Table 2: Accuracy and efficiency comparison for different methods.

	MC	MixIS [4]	SS [6]	Proposed
prob.(failure)	4.675E-4	4.332E-4	4.208E-4	4.7E-4
accuracy	88%	90%	90%	90%
#runs	1.2E+6	2.85E+4	9.771E+3	231
speedup	1X	42X	123X	5200X

- For the same number of samples, the proposed method outperforms existing approaches in terms of the estimator accuracy. For instance, we compare accuracy level of all different methods in Table(1) with only 10,000 samples. In this table, the proposed method can provide 98.2% accuracy while other methods can only reach up to 90.42%, which is attributed to the choice of the optimal sampling distribution.

4.3 Efficiency Comparison

4.3.1 Comparison of Convergence Speed

Fig.2(b) illustrates the efficiency of proposed algorithm, which is shown to have the fastest speed of convergence among all the different methods. In this figure, the proposed method chooses more failed samples and increasingly improves the accuracy to an extremely high level due to the optimal sampling distribution.

Similar observations can be made from Fig.2(a): the proposed method starts with the estimation that is very close to the final accurate results and quickly converges to the 95% confidence interval of the final Monte Carlo result (denoted by two dashed lines). Meanwhile, the estimations of other methods keep fluctuating before asymptotically approaching the final accurate results.

In fact, the proposed method can achieve 90% accuracy and 90% confidence interval with only 231 samples. In the contrast, MixIS and SS need 2.85e+4 and 9.77e+3 samples to reach the same accuracy level, respectively. Monte Carlo method cannot even reach 90% accuracy with up to 1.2e+6 samples. In other words, the proposed method can achieve $5200X$ speedup over Monte Carlo, $123X$ speedup over MixIS [4] and $42X$ speedup over SS [6].

4.3.2 Other Efficiency Comparison

It should be noted that all importance sampling based methods require some "extra" samples to find the new sampling distribution, called "extra" because Monte Carlo method does not need these extra samples in simulations. For example, the stage 1 and stage 2 in Algorithm (2) need some "extra" samples to construct the optimal sampling distribution before the failure probability can be estimated in stage 3.

Specifically, in our experiments, the MixIS needs 3000 samples to find the sampling distribution, because it mixes the uniform distribution, given sampling distribution and mean-shifted distribution together and requires more samples. The SS method needs 2000 samples to locate the failed samples with a minimum L_2-norm in a spherical manner. The proposed method also needs 2000 samples to find the optimal sampling distribution. However, these "extra" samples turn out to be negligible when compared to the Monte Carlo method.

5. CONCLUSION

In this paper, we presented an improved importance sampling algorithm based on probability collectives method to efficiently estimate the rare event failures of SRAM cells. This method adopts the "Kullback-Leibler (KL) distance" to represent the distance between the optimal sampling distribution and a given sampling distribution. The KL distance is further analytically minimized using immediate sampling based probability collectives method and a set of parameterized Gaussian distributions are obtained as the optimal sampling distribution. The experiments demonstrate that proposed algorithm can provide extremely high accuracy and dramatically improve the convergence of importance sampling. For instance, the proposed method can be $5200X$ faster than Monte Carlo method and offer more than $40X$ speedup over other existing state-of-the-art techniques (e.g., mixture importance sampling [4] and spherical sampling [6]) with the same accuracy.

6. REFERENCES

[1] R. Heald and P. Wang, "Variability in sub-100nm SRAM designs," in *Computer Aided Design, 2004. ICCAD-2004. IEEE/ACM International Conference on*, pp. 347–352, 2004.

[2] P. Girard, A. Bosio, L. Dilillo, P. S., and A. Virazel, "Advanced test methods for SRAMs: Effective solutions for dynamic fault detection in nanoscaled technologies," 2009.

[3] K. Agarwal and S. Nassif, "Statistical analysis of SRAM cell stability," in *Proceedings of the 43rd annual Design Automation Conference*, DAC '06, pp. 57–62, 2006.

[4] R. Kanj, R. Joshi, and S. Nassif, "Mixture importance sampling and its application to the analysis of SRAM designs in the presence of rare failure events," in *Proceedings of the 43rd annual Design Automation Conference*, DAC'06, pp. 69–72, 2006.

[5] L. Dolecek, M. Qazi, D. Shah, and A. Chandrakasan, "Breaking the simulation barrier: SRAM evaluation through norm minimization," in *Proceedings of the 2008 IEEE/ACM International Conference on Computer-Aided Design*, ICCAD '08, pp. 322–329, 2008.

[6] M. Qazi, M. Tikekar, L. Dolecek, D. Shah, and A. Chandrakasan, "Loop flattening and spherical sampling: Highly efficient model reduction techniques for SRAM yield analysis," in *Design, Automation Test in Europe Conference Exhibition (DATE), 2010*, pp. 801 –806, 2010.

[7] A. Singhee and R. Rutenbar, "Statistical Blockade: A novel method for very fast monte carlo simulation of rare circuit events, and its application," in *Design, Automation Test in Europe Conference Exhibition, 2007. DATE '07*, pp. 1–6, 2007.

[8] K. Katayama, S. Hagiwara, H. Tsutsui, H. Ochi, and T. Sato, "Sequential importance sampling for low-probability and high-dimensional SRAM yield analysis," in *IEEE/ACM International Conference on Computer-Aided Design*, ICCAD '10, 2010.

[9] C. Dong and X. Li, "Efficient SRAM failure rateprediction via Gibbs sampling," in *Proceedings of the 43rd annual Design Automation Conference*, DAC'11, 2011.

[10] V. K. Rohatgi and A. K. M. Ehsanes Saleh, "An introduction to probability and statistics," *Wiley-Interscience*, 2000.

[11] T. M. Cover and J. A. Thomas, "Elements of information theory," *John Wiley and Sons*, 1991.

[12] D. Rajnarayan, D. H. Wolpert, and I. Kroo, "Optimization under uncertainty using probability collectives," *10th AIAA/ISSMO Multidisciplinary Analysis and Optimization Conference*, 2006.

[13] D. Rajnarayan, I. Kroo, and D. H. Wolpert, "Probability collectives for optimization of computer simulations," *AIAA/ASME/ASCE/AHS/ASC Structures, Structural Dynamics, and Materials Conference*, 2007.

[14] A. Ridder and R. Y. Rubinstein, "Minimum cross-entropy methods for rare-event simulation," *Simulation: Transactions of the Society for Modeling and Simulation International*, vol. 83, pp. 769–784, 2007.

[15] T. H. de Mello, "A study on the cross-entropy method for rare event probability estimation," *INFORMS Journal on Computing*, vol. 19, no. 3, pp. 381–394, 2007.

[16] P. T. de Boer, D. P. Kroese, S. Mannor, and R. Y. Rubinstein, "A tutorial on the cross entropy method," *Annals of Operations Research*, vol. 134, pp. 19–67, 2005.

[17] H. Niederreiter, "Random number generation and quasi-monte carlo methods," *Society for Industrial and Applied Mathematics*, 1992.

[18] D. Mukherjee, H. K. Mondal, and B. Reddy, "Static noise margin analysis of SRAM cell for high speed application," *International Journal of Computer Science Issues*, 2010.

Integrated Fluidic-Chip Co-Design Methodology for Digital Microfluidic Biochips

Tsung-Wei Huang, Jia-Wen Chang, Tsung-Yi Ho
Department of Computer Science and Information Engineering
National Cheng Kung University, Tainan, Taiwan
{electron, jwchang}@eda.csie.ncku.edu.tw
tyho@csie.ncku.edu.tw

ABSTRACT

Recently, digital microfluidic biochips (DMFBs) have revolutionized many biochemical laboratory procedures and received much attention due to many advantages such as high throughput, automatic control, and low cost. To meet the challenges of increasing design complexity, computer-aided-design (CAD) tools have been involved to build DMFBs efficiently. Current CAD tools generally conduct a two-stage based design flow of fluidic-level synthesis followed by chip-level design to optimize fluidic behaviors and chip architecture separately. Nevertheless, existing *fluidic-chip* design gap will become even wider with a rapid escalation in the number of assay operations incorporated into a single DMFB. As more and more large-scale assay protocols are delivered in current emerging marketplace, this problem may potentially restrict the effectiveness and feasibility of the entire DMFB realization and thus needs to be solved quickly. In this paper, we propose the *first* fluidic-chip co-design methodology for DMFBs to effectively bridge the fluidic-chip design gap. Our work provides a comprehensive integration throughout fluidic-operation scheduling, chip layout generation, control pin assignment, and wiring solution to achieve higher design performance and feasibility. Experimental results show the effectiveness, robustness, and scalability of our co-design methodology on a set of real-life assay applications.

Categories and Subject Descriptors

B.7.2 [**Integrated Circuits**]: Design Aids - Layout, Place and Route

General Terms

Algorithms, Designs

Keywords

Biochip, microfluidics, Co-design

1. INTRODUCTION

Digital microfluidic biochips (DMFBs), a more versatile category of microfluidic technology, have recently emerged

as a popular alternative for laboratory experiments. Compared to conventional bench-top procedures, DMFB technology offers advantages of low sample and reagent consumption, less likelihood of error due to minimal human intervention, high throughput and sensitivity, automatic control, and low cost. With these advantages, DMFBs are gaining increasing applications including DNA analysis, proteomic analysis, immunoassay, and point-of-care diagnosis [8, 10].

Generally, a DMFB consists of a two-dimensional (2D) electrode array and peripheral devices (optical detector, dispensing ports, etc.) [3, 10]. On a DMFB, the sample carriers (i.e., *droplets*) are controlled by underlying electrodes using electrical actuation (a principle called electrowetting-on-dielectric or EWOD) [8]. By assigning time-varying voltage values to turn on/off the electrodes, droplets can be moved around the entire 2D array to perform fundamental operations. For instance, we can transport a droplet from a source location to a target location for a detecting operation or cluster adjacent electrodes to form a mixing device. These operations are carried out in a *reconfigurable* manner due to their flexibility in area and time domain. That is, we can perform these operations anywhere on the 2D plane during different time steps [12, 15].

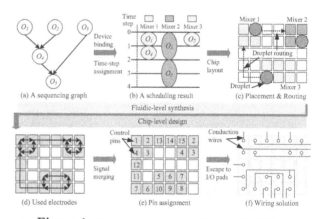

Figure 1: Conventional design flow of DMFBs.

As design complexity increases, CAD tools have been developed to build DMFBs efficiently. Traditionally, designing and realizing DMFBs consist of two major stages, *fluidic-level synthesis* and *chip-level design* [3]. In fluidic-level synthesis illustrated in Fig. 1(a)-(c), different assay operations (e.g., mixing, dilution, etc.) and their mutual dependences are first represented as a sequencing graph. Next, scheduling assigns time-multiplexed steps to these assay operations and binds them to a given number of devices so as to maximize parallelism [9]. Based on the scheduling result, device placement and droplet routing are conducted to generate a chip layout and establish droplet routing connections between devices in a reconfigurable manner [4, 11, 12]. On the other hand, chip-level design determines the required control pins and corresponding wiring connections for the underlying electrodes to execute the synthesis result. As illustrated in Fig. 1(d)-(f), fluidic-control information of used electrodes is obtained from the previous synthesis. Then, control pins must be *minimally* and appropriately assigned to electrodes for minimizing the bandwidth of input signals. Several signal merging strategies are proposed to facilitate the pin assignment with minimum control pin usage [6, 13, 14, 16]. Finally, conduction wires must be routed to establish correspondence among control pins and external driving ports [5].

Regarding this conventional design flow, a number of high-quality CAD tools have been developed to solve several associated combinational optimization problems [4, 5, 6, 9, 11, 12, 13, 14, 15, 16]. Due to the distinct nature between fluidic-level synthesis and chip-level design, most CAD tools are separately developed for the two design stages to simplify the complexity. For example, the works in [4, 9, 11, 12, 15] focus on specific stages in fluidic-level synthesis and works in [5, 13, 14] deal with the chip-level design. However, existing *fluidic-chip* (fluidic-level and chip-level) design gap may restrict the effectiveness and feasibility of the entire DMFB realization. Specifically, a successful fluidic-level-synthesis result cannot guarantee a successful chip-level-design solution. Fluidic-chip design gap is not concerned in previous works and thus the feasibility to realize the entire DMFBs is restricted. Even though there are researches [6, 16] take the pin-count issue of chip-level design into consideration in earlier stage, the wiring problem, which is a critical step for chip fabrication, is not referred. Therefore, it may need additional processes such as pin reassignment, rerouting, extra pin-count demand, or even a multi-layer routing structure to obtain a feasible wiring and successful chip-level design solution. Such processes add additional cost to chip fabrication and are not desirable for low-cost DMFBs. Fluidic-chip design gap problem will become even critical with a rapid escalation in the number of assay operations incorporated into a single large-scale DMFB. Unfortunately, to the best knowledge of the authors, there is still no work in the literature that provides a solution to deal with this concern.

1.1 Our Contributions

In this paper, we propose a fluidic-chip co-design methodology for DMFBs to effectively bridge the fluidic-chip design gap. Fig. 2 shows the comparison of conventional design flow and our co-design methodology. Compared with the conventional design flow, our co-design methodology is the first work in the literature that converges the fluidic-level synthesis and chip-level design. We provide a comprehensive integration throughout fluidic-operation scheduling, chip layout generation, control pin assignment, and wiring solution to achieve high design performance and feasibility. In realizing the co-design methodology efficiently and effectively, we identify three major design concerns that must be solved to meet the design convergence.

1. Existing scheduling methods focus on parallel controls of assay operations to increase throughput. However, device count is not restricted and thus much latter design effort is required especially for pin assignment and wiring.

2. Existing placement and routing methods allow devices and droplets to be arbitrarily moved in area and time domain. Although this scheme utilizes the reconfig-

urable properties, the entire solution incurs a lot of used electrodes with high pin-count demand and wiring problems.

3. Existing pin-assignment methods conduct the signal merging throughout the entire electrode set without careful arrangement. Despite achieving low pin-count, the underlying wiring situation may confront congestion, detour, or even infeasible problems.

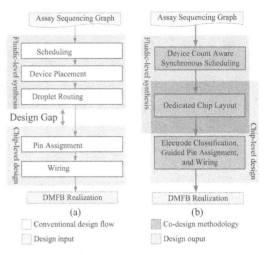

Figure 2: Comparison of conventional design flow and proposed fluidic-chip co-design methodology.

In order to handle these issues, our co-design methodology, as presented in Fig. 2(b), consists of three major stages: (1) Device count aware synchronous scheduling indeed reduces the required pin-count and wiring effort; (2) Dedicated chip layout simultaneously determines the 2D orientation of devices and routing paths, while simplifying and facilitating the designs of device placement, droplet routing, pin assignment, and wiring; (3) Electrode classification allows control pins to be orderly assigned with minimum wiring interference around different control signals. Guided pin assignment and wiring reduce the demanded pin-count and wirelength for signal connections.

Along with the proposed co-design methodology, our contributions include the followings.

1. We propose the *first* fluidic-chip co-design methodology for DMFBs that provides an integration throughout fluidic-level synthesis and chip-level design and achieve high success rate of the entire DMFB realization.

2. For a given assay, we derive an exact ILP formulation to optimally minimize the device count which impacts on pin-count and wiring issue for chip-level design. An effective stage-count reduction scheme is also provided to reduce the problem size, while keeping assay completion time satisfied.

3. We propose a novel chip layout to avoid wiring congestion induced by arbitrary device placement and control signal merging in conventional design flow. With the proposed chip layout, following pin-assignment and wiring guidelines maintain the low pin-count and short wirelength.

Experimental results show the effectiveness and robustness of our co-design methodology for different real-life assays. Our co-design methodology successfully realizes all assays on DMFBs while the conventional design flow can only pass one case. We also explore how the success rate of DMFB realization changes when the problem size becomes large to demonstrate the scalability of the proposed co-design methodology. Evaluation shows that our co-design methodology always maintains high success rate, while the success rate realized by conventional design flow continues to decrease dramatically as the problem size increases.

The remainder of this paper is organized as follows. Section 2 details the proposed co-design methodology. Section 3 shows our experimental results. Section 4 gives the conclusion.

2. FLUIDIC-CHIP CO-DESIGN METHODOLOGY

In this section, we first describe the idea of synchronous reactions and then detail each phase of the proposed fluidic-chip co-design methodology for DMFBs.

2.1 Synchronous Reactions

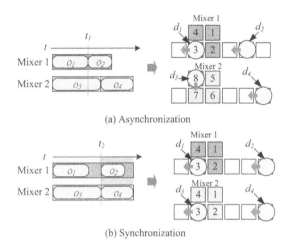

(a) Asynchronization

(b) Synchronization

Figure 3: Comparison of asynchronous reactions and synchronous reactions. The numbers in boxes represent pin numbers assigned to electrodes.

The objective of synchronous reactions is to facilitate the control signal merging for pin-count reduction by deriving a series of execution stages for synchronous controls. The differences between asynchronous and synchronous control can be explained by Fig. 3. Let us consider four operations o_1, o_2, o_3, and o_4, which are carried out by droplets d_1, d_2, d_3, and d_4, respectively. As shown in Fig. 3(a), asynchronous control methods in most previous works [9, 10] make each operation seamlessly enter into the subsequent operation in the end of its execution. Although this scheme maximizes the flexibility and independence of droplet controls, it may restrict the solution quality of pin-count reduction. Take the ending time step of o_1 (t_1 in Fig. 3(a)) for example, mixer 1 is ceased to let d_1 leave, then followed by letting d_2 in for o_2. Meanwhile, mixer 2 is still running for o_3, so d_4 cannot enter into mixer 2 for o_4. To realize the scheduling result without any fluidic error, mixers 1 and 2 cannot share the same control pins. Thus, mixer 1 needs 4 control pins (pin 1, 2, 3, 4) and mixer 2 needs another 4 control pins (pin 5, 6, 7, 8). This design method may incur a high pin-count demand and is not suitable for low-cost DMFBs [6, 14]. On the other hand, in Fig. 3(b), mixers 1 and 2 are controlled together, meaning that the mixers must begin and cease the executions synchronously. That is, o_1 and o_3 are simultaneously begun and ceased when the slower operation (o_3 in Fig. 3(b)) is completed. Then, o_2 and o_4 simultaneously

begin after o_1 and o_3 are both completed (t_2 in Fig. 3(b)) and then ceased in the same feature. In this manner, mixers 1 and 2 can share their control signals (pin 1, 2, 3, 4 in Fig. 3(b)), which is more favorable for pin-count reduction.

To achieve synchronous reactions, the work in [6] expands the scheduling to a stage assignment problem by deriving a series of execution stages and align operations to these stages for synchronous controls. As an example shown in Fig. 4, to enable synchronous controls, o_1, o_2, and o_3 can be aligned to stage 1 and o_4 and o_5 can be aligned to stages 2 and 3, respectively. After binding all operations with appropriate devices, the synchronized control signals will *simultaneously* begin (cease) the execution in the start (end) of each stage.

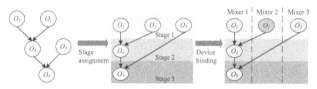

Figure 4: Synchronous scheduling by stage assignment.

Although stage assignment provides a good solution for synchronous control, the followed chip-level design issue is not addressed. To realize the synchronous reaction, there must be conduction wires to connect synchronized devices together in wiring step. That is, introducing the synchronous reactions will inevitability incur wire congestion, detour, thereby increasing the complexity of wire routing if device placement and droplet routing in fluidic-level synthesis are not carefully designed. As a result, a novel methodology to deal with this concern is needed. In following subsections, we propose a comprehensive co-design methodology to bridge the design gap between fluidic-level synthesis and chip-level design.

2.2 Device Count Aware Synchronous Scheduling

Based on the concept of synchronous reactions, we observe that different stage-assignment results can lead to different device counts. As explained in Fig. 5, o_3 can be aligned to two stages, stage 1 or stage 2, without violating the precedence on o_5. If we align o_3 to stage 1 as shown in (a), additional mixing device, mixer 3, must be included for execution, implying extra pin count and wiring effort must be added to drive it. On the other hand, if we align o_3 to stage 2 as shown in (b), o_3 can share the mixer 2 with o_2 in different execution stages and thus the required number of devices can be decremented. In this case, we achieve a fewer number of control pins and simple wiring solution to realize the DMFB. As long as a device is placed on the microfluidic array, associated control pins and wiring must be included for this device and its connections to other devices. To avoid design overhead, it is consequently desirable to minimize the required device count for fluidic-operation scheduling.

2.2.1 Problem Formulation

Concerning those design issues discussed above, the scheduling problem can be formulated as follows.

Given: *A sequencing graph of operations and a device library.*

Objective: *Derive a set of execution stages for synchronous execution and bind all operations with appropriate devices to these stages, while minimizing the device count.*

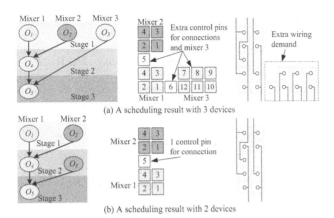

(a) A scheduling result with 3 devices

(b) A scheduling result with 2 devices

Figure 5: A scheduling result with higher device count in (a) results in more control pins and wiring overhead than (b).

2.2.2 ILP Formulation

To optimally solve the scheduling problem while minimizing the device count, we present an ILP formulation with the following notations, objective function, and constraints.

Notations:

1. An operation set O with each operation indexed by i.
2. A device set D with each device indexed by j.
3. A stage set S with each stage indexed by k.
4. Available device set D_i for operation i, $\forall i \in O$, $D_i \subseteq D$.
5. Execution time T_j for device j, $\forall j \in D$.
6. A 0-1 variable $o_{i,j,k}$ to represent that operation i is bound with device j and executed in stage k, $\forall i \in O$, $\forall j \in D_i$, $\forall k \in S$.
7. An integer variable b_k (f_k) to represent the beginning (finishing) time for stage k, $\forall k \in S$.
8. An integer variable r_j to represent the required number for device j, $\forall j \in D$.

Objective: The objective is to minimize the total device count as follows.

$$Minimize : \sum_{j \in D} r_j \qquad (1)$$

Subject to:

1. *Exclusivity constraints:* An operation is bound with one device and executed in one stage.

$$\sum_{j \in D_i} \sum_{k \in S} o_{i,j,k} = 1, \forall i \in O \qquad (2)$$

2. *Timing constraints:* For each stage, the beginning time should be less than the finishing time and all stages should be finished by a maximum timing value.

$$0 \leq b_k \leq f_k \leq T_{max}, \forall k \in S \qquad (3)$$

Note that T_{max} can be set to an upper bound of the minimum assay completion time obtained by greedy assignment or a user-defined value [9].

3. *Synchronization constraints:* In realizing the synchronization, all stages must be orderly executed without overlapping.

$$f_{k_1} \leq b_{k_2}, \forall k_1 \in S, \forall k_2 \in S, k_1 < k_2 \qquad (4)$$

Besides, a stage will not enter into the subsequent stage until all included operations have been finished. In other words, the duration of a stage is lower-bounded by the slowest operation assigned to the stage.

$$f_k - b_k - T_j o_{i,j,k} \geq 0, \forall k \in S, \forall i \in O, \forall j \in D_i \qquad (5)$$

4. *Dependency constraints:* If there is a dependency "→" between operations i_1 and i_2 (i.e., operation i_2 should be executed after operation i_1 is finished), the stage k_2 which includes operation i_2 must appear after the stage k_1 which includes operation i_1.

$$f_{k_1} - b_{k_2} + M(o_{i_1,j_1,k_1} + o_{i_2,j_2,k_2} - 2) \leq 0,$$
$$\forall i_1 \rightarrow i_2, j_1 \in D_{i_1}, j_2 \in D_{i_2}, k_1 \in S, k_2 \in S \qquad (6)$$

Note that M is a big number and must be larger than T_{max} for correct formulations.

5. *Device requirement:* The inherent reconfigurability of DMFBs allows operations to share the same device in different execution stages. Therefore, the required number for each device should be more than its requirement in any stage.

$$r_j \geq \sum_{i \in O} o_{i,j,k}, \forall j \in D, \forall k \in S \qquad (7)$$

Note that the summation of $o_{i,j,k}$ is only calculated when $j \in D_i$ to avoid the boundary overflow.

2.2.3 Solution Space Reduction

Although the basic ILP formulation presented in the previous subsection can optimally solve the scheduling problem, it is complicated for large-scale assays and not efficient to be solved without reductions. The major difficulty is the number of execution stages, which dominates the major complexity of constraints and variables. Thus, the stage count must be conditionally bounded to avoid runtime overhead.

To handle such an issue, we conduct explicit stage assignments for all operations. We assign a stage interval to each operation o_i, denoted as a set S_i of a consecutive stage sequence, to make operation o_i only be executed in these stages. The objective is to minimally bound the sizes of stage intervals while considering the dependencies around operations. Since the sequencing graph is a directed acyclic graph, the number of stages can be bounded by the length of a critical path on the graph [3, 9, 10]. With respect to the critical path, we can easily obtain the stage interval (i.e., earliest and latest available stages) in which an operation should be executed.

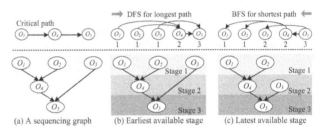

(a) A sequencing graph   (b) Earliest available stage   (c) Latest available stage

Figure 6: Finding the stage intervals (earliest and latest available stages with respect to the critical path) for operations to reduce the ILP complexity.

For example, suppose we have a sequencing graph with 5 operations as shown in Fig. 6(a). With respect to the critical path, $o_1 \rightarrow o_4 \rightarrow o_5$, we conduct DFS (depth-first search) to trace the longest paths from beginning nodes $\{o_1, o_2, o_3\}$ (i.e., zero indegree) to all other nodes, as shown in Fig. 6(b). Note that the distance values of beginning nodes are initialized as 1. Then, we reverse the sequencing graph and conduct BFS (breadth-first search) to backtrace the shortest paths from the ending node $\{o_5\}$ (i.e., zero outdegree) to all other nodes. Note that the distances of ending nodes are initialized as their longest distance values and will decrease level by level in BFS routine. Since all executions should follow the dependencies, the longest (shortest) distances thus represent the earliest (latest) stages in which operations can be executed. For example, since the longest and shortest distances for o_3 are 1 and 2, the earliest and latest available stages for o_3 are stage 1 and stage 2, respectively.

After finding all stage intervals S_i, we replace S with S_i for constraints (2), (5), and (6) in our ILP formulation. Since S_i is relatively small compared with S, the entire solution space and complexity of constraints and variables can be significantly reduced. Besides, the proposed reduction method follows the dependencies when assigning stage intervals for operations. The dependency constraints around those operations with only *one* assigned stage become redundant and thus can be removed. Take Fig. 6(b)-(c) for example, since both earliest and latest stages of operations o_1 (o_4) are assigned by stage 1 (stage 2), the dependency constraint for $o_1 \rightarrow o_4$ can be removed.

The advantages of the proposed reduction methods are threefold. First, since the stage count is tightly bounded by the length of the critical path, the assay completion time can be also minimally bounded without excessively suspending operations between stages. On the other hand, minimally assigning the stage intervals can also avoid redundant solution search and restrict unnecessary formulations of constraints and variables. Furthermore, our reduction method also provides flexibility. If the initial assignments are too tight, we can easily and incrementally expand the stage intervals to find a better solution without numerous modification of the entire routine. That is, we can expand all latest available stages for all operations by a user-defined range. Since such expansion may result in longer assay completion time and ILP solving time, we empirically set the expansion to be 1.

2.3 Chip Layout Construction

After all operations are scheduled with minimum device count, these devices must be well-placed to generate a chip layout. Most previous device placement methods focus on determining the device locations during different time steps to guarantee correct execution and feasible routing solution. Strategies such as segregation between devices to avoid unexpected mixing and dynamically moving devices to free more cells for routing are also considered [3, 12, 15].

Pin	Cell	time		
1	c_1	Mixer	Dilutor	Mixer
2	c_2	Mixer	Dilutor	Mixer
3	c_3	Mixer	Mixer	Dilutor
4	c_4	Dilutor	Mixer	Dilutor
5	c_5	Dilutor	Mixer	Mixer

(a) Pin count: 5

Pin	Cell	time		
1	c_1	Mixer	Mixer	Mixer
1	c_2	Mixer	Mixer	Mixer
1	c_3	Mixer	Mixer	Mixer
2	c_4	Dilutor	Dilutor	Dilutor
2	c_5	Dilutor	Dilutor	Dilutor

(b) Pin count: 2

Figure 7: (a) Cells are arbitrarily functioned. (b) Cells are dedicated to specific functions.

Although these works can solve the device placement and droplet routing, they may suffer from chip-level design problems. An essential problem of previous device placement methods is the excessive allowance of a cell (i.e., electrode) to be arbitrarily functioned, such as mixing, diluting and detecting devices in different time steps. However, different devices are always associated with different control signals, implying such methods require high pin-count demand to maximize the signal bandwidth and freedom of electrode controls, as shown in Fig. 7(a). Thus, to improve this deficiency, it is desirable to make cells be dedicated to unique devices or specific functions. As shown in Fig. 7(b), by dedicating cells to specific functions, cells with same action can be controlled synchronously. What we concern now is determining the *fixed* 2D location on specific cells for each device, while achieving feasible droplet routing solutions.

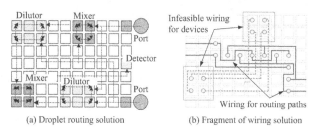

(a) Droplet routing solution (b) Fragment of wiring solution

Figure 8: Spreading out the droplet routing paths around devices may lead to wiring problems.

On the other hand, the major problem of previous droplet routing methods is that they spread out the routing paths between devices for establishing connections [4, 11], as illustrated in Fig. 8(a). Although such methods utilize the temporal flexibility of DMFBs, the induced wiring complexity may trigger many blocking or detouring problems. Take Fig. 8(b) for example, the wiring solution for droplet routing paths significantly blocks the wiring for devices, which makes an infeasible design outcome especially for connecting synchronous devices. Hence, it is necessary to conduct a specific orientation for routing paths with device connections, while taking the chip-level design issues into consideration.

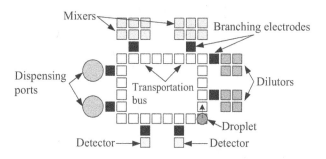

Figure 9: The proposed chip layout.

To deal with these issues, the device placement and droplet routing for the entire assay should be decided simultaneously on a 2D plane. In this concern, we propose a dedicated chip layout to deal with these issues and support our scheduling method. As presented in Fig. 9, the architectural layout consists of three major components, *surrounding devices* for executing operations, *central transportation bus* for droplet routing and device connections, and *branching electrodes* for accessing the passages between bus and devices. With the proposed chip layout, we achieve three major advantages as follows.

1. Since the outside surrounding devices are fixed without arbitrary movements and the inside bus is independently oriented without significant interference with devices, the aforementioned wiring problem can be avoided.
2. In addition to routing droplets, the central transportation bus can serve as storages for intermediate droplets. Thus, extra storing devices for holding droplets are omitted and thus the design effort can be reduced.
3. Instead of directly constructing the chip layout in 3D configuration (z-axis for time), the proposed layout focuses on determining the 2D orientation of devices and transportation bus, which greatly reduces the design complexity.

Regarding this, different device orderings and bus lengths may produce different droplet routing distances and realizations of the previous scheduling results. Consider the sequencing graph in Fig. 5(a). Each edge in the sequencing graph implies the droplet transportation if the incident two operations are bound with different devices. We refer to this droplet transportation as *connection* between these devices. The connections can be used to represent the relationships between devices. We can get a shorter transportation length if we put the device of o_3 and o_5 closer. That is, a means to determine the ordering of surrounding devices and the length of transportation bus is the major concern of constructing the chip layout.

2.3.1 Problem Formulation

Regarding those issues discussed in the previous subsection, the problem of constructing a chip layout can be formulated as follows.

Given: *A set of devices and their connection relations.*

Objective: *Derive a device ordering to minimize the total transportation distances for droplets, while minimally determining the bus length to supply all device connections and required droplets.*

2.3.2 Linear Assignment for Device Ordering

As droplets are routed between devices, if device orderings are not well-oriented, a longer droplet routing time is required, which may potentially affect the assay throughput [4, 10, 11, 16]. Since it is not straightforward to recognize the device ordering on a 2D plane, we transform the surrounding devices to an 1D ordering, as demonstrated in Fig. 10. Referring to the scheduling result, we can obtain the connections between devices due to original operation dependencies. To generally simplify the problem, adjacent devices are associated with one unit distance. Therefore, the problem now is deriving an 1D ordering such that the total induced distances of device connections are minimized.

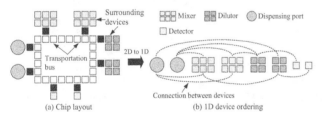

Figure 10: An example of transforming surrounding devices to an 1D ordering.

The problem we formulate above is actually a classic linear assignment problem, which is well-known as NP-complete [2]. This feature justifies the necessity of using heuristic approaches to solve our problem. And fortunately, there are high-quality heuristics and approximation algorithms for solving the linear assignment problem. In this paper, we conduct the typical method by iteratively swapping pairs of the devices as long as the total distances can be reduced. The swaps are repeatedly conducted until an equilibrium is reached. That is, the amount of distances can no longer be reduced by swapping any pair of devices.

Besides the minimization of the droplet routing time, chip-level design issues must be also considered. An essential concern is the devices with the same type (e.g., mixing devices or detecting devices) that are synchronized and controlled together. These synchronized devices share the same control signals to reduce the pin count. Considering the wiring issues, it is desirable to place these devices closely so as to

minimize the wirelength and to simplify the wiring problem. Therefore, for synchronized devices, we contract them as a single device in formulating the linear assignment problem to guarantee they are placed in adjacent positions.

2.3.3 Transportation Bus Length Determination

Another key issue is to determine enough length of transportation bus to supply droplets for realizing the scheduling result. In other words, following pre-assigned execution stages, the bus length must be minimally bounded while accommodating the maximum number of droplets that are delivered between stages.

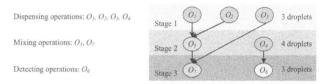

Figure 11: Determining the bus length to supply enough droplets.

The droplet accommodation of the bus can be determined by examining the droplet requirement for each stage while picking a maximum one as the solution. For each stage, the droplet requirement comes from two parts, its own dispensing operations and predecessor operations. As explained by Fig. 11, since there are only 3 dispensing operations in stage 1, the bus must supply 3 droplets. In stage 2, there are 1 dispensing operation o_4 and 3 predecessor operations, o_1, o_2, and o_3 with edges passing through this stage, and thus we need 4 droplets on the bus for this stage. In the same way, stage 3 has only 3 predecessor operations, o_3, o_4, and o_5 that will deliver droplets to this stage and thus we need 3 droplets on the bus for this stage. In these concerns, a transportation bus which can supply simultaneous movements of at least 4 droplets is required. To avoid unexpected mixing between droplets, a minimum segregation of 4 cells between two droplets must be included (we will discuss it in the next subsection). Therefore, the bus length can be determined as a multiple of 4. That is, total $4 \times 4 = 16$ cells (i.e., electrodes) are required for constructing the transportation bus for this example. Then, based on pre-determined device ordering, we can circularly connect devices to the bus by adding branching cells to generate the entire 2D chip layout. Note that there are three detailed implementation concerns as follows. (1) Both square and rectangular orientations of the bus are allowed. (2) Segregation between devices is also included to avoid unexpected mixing. (3) If devices are too large to be connected by current bus length, we incrementally increase the bus length by 4.

2.4 Pin Assignment and Wiring Solution

To realize the scheduled fluidic functions, control pins must be appropriately and minimally assigned on the chip for low-cost fabrication [13, 14]. Besides synchronizing the devices in each stage, we also synchronously control the droplets to achieve pin sharing on the transportation bus. A primary issue is the avoidance of fluidic errors between droplets during control signal sharing. As explained in Fig. 12, to avoid undesirable fluidic errors (e.g., mixing or splitting) caused by neighboring actuations, we require at least 4 control pins on the transportation bus, such that a safe spacing can be maintained between droplet movements [11].

After control pins are assigned on the chip, wires must be routed to establish the correspondence between control pins and then escape to outside I/O pads. Since multiple electrodes may share the same control pins, the wiring problem

(a) 3 pins for droplet routing   (b) 4 pins for droplet routing

Figure 12: Synchronized droplet movements require at least four control pins to prevent the fluidic error.

thus can be viewed as an escape routing with multi-terminal pins. However, such a wiring problem is well-known as NP-complete and is computationally expensive if we directly solve it [3, 5]. Therefore, a specific method must be developed to handle this issue.

2.4.1 Problem Formulation

The pin-assignment and wiring problems can be formulated as follows.

Given: *A scheduling result and the corresponding chip layout.*

Objective: *Derive a pin-assignment result and establish wiring connections between control pins to realize the fluidic-level synthesis result without introducing erroneous fluidic behaviors.*

2.4.2 Classifications of Pins and Wires

The major reason that the previous pin-assignment method suffers from wiring problems is the excessive signal merging without any restriction [5, 14, 16]. Since the pins with the same control signal must be wired together to receive the same input signal, overly merging the control signals around all electrodes may potentially incur detours, blocking, or even deadlock between wires. Therefore, it is desirable to make a specialized plan throughout the entire electrode set to avoid this problem.

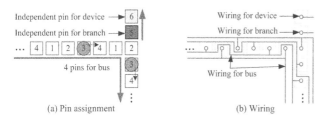

(a) Pin assignment   (b) Wiring

Figure 13: Pin assignment and wiring for transportation bus, branch, and device.

To this end, we classify all electrodes to three categories, bus, branch, and device by their functionality and conduct pin assignment and wiring individually for the three categories, as explained in Fig. 13. Regarding the pin-constrained design issue [3, 10, 14], we propose the following pin-assignment guidelines for pin-count reduction.

1. Since droplets are synchronously moved, four independent control pins 1, 2, 3, and 4 are sequentially and repeatedly assigned to the transportation bus.
2. Branching electrodes are assigned by dedicated control pins for independent accesses of the passages between transportation bus and devices.
3. Devices with the same type are synchronously controlled and shared the same control pins.

As we comprehensively address the wiring issues in the proposed co-design methodology, wiring connections between control pins can be systematically established based on the following guidelines.

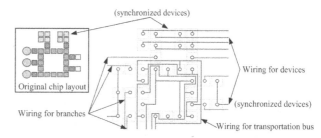

Figure 14: Realization of wiring solution.

1. Since there are always four control pins assigned on the well-oriented transportation bus, the associated wiring solution can be sequentially constructed from innermost region to outermost region, following the pin order 1, 2, 3, 4 (see the wiring for transportation bus in Fig. 14).
2. Since synchronized devices are placed adjacent to each other, shared control pins can be wired together without lots of detours so as to minimize the wirelength (see the wiring for devices in Fig. 14).
3. For branching electrodes that are assigned by dedicated control pins, the preserved segregation between devices provides region for wiring these pins (see the wiring for branches in Fig. 14).

Note that since we have minimized the required device count in previous scheduling, the transportation bus length and branching electrodes can be also minimally bounded. Thus, the entire pin-assignment and wiring effort is also reduced.

3. EXPERIMENTAL RESULTS

We implement our fluidic-chip co-design methodology in C++ language on a 2-GHz 64-bit linux machine with 16GB memory and use GLPK [1] as our ILP solver. Evaluations are based on a set of real-life chip applications for multiplexed assay, DNA sequencing assay, in-vitro diagnostic assay, and PCR amplification assay [7, 9, 14]. We use the real chip specification of the wiring model in [5]. To show the effectiveness of our integrated design method, we implement the conventional design flow for DMFB realization by five stages of scheduling, device placement, droplet routing, pin assignment, followed by wiring. For purpose of comparison, we choose four state-of-the-art works from the same group at Duke University as our conventional design flow algorithms. Works of [9, 11, 12, 14] are conducted as the solvers for scheduling, device placement, droplet routing, and pin assignment, respectively. Then, since the associated wiring problem is NP-complete, we conduct the approach in [5] which is based on maze routing to sequentially and iteratively route a nearest electrode pair with the same control pin until all electrodes are routed.

Table 1: Comparison between the conventional design flow and our fluidic-chip co-design methodology.

Assay	Conventional design				Ours			
	#D	T	#P	#WL	#D	T	#P	#WL
Multiplexed	7	63	25	N/A	6	67	14	562
DNA	11	67	38	1094	9	79	22	610
Diagnostic	11	50	26	N/A	9	63	18	748
PCR	14	26	14	N/A	10	47	18	820
Total	43	206	95		34	256	72	

#T: assay completion time (sec.)   #D: device count   #P: pin-count
#WL: wirelength (measured by the number of used grids)

As listed in Table 1, the overall comparison results show that our integrated fluidic-chip co-design methodology successfully accomplishes all assays by correctly deriving the result from fluidic-level synthesis to chip-level design. However, the conventional design flow can only accomplish one assay of DNA sequencing. The major reason is that we provide an integration between fluidic-level synthesis and chip-level design, which is different from the conventional one without design convergence. In our synchronous scheduling, we first minimize the required device count by 21% for reducing the effort in later design, especially in chip level. The increase of assay completion time is expected, since synchronization may delay some operations in an execution stage until the slowest one is finished. Based on the scheduling result, we propose a dedicated chip layout for realizing the scheduling result while considering the impact on chip-level designs. With appropriate electrode classification to avoid the wiring congestion followed by pin-assignment and wiring guidelines, we reduce the required pin count by 30%, while maintaining successful wiring solutions for all assays. Even compared with the only DNA sequencing assay which can be successfully realized by the conventional design flow, we also reduce the wirelength by 44%. Note that in CPU runtime all assays can be solved within 1 second by conventional design flow and our co-design methodology.

Table 2: Run time results of the optimal ILP formulation and reduced ILP formulation

Assay	Optimal ILP		Reduced ILP		
	#D	CPU	#D	Error	CPU
Multiplexed	6	> 1 day	6	0%	0.04
DNA	11	> 1 day	11	0%	0.07
Diagnostic	7	> 1 day	9	29%	0.12
PCR	9	> 1 day	10	11%	0.10

#D: device count   CPU: cpu run time in seconds

In runtime evaluation, solving the ILP formulation for synchronous scheduling with minimum device count dominates the entire runtime performance by about $90 \sim 95\%$. We list the runtime results of the optimal ILP formulation and the proposed reduced ILP formulation in Table 2. Compared with the optimal ILP formulation, our reduced ILP formulation can efficiently solve all the assays in much reasonable CPU runtime by all less than 1 second. We also achieve two optimal results in multiplexed and DNA assays.

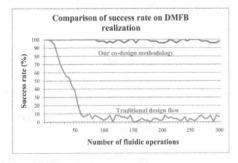

Figure 15: Comparison of the success rate on DMFB realization between our co-design methodology and conventional design flow.

Then, we randomly generate a set of assays with large problem size to demonstrate the scalability of our co-design methodology. The problem size is evaluated as the number of operations in an assay. For each problem size, we simulate the fluidic operations and generate 50 assays to test the success rate of DMFB realization (from fluidic-level synthesis to chip-level design without any failure) of our co-design

methodology and conventional design flow. As illustrated in Fig. 15, our co-design methodology always maintains high success rate of average 99% for all test assays (note that our fluidic-chip co-design methodology can efficiently solve the largest assay with 300 operations by taking only 23 seconds). Nevertheless, the success rate by conventional design flow decreases dramatically as the problem size increases. We find out that when the problem size increases to about 60 operations, the success rate is always lower than 10%, which is not strong enough to realize many large-scale DMFBs.

4. CONCLUSION

In this paper, we have presented the *first* integrated fluidic-level and chip-level co-design methodology for digital microfluidic biochips (DMFBs) that provides an integration throughout fluidic-level synthesis and chip-level design. We also have comprehensively identified the factors that would affect DMFB realization and explore properties that are favorable for bridging the fluidic-chip performance gap. Experimental results have demonstrated that our co-design methodology achieves higher success rate of DMFB realization over the conventional design flow.

5. REFERENCES

[1] http://www.gnu.org/software/glpk/.
[2] M. R. Garey and D. S. Johnson, "Computers and Intractibility: A Guide to the Theory of NP-Completeness," 1979.
[3] T.-Y. Ho, J. Zeng, and K. Chakrabarty, "Digital microfluidic biochips: A vision for functional diversity and more than Moore," *IEEE/ACM ICCAD*, pp. 578–585, 2010.
[4] T.-W. Huang and T.-Y. Ho, "A fast routability- and performance-driven droplet routing algorithm for digital microfluidic biochips," *IEEE ICCD*, pp. 445–450, 2009.
[5] T.-W. Huang, S.-Y. Yeh, and T.-Y. Ho, "A network-flow based pin-count aware routing algorithm for broadcast electrode-addressing EWOD chips," *IEEE/ACM ICCAD*, pp. 425–431, 2010.
[6] C. C.-Y. Lin and Y.-W. Chang, "ILP-based pin-count aware design methodology for microfluidic biochips," *Proc. IEEE/ACM DAC*, pp. 258–263, 2009.
[7] R.-W. Liao and T.-H. Yang, "Design and fabrication of biochip for in-situ protein synthesis," Master's Thesis, Department of Optics and Photonics, National Central University, Taiwan, 2008.
[8] M. G. Pollack, A. D. Shenderov, and R. B. Fair, "Electrowetting-based actuation of droplets for integrated microfluidics," *LOC*, pp. 96–101, 2002.
[9] F. Su and K. Chakrabarty, "Architectural-level synthesis of digital microfluidics-based biochips," *IEEE/ACM ICCAD*, pp. 223–228, 2004.
[10] F. Su, K. Chakrabarty, and R. B. Fair, "Microfluidics based biochips: Technology issues, implementation platforms, and design-automation challenges," *IEEE Trans. on CAD*, pp. 211–223, 2006.
[11] F. Su, W. Hwang, and K. Chakrabarty, "Droplet routing in the synthesis of digital microfluidic biochips," *IEEE/ACM DATE*, pp. 1–6, 2006.
[12] F. Su and K. Chakrabarty, "Module placement for fault-tolerant microfluidics-based biochips," *ACM TODAES*, vol. 11, pp. 682–710, 2006.
[13] T. Xu and K. Chakrabarty, "Droplet-trace-based array partitioning and a pin assignment algorithm for the automated design of digital microfluidic biochips," *Proc. CODES+ISSS*, pp. 112–117, 2006.
[14] T. Xu and K. Chakrabarty, "Broadcast electrode-addressing for pin-constrained multi-functional digital microfluidic biochips," *Proc. IEEE/ACM DAC*, pp. 173–178, 2008.
[15] P.-H. Yuh, C.-L. Yang, and Y.-W. Chang, "Placement of defect-tolerant digital microfluidic biochips using the T-tree formulation," *ACM JETC*, vol. 3, no. 3, 2007.
[16] Y. Zhao, R. Sturmer, K. Chakrabarty and V. K. Pamula, "Synchronization of concurrently-implemented fluidic operations in pin-constrained digital microfluidic biochips," *IEEE Int. Conf. on VLSI Design*, pp. 69-74, 2010.

Transformation from Ad Hoc EDA to Algorithmic EDA

Jason Cong
Department of Computer Science
University of California, Los Angeles
Los Angeles, CA 90095, USA
cong@cs.ucla.edu

ABSTRACT

In this paper I will attempt to provide an overview of Professor C. L. (Dave) Liu's contributions to electronic design automation (EDA). I will start with his early work as a pioneering researcher in combinatorial mathematics and algorithms, and then highlight several of his algorithmic contributions to a number of EDA problems—including floorplanning, placement, routing, FPGA synthesis, high-level synthesis, and fault-tolerant designs. Liu's studies of these problems were very timely, right at the time when they became important to the integrated circuit (IC) design industry. Many of his solution techniques are also timeless, as they have been applied to other EDA problems. Finally, I discuss his contributions as a great educator—one who has trained multiple generations of leaders, researchers and entrepreneurs in EDA and beyond. These people carry on Liu's vision and efforts in the transformation of ad hoc EDA to algorithmic EDA.

Categories and Subject Descriptors
B.7.2 [**Hardware**]: Design aids.

General Terms
Algorithms, design, theory.

Keywords
Electronic design automation, algorithm, floorplanning, placement, routing, synthesis, fault-tolerant designs.

1. INTRODUCTION

Professor C. L. Liu led the trend and transformation from ad hoc EDA to algorithmic EDA. His vision, leadership, and sustained effort over several decades demonstrated the effectiveness of "the algorithmic" approach, which is to say, bringing what were once considered to be esoteric techniques to the mainstream of EDA tools. He was one of the early advocates for a more rigorous mindset to design automation, arguing that powerful, formal algorithmic techniques were essential to the effective solution of complex design automation problems. In contrast to the ad hoc solution methods prevalent during the early days of EDA, techniques such as mathematical programming, dynamic programming, simulated annealing, network flow algorithms, branch and bound, and graph theoretic algorithms are now applied pervasively to all aspects of VLSI design. Their widespread use is due in part to Liu's innovative application of these techniques to numerous problems arising in electronic design. His technical contributions are embedded in the foundations of a multitude of current EDA tools within several disciplines—including behavioral synthesis, logic synthesis, and physical design. To recognize his distinguished contributions, the EDA Consortium and IEEE Council on Electronic Design Automation (CEDA) selected Prof. Liu for the prestigious Phil Kaufman Award last year. I had the honor and pleasure to be the award presenter to introduce Prof. Liu at the award banquet held on November 8, 2011. This gave me a unique opportunity to learn even more about the breadth and magnitude of Liu's professional contributions during his career. In this article I share some of the information that I collected, with an emphasis on Prof. Liu's contributions to EDA.

Permission to make digital or hard copies of all or part of this work for personal or classroom use is granted without fee provided that copies are not made or distributed for profit or commercial advantage, and that copies bear this notice and the full citation on the first page. To copy otherwise, or republish, to post on servers or to redistribute to lists, requires prior specific permission and/or a fee.
ISPD'12, February 8–12, 2011, Napa, California, USA.
Copyright 2012 ACM 978-1-4503-1167-0/12/03...$10.00.

2. A PIONEER IN COMPUTER ALGORITHM RESEARCH

Prof. Liu received his BS degree from the National Cheng-Kung University (NCKU) in Taiwan in 1956. In fact, when he entered the university, it was still called the Tainan College of Engineering, and he was among the first to graduate from NCKU under its new name. Liu then began his graduate study at MIT. His master's thesis was on "A Study in Machine Aided Learning," completed in 1960 under Ronald Howard. This thesis investigated "the advantages and the possibilities of using a computer as a teaching machine," and "A program using an IBM 704 computer to teach arithmetic and a program using an IBM 704 to teach matrix algebra are suggested."[1] Clearly, this was a pioneering work on distant learning. However, a better time-sharing operating system was needed to support such distant learning capabilities.

While waiting for that operating system to be a reality, Prof. Liu's interest turned to a subject much more closer to EDA. His Ph.D. thesis was on "Some Memory Aspects of Finite Automata," which investigated several special cases where "the unique determination of the behavior of an automation is possible, even when a portion of its past history is unknown."[2] The thesis work was completed in 1962 under Dean Arden. It is worthwhile to mention that Arden was also the Ph.D. advisor of several other prominent computer scientists, including Frederick Hennie and

[1] From the abstract of Liu's M.S. thesis.

[2] From the abstract of Liu's Ph.D. thesis.

Jack Dennis, both faculty members at MIT (and Dennis was the Ph.D. advisor of Randy Bryant, another Phil Kaufman Award recipient).

After his Ph.D. study, Liu stayed at MIT as a faculty member, where he made a number of important and pioneering contributions to combinatorial mathematics and computer algorithms. One landmark contribution was his book, *Introduction to Combinatorial Mathematics,* published in 1968 [1], which covered a number of important techniques critical to the development of the computer algorithm field. Some sample chapters include

- Chapter 3: Recurrence Relations
- Chapter 6: Fundamental Concepts in the Theory of Graphs
- Chapter 7: Trees, Circuits, and Cut-sets
- Chapter 10: Transport Networks
- Chapter 11: Matching Theory
- Chapter 12: Linear Programming
- Chapter 13: Dynamic Programming

Many of the concepts and techniques covered in the book were widely used later on in the EDA algorithms. It is worth noting that in the same year, another important book was published: *The Art of Computer Programming, Vol. 1, Fundamental Algorithms* by Donald Knuth. Both books had a profound impact on computer science.

Liu was also among the first in extending and applying these algorithmic techniques to solve real-life problems in other disciplines of computer science. One best-known example is his work on real-time scheduling. His classic paper, "Scheduling Algorithms for Multiprogramming in a Hard-Real-Time Environment," with J. W. Layland published in the Journal of ACM in 1973 [2], is a cornerstone of scheduling theory for real-time systems. It proved that for static scheduling, rate-monotonic scheduling (or RMS in short) gives the best solution (that is, one should schedule events according to the decreasing order of their occurrence rates), and for dynamic scheduling, the deadline-driven scheduling guarantees full utilization. This is the most cited paper in the real-time system community, with a total citation count over 7500 (according to Google Scholar and as of January 2012). This number also matches the best-cited paper in EDA, which is by Randy Bryant on binary-decision diagrams [3] (about 7600 citations according to Google Scholar). Liu's work on the RMS algorithm had a significant impact on the embedded systems world, the market size of which is measured in billions of dollars. One good example in this market segment is WindRiver whose real-time operating system has served a wide range of semiconductor and embedded system companies (acquired by Intel in 2009).

1973 was also the year that Liu made the transition from MIT to the University of Illinois at Urbana-Champaign (UIUC), together with his wife Jane Liu who also served on the faculty of the Computer Science Department. In the first a few years at UIUC, Liu continued his research on combinatorial mathematics and algorithms, and published another landmark book, *Elements of Discrete Mathematics*, in 1977 [4]. It covers topics such as

- Sets and propositions
- Relations and functions
- Graphs and trees
- Recurrence relations and generating functions
- Groups and rings
- Boolean algebra

These topics became part of the standard curriculum for any undergraduate computer science program worldwide, and Liu's book was adopted by many universities as the standard textbook on discrete mathematics. It was translated into Japanese in 1980 and into Chinese in 1981. In fact, it was used in my undergraduate course on discrete mathematics at Peking University in the early 1980s, and greatly influenced my decision to apply to UIUC to work with Liu for my Ph.D. studies. I have always been grateful that he gave me that opportunity.

3. TRANSFORMATION FROM AD HOC EDA TO ALGORITHMIC EDA

Prof. Liu branched into EDA at the beginning of the 1980s. His first paper in the Design Automation Conference (DAC) was on "Optimal Bipartite Folding of PLA" [5] published in 1982 DAC. The timing was perfect, as this was the beginning of the long exponential growth of the IC industry. As a point of reference, in 1981, IBM introduced the first IBM PC, which ran on the Intel 8088 microprocessor with a clock frequency of 4.77 MHz and transistor count of 29,000. Over the next thirty years, the transistor count followed Moore's Law faithfully, doubling roughly every two years, and now with several billions of transistors integrated on a single chip. The design complexity quickly outgrew the simple ad hoc EDA algorithms used in the 1970s (e.g., iterative pairwise exchanges for circuit partitioning, greedy construction or min-cut based circuit placement, line-probe + maze routing). The industry needed more efficient EDA algorithms to cope with the exponentially growing design complexity. Liu and his research group provided timely solutions to many important EDA problems as the IC technology and complexity advanced. Here, I would like to highlight a few examples.

3.1 Automatic Floorplanning

One of the earliest IC physical design problems was floorplanning, where one needs to determine positions, orientations, and dimensions of a set of interconnected logic blocks to optimize certain metrics (such as the total wirelength). Prior to Prof. Liu's work in 1986, there was really *no practical way* to quickly pack rectangular shapes, let alone minimize their wire-length, or deal with the fact that some logic blocks may have flexible geometric shapes. He and his former student Martin developed an elegant string-based encoding for topological representation of various slicing floorplan solutions [6]. They combined the newly invented simulated annealing method [7] for searching the topologic solution space and the dynamic programming based approach by Stockmeyer [8] for computing the optimal module orientation and sizing. The work was an immediate hit and received the Best Paper Award of 1986 DAC. All modern discrete-optimizer floorplanning approaches stem from Liu's pioneering work in this area. His method was widely used in both in-house floorplanning systems (such as those in Bell Labs and National Semiconductors) back in 1980's, and latest physical synthesis and design planning tools (such as IC Compiler from Synopsys and SpyGlass Physical from Atrenta).

3.2 Multi-Layer Routing

Early ICs were designed with two routing layers (one layer of polysilicon and another layer of metal) and based mainly on using channel routing. IC manufacturers started introducing two and more metal layers in the 1980s. So, multi-layer routing became a topic of interest to Prof. Liu's group. I was fortunate to participate in this work as part of my Ph.D. study. In 1986, Cong, Wong, and Liu developed a novel approach to three-layer channel routing based on transformation from good two-layer channel routing solutions [9]. This work showed that optimal track permutation can be modeled as a two-processor scheduling problem, and optimal track grouping can be modeled as a shortest-path problem. Both can be solved optimally in polynomial time. This approach led to the first optimal three-layer routing solution to the then-famous "Deutsch's difficulty example." Later, Cong and Liu also investigated the over-the-cell channel routing problem, where part of the routing can be done outside of channel over the standard cells [10]. This work showed that the problem of selecting a maximum subset of wire segments to be routed over the cells could be modeled as the problem of finding a maximum independent set in a circle graph, which can be solved optimally in polynomial time. Over-the-cell routing helped to reduce the channel routing density considerably, and was used in practice before the availability of more metal routing layers and the adoption of general area routing (or channel-less routing).

3.3 Timing-Driven Placement for FPGAs

FPGAs became a fast growing IC implementation alternative in the 1990s, and it was then that Prof. Liu's group started its research on FPGA design automation. They first investigated the timing-driven placement problem for FPGAs, which is more specialized than that for standard-cell based ASICs due to the regularity of the FPGA architecture and the fact that any logic block can implement the same set of functionality. Liu and former student Anmal Mather exploited this symmetry to devise a new approach that can be generalized to any placement problem where the elements being placed are regular and can be moved to predefined discrete locations [11]. They presented a novel iterative approach to placement, with each iteration comprising two phases—compression and relaxation. The compression phase identifies paths in the current placement that violate timing constraints (critical paths) and attempts to fix these by moving the clusters involved in the critical path closer to each other. In the process, it may generate an illegal placement with multiple clusters placed on the same physical logic block. The relaxation phase uses the notion of slack neighborhood graphs to regenerate a legal placement by moving overlapping clusters to neighboring logic blocks in their "slack neighborhood." The combination of compression and relaxation allows this methodology to explore a much larger, more global neighborhood structure in local search. In fact, this approach is not only effective for timing-driven FPGA placement, it also has been used to solve several other problems, including incremental placement for ECO (engineering change order) support [12] and fault-tolerant placement for FPGAs [13][14].

3.4 FPGA Technology Mapping for Sequential Circuits

Prof. Liu's group also studied the logic synthesis problem for FPGA designs. In particular, he and former student Peichen Pan had a breakthrough result on optimal FPGA mapping for sequential circuits [15]. Technology mapping is an important step in design automation. Retiming is a powerful technique for optimizing sequential designs. Although technology mapping [16] and retiming [17] can be solved optimally individually, only

heuristic algorithms existed for the combined problem due to the large solution space. In their work, Liu and Pan proposed a novel approach that attempts to find the best mapping solutions in the combined large solution space. They introduced two important concepts: sequential cuts, which extend mapping across register boundaries, and l-values, which generalize the traditional arrival times. The concepts proposed in their work have since been applied to many other design automation problems—such as circuit clustering, logic optimizations, and layout optimizations. Their proposed technology mapping algorithm uses dynamic programming and successive approximation, and guarantees to find optimal solutions in polynomial time—an elegant solution to a very difficult problem!

3.5 Conditional Scheduling in High-Level Synthesis

In the 1990s, as the design complexity continued to grow, the industry and the research communities became very interested in high-level synthesis (HLS) in order to further raise the level of design abstraction. Prof. Liu and his research group actively participated in this effort. As an example, one problem addressed by Liu and former student Taewhan Kim was that of scheduling operations in data flow graphs with conditional branches. At that time, many good scheduling algorithms had been proposed for scheduling operations in data flow graphs with no conditional branches, but there were few algorithms for scheduling data flow graphs with conditional branches. Camposano's path-based scheduling [18] and Wakabayashi's condition vector-based scheduling [19] were two noticeable works in the early 1990s. However, Camposano's work enumerates all execution paths, which could be exponential in number, while Wakabayashi's work involves a lot of complication in scheduling due to the simultaneous consideration of conditional and unconditional resource sharing by operations. Kim and Liu's idea was to reduce the problem to an equivalent one without conditional branches [20]. The proposed algorithm goes through three steps: first, the dataflow graph with conditional branches is transformed into one without conditional branches; then an optimized conventional HLS scheduling algorithm is computed for the transformed data flow graph; and finally, the schedule of the original data flow graph is derived. The approach is algorithmically simple, but effective for data flow graphs with many nested conditional blocks.

3.6 Fault-Tolerant Design of Random-Access Memory Systems

Fault-tolerant design was another theme of Prof. Liu's group in the 1990s, where the graph-theoretical approach was used extensively. As the degree of IC integration increases, so does the likelihood of having faults in the circuits. Redundancy is introduced to improve the yield and reliability. Liu's group studied a collection of problems related to reconfiguring random access memories (RAM) with redundant rows and columns. For example, Liu and former student Ran Libeskind-Hadas used graph theoretic techniques to find new effective heuristics for the NP-complete problem of reconfiguring faulty RAMs [21]. Additionally, Liu's group explored generalizations of the RAM reconfiguration problem that permit multiple arrays sharing the same spare rows and columns as well as arrays with heterogeneous elements. Their work gave a sharp characterization of the cases that could be solved in polynomial time and gave heuristics for the NP-complete cases [21]. This collection of work led to a research monograph "Fault Covering Problems in Reconfigurable VLSI Systems" by Libeskind-Hadas, Hasan, Cong, McKinley, and Liu [23].

It is clear from these examples that Liu's group provided much-needed algorithmic innovation to the EDA industry to address the rapidly growing demand and complexity of the IC designs. Not only were his studies of these problems very timely, right at the time when they became critical to the integrated circuit (IC) design industry, but many of his solution techniques are also timeless, as they have been applied to other EDA problems. One example is that the idea of transforming a two-layer channel routing solution into a three-layer solution developed in 1986 was used again twenty years later — this time for three-dimensional circuit placement [24] where a three-dimensional circuit placement solution was constructed from a two-dimensional placement.

4. A GREAT EDUCATOR

Equally important, if not more, Prof. Liu is a great educator who has trained multiple generations of leaders, researchers, and entrepreneurs in computer science and especially in EDA. Collectively, they carried out the effort of transforming ad hoc EDAs to algorithmic EDAs. A number of Liu's students have built very successful careers in academia:

- Jason Cong, chancellor's professor at UCLA and former chair of the Computer Science Department (2005-2008)
- Taewhan Kim, professor at the Seoul National University
- H. W. Leong, professor at the National University of Singapore
- Pravin M. Vaidya, formerly with the University of Illinois at Urbana-Champaign
- Martin D. F. Wong, professor at the University of Illinois at Urbana-Champaign
- Andrew C. Yao, Tsinghua University in China (formerly with Princeton University), and the winner of the 2000 ACM Turing Award, usually referred to as "the Nobel Prize of Computer Science."

Also, many of Liu's students became business and technology leaders in the EDA and semiconductor industries:

- Jason Cong, founder and president of Aplus Design Technologies (acquired by Magma); also co-founder and chief technology advisor of AutoESL Design Technologies (acquired by Xilinx)
- Tong Gao, Synopsys Fellow
- Anmol Mathur, CTO and co-founder of Calypto Design Systems
- Peichen Pan, director of engineering at Aplus (acquired by Magma, and head of R&D at AutoESL (acquired by Xilinx)

These former students have been leaders in the research and development of a number of successful EDA commercial products, where the algorithmic approach advocated by Prof. Liu was the guiding principle for innovations.

As a great educator, another significant contribution that Prof. Liu made was that of his leadership as the president (1998-2002) of National Tsing Hua University, one of Taiwan's (and the world's) great technical universities. Liu made monumental contributions to push NTHU to new heights. In 2000 he established a university venture capital firm to facilitate technology transfer, and established the College of Technology Management with a $150

million (NT) donation form TSMC. He also raised the funding for construction of the new library building that started in 2001 with a $300 million (NT) donation from Macronix. Total fundraising during his four-year tenure was up to $700 million (NT). The student population also increased from 6,400 to 8,000.

Prof. Liu's move to Taiwan was also a true blessing to the Taiwan EDA community. he has been instrumental in setting up funding programs and also special faculty positions in EDA since 2000. This had a great impact on Taiwan and put Taiwan's EDA on the map. As a example, in the 2009 Design Automation Conference, the number of papers from USA was 80 (ranked the highest) and the number of papers from Taiwan was 15 (ranked the second highest). This was almost 20% of USA, surpassed much larger players such as Canada, Germany, Japan, and Mainland China (note that Taiwan's GPD is less than 3% than that of the USA).

Finally, Prof. Liu's contributions as a great educator are not limited to only higher education. He proactively reaches out to the general public to share his knowledge, experience, and wisdom. After he retired from the presidency of NTHU, Liu created his own radio talk show series called "Let my tattling make you laugh" (available from http://www.ic975.com but in Chinese). I listened to those topics that were selected for inclusion into a three-CD collection. They are highly educational, yet at the same time very entertaining. Some example topics follow:

- Start of Google
- Short stories by famous writers (such as "The Old man and the Sea" by Hemingway)
- Obama's inaugural speech
- Steve Job's commencement speech at Stanford

Prof. Liu is a highly sought-after invited speaker for many events and occasions. For example, he has been a keynote speaker for every annual symposium of the International Center on System-on-a-Chip (IC-SOC) that I co-directed with Tim Cheng at UC Santa Barbara since 2001. It now becomes the Pacific-Rim Outlook Forum for IC Technology (PROFIT) [25]. The following is a list of Liu's keynote speeches in the past eight years at this event.

- There is Mathematics in Poetry, and Poetry in Mathematics (2004)
- To See the World in a Grain of Sand (2005)
- The Joy of Speech Making (2006)
- Representation of Information: Coding Theory and Cryptography (2006)
- Ten Commandments for Technical Managers (2007)
- What is a Professor? (2008)
- Heaven and Literature (2009)
- The 19th Design Automation Conference (2010)
- Trinity of Discovery, Invention, and Creativity: From Computing to EDA (2011)

These speeches are available for download from PROFIT's website. They are highly representative of Liu's speeches—educational, stimulating, humorous, and highly accessible to the general public. Through his radio show and these invited talks, Prof. Liu is setting up an excellent example for lifetime learning.

5. A HIGHLY-RESPECTED BUSINESS LEADER

In the last twelve years since his return to Taiwan, Prof. Liu's contribution to Taiwan's semiconductor and EDA industry has been both broad and significant. His vision, experience, wisdom and leadership are highly sought after by world-class companies as well as promising startups. An abbreviated list of his corporate involvements (as of 2011) demonstrates his reach and impact:

- Chairman of the board: TrendForce (formerly DramExchange), a leading market intelligence provider in the DRAM and LED technical segments.
- Member of the board, large companies: Powerchip Semiconductor Corp., United Microelectronics Corp. (one of the world's leading foundries), MediaTek (perhaps the fastest-growing fabless company in the past five years), and Macronix International Co., Ltd. All are listed on the Taiwan stock exchange; each has an annual revenue exceeding $1 billion.
- Member of the board, small startups: Anpec Electronics Corporation, Andes Corporation, Cadence Methodology Service Company, and Mototech Technology Corp. All are start-ups in Taiwan.

Clearly, Prof. Liu's impact and contributions have gone much beyond the EDA industry.

6. CONCLUDING REMARKS

It is almost an impossible task to give an overview of Prof. Liu's professional contributions, given the breadth and depth of his contributions in so many dimensions. This article covers only a small subset, but I hope it helps the reader to appreciate Liu's impact on EDA as a distinguished researcher and educator.

The 1980s decade was an exciting time for the EDA research community and industry. The PC revolution started, and the simplified design rules by Mead and Convey made it possible for every graduate student to design integrated circuits. The success and impact of the VLSI technology and the challenge of coping with the rapid increase in IC design complexity attracted many top theoreticians, including Dave Liu, to work on various VLSI CAD problems. For several years, the theoretical computer science conferences and journals published research papers on problems such as channel routing and circuit placement. However, most theoreticians moved out of this field by the end of 1980s. The EDA field was very fortunate that Liu stayed and continued to carry out a highly impactful mission with his students in bringing systematic, efficient algorithmic techniques to EDA that helped to transform the industry. Thirty years later, the EDA industry is once again facing a design complexity crisis, with an over 100,000X increase in transistor count (compared to that of the 1980s), coupled with many nano-meter design challenges, such as variability, reliability, power and thermal problems. I would like to raise the question: Is there value to partnering with the theoretical computer science community again, and, if so, what shall we do to attract and engage a future Dave Liu equivalent for a career-long association and devotion to EDA?

7. ACKNOWLEDGEMENTS

I had great pleasure working with Rob Rutenbar to put together the nomination package of Dave Liu for the Phil Kaufman Award. Rutenbar served as the official nominator and I served as the award presenter. Some of the materials in this paper are taken from the award nomination package and award presentation slides. I would also like to thank Liu's former students, Taewhan Kim, Ran Libeskind-Hadas, Anmal Mather, and Peichen Pan, for providing highlights of their research work with Dave.

8. REFERENCES

[1] C. L. Liu, "Introduction to Combinatorial Mathematics," *McGraw-Hill Book Company*, 1968, (Japanese translation, 1972, Chinese translation, 1982).

[2] C. L. Liu and J. W. Layland, "Scheduling algorithms for multiprogramming in a hard real time environment," *J. ACM*, vol. 20, no. 1, pp. 46-61, 1973.

[3] R. E. Bryant, "Graph-based algorithm for Boolean function manipulation," *IEEE Trans. on Computers*, vol. C-35, no.8, pp. 677-691, 1986.

[4] C. L. Liu, "Elements of Discrete Mathematics," *McGraw-Hill Book Company*, 1977, (Japanese translation, 1972, Chinese translation, 1982).

[5] J. R. Egan and C. L. Liu, "Optimal bipartite folding of a PLA," *Design Automation Conference*, 1982, pp. 141-146.

[6] D. F. Wong and C. L. Liu, "A new algorithm for floorplan designs," *Design Automation Conference*, 1986, pp. 101-107.

[7] S. Kirkpatrick, C. D. Gelatt and M. P. Vecchi, "Optimization by simulated annealing," *Science*, vol. 220, pp. 671-680 1983.

[8] L. Stockmeyer, "Optimal orientations of cells in slicing floorplan designs," *Information and Control*, vol. 59, pp. 91-101, 1983.

[9] J. Cong, D. F. Wong and C. L. Liu, "A new approach to the three layer channel routing," *Proc. Int'l Conf. Computer-Aided Design*, November 1987, pp. 378-381.

[10] J. Cong and C. L. Liu, "Over-the-cell channel routing," *Proc. Int'l Conf. Computer-Aided Design*, November 1988, pp. 80-83.

[11] Anmol Mathur and C. L. Liu, "Compression-relaxation: a new approach to performance driven placement for regular architectures," *IEEE Trans. on Computer-Aided Design of Integrated Circuits and Systems*, July, 1997.

[12] Anmol Mathur, K. C. Chen and C. L. Liu, "Re-engineering of Timing Constrained Placements for Regular Architectures," *Proc. Int'l Conf. Computer-Aided Design*, 1995.

[13] Anmol Mathur and C. L. Liu, "Timing driven placement reconfiguration for fault tolerance and yield enhancement in FPGAs," *Proceedings of European Design and Test Conference*, 1996.

[14] A. Agarwal, J. Cong, and B. Tagiku, "Fault tolerant placement and defect reconfiguration for nano-FPGAs," *Proc. Int'l Conf. Computer-Aided Design*, pp. 714-721, 2008.

[15] P. Pan and C. L. Liu, "Optimal clock period FPGA technology mapping for sequential circuits," *Design Automation Conference*, June 1996.

[16] J. Cong and Y. Ding, "FlowMap: An Optimal Technology Mapping Algorithm for Delay Optimization in Lookup-Table Based FPGA Designs," *IEEE Trans. on Computer-Aided Design*, vol. 13, no. 1, pp. 1-12, January 1994.

[17] C. E. Leiserson and J. B. Saxe, "Retiming synchronous circuitry," *Algorithmica*, vol. 6, pp. 5–35, 1991.

[18] R. Camposano, "Path based scheduling for synthesis," *IEEE Trans. on Computer-Aided Design of Integrated Circuits and Systems*, vol. 10, no. 1, pp. 85-93, Jan. 1991.

[19] K. Wakabayashi and T. Yoshimura, "A resource sharing control synthesis method for conditional branches," *Proc. Int'l Conf. Computer-Aided Design*, pp. 62-65, 1989.

[20] T. Kim, N. Yonezawa, J. W. S. Liu, and C. L. Liu, "A scheduling algorithm for conditional resource sharing – a hierarchical reduction approach," *IEEE Trans. on Computer-Aided Design of Integrated Circuits and Systems*, vol. 14, no. 4, pp. 425-438, Apr. 1994.

[21] R. Libeskind-Hadas and C. L. Liu, "Fast Search Algorithms for Reconfiguration Problems," *IEEE International Workshop on Defect and Fault Tolerance in VLSI Systems*, 1991.

[22] P. K. McKinley, N. Hasan, R. Libeskind-Hadas and C. L. Liu, "Disjoint covers in replicated heterogeneous arrays," *SIAM Journal on Discrete Mathematics*, vol. 4, no. 2, 1991.

[23] R. Libeskind-Hadas, N. Hasan, J. Cong, Philip Mckinley and C. L. Liu, "Fault covering problems in reconfigurable VLSI Systems," *Kluwer Academic Publishers*, 1992.

[24] J. Cong, G. Luo, J. Wei, and Y. Zhang, "Thermal-aware 3D IC placement via transformation," *Asia and South Pacific Design Automation Conference*, pp. 780-785, January, 2007.

[25] http://www.profitforum.org

On Simulated Annealing in EDA

Martin D. F. Wong
Department of Electrical and Computer Engineering
University of Illinois at Urbana-Champaign
mdfwong@illinois.edu

ABSTRACT

Simulated annealing was first introduced in 1983 as a generic stochastic algorithmic approach to solve optimization problems. Prof. C. L. Liu and his students H. W. Leong and D. F. Wong were among the earliest EDA researchers who applied simulated annealing to EDA. They solved a wide range of EDA problems with successes and reported their results in a series of papers at premier EDA conferences such DAC and ICCAD. Liu, Leong, and Wong later summarized their works in a research monograph entitled Simulated Annealing for VLSI Design, published in 1988 by Kluwer Academic Publishers. In the Preface of their book, the authors wrote "We hope that our experiences with the techniques we employed, some of which indeed bear certain similarities for different problems, could be useful as hints and guides for other researchers in applying the method to the solutions of other problems". Indeed, there were "similarities in techniques" among Liu's works that had influenced the design of simulated annealing EDA algorithms in the past 20 some years. To better understand Liu's contributions in simulated annealing, one should note that a typical simulated annealing algorithm uses a solution space and a cost function that come directly with the problem. Although computation time may be high, the algorithm is straightforward to design, making simulated annealing an attractive option for difficult problems where clever algorithms are hard to come by. On the contrary, Liu's simulated annealing algorithms have a completely different style: They resemble clever algorithms that attempt to solve difficult problems in polynomial time, and they often have a strong algorithmic flavor with a significant effort spent on optimizing the solution space and cost function. In this talk, we will discuss Prof. Liu's pioneering contributions in simulated annealing for EDA.

Categories and Subject Descriptors

D.7.2 [Integrated Circuits]: Design Aids

General Terms

Algorithms, Design, Theory

Keywords

Simulated annealing, electronic design automation, EDA, computer-aided design, CAD, VLSI

On Pioneering Nanometer-Era Routing Problems

Tong Gao
Synopsys, Inc.,
445 N Mary Ave
Sunnyvale, CA 94085 USA

tonggao@synopsys.com

Prashant Saxena
Synopsys, Inc.,
2025 NW Cornelius Pass Rd
Hillsboro, OR 97124 USA

saxena@synopsys.com

ABSTRACT

In this paper, we present a tribute to Professor C. L. (David) Liu for his numerous contributions in the field of the physical design of VLSI circuits by highlighting some of his work in the area of routing in general, and performance-driven routing in particular. We point out how he pioneered several important problem formulations along with elegant algorithmic solutions for them, often 5-10 years ahead of the time when they would become important in the semiconductor industry. More specifically, we present a brief discussion of his work on the problems of interconnect crosstalk optimization and performance-driven layer assignment, showing how it influenced subsequent academic research as well as the evolution of industrial layout tools, as an illustration of his visionary and transformative approach to physical design.

Categories and Subject Descriptors

B.7.2 [**Integrated Circuits**]: Design Aids – *Placement and Routing.*

General Terms

Algorithms, Performance, Design, Experimentation, Theory.

Keywords

Routing, VLSI routing, performance-driven routing, coupling capacitance, wire delay, crosstalk optimization, layer assignment, signal integrity, FPGA routing, channel routing, area routing, negotiation-based routing, scaling.

1. INTRODUCTION

The techniques used for the routing of electronic circuits have evolved from the manual placement of stretches of colored tapes on large drafting boards having tens of circuit components drawn on them (in the 1960s) to sophisticated routing engines that can rapidly generate high quality layouts for circuits containing tens of millions of standard cells and hundreds of macro blockages while balancing multiple objectives pertaining to route completion, electrical performance, manufacturability, and variation tolerance. This evolution of automated routing engines was powered by a handful of research groups that pioneered novel

formulations and heuristics that could model the evolving routing architectures and generate optimized routings for these models. One of these premier research groups was the one led by Liu; this group was characterized by its emphasis on clean combinatorial abstractions that were amenable to powerful algorithmic techniques to generate efficient routing solutions. Furthermore, Liu distinguished himself with a track-record of the very early identification and formulation of several problems that were subsequently to become crucial to the semiconductor and electronic design automation (EDA) industries; examples of such problems in the routing space include interconnect crosstalk optimization and performance-driven layer assignment during area routing. Routing engines interact closely with interconnect process technology, whose evolution does not always follow ideal scaling principles. Therefore, it is difficult to consistently prognosticate successfully on the future of routing technology. This makes Liu's visionary forays into modern routing concerns even more creditable.

Liu's earliest work with applications to EDA (specifically, sequential logic synthesis) was carried out in the early 1960s; this work (such as [20,21,22]) focused on the theory of finite state machines. In the early 1980s, Liu started looking at several physical design problems arising from the layout of electronic circuits, and soon came up with several pioneering results in the areas of floorplanning, placement, and routing.

2. EARLY WORK IN ROUTING

During the early 1980s, research in VLSI routing focused largely on the Channel Routing model. This model had been first formulated in 1971 by Hashimoto and Stevens [13], with much research in this space being motivated by the so-called "Deutsch difficult example" [7]. All through the 1980s, Liu's group was one of the most active research groups working in this area and kept coming up with a series of elegant algorithms that relied on combinatorial methods, simulated annealing and force-directed methods to solve channel routing problems efficiently [17][18][19][37][12] that improved significantly on the state-of-the-art [36]. During this period, manufacturing technology was also evolving to allow additional routing layers as well as over-the-cell routing, and Liu's team kept pace with these developments with influential works in multi-layer channel routing [6] as well as over-the-cell channel routing [3][5].

This was also the time when reconfigurable architectures such as field-programmable gate arrays (FPGAs) were gaining in popularity and designers were increasingly using such platforms for the prototyping and low-volume manufacture of larger and larger circuit designs. This in turn led to the development of design automation tools for the technology mapping and layout of circuits using such architectures. Liu's team was again prominent in exploring these problems; in the routing space, they came up with several elegant algorithms targeted towards generating high-quality routes for such constrained routing architectures [29][8][32][30][25].

3. INTERCONNECT CROSSTALK OPTIMIZATION

Routing algorithms had traditionally focused solely on route completion and wirelength minimization. However, with the continual shrinking of device geometries due to process scaling, several electrical effects that could earlier be safely ignored in interconnects started becoming non-trivial. These included the signal integrity and delay variation impact of the coupling capacitance between wires routed in adjacent tracks, as well as the increasing resistance of the wires that degraded wire delays.

As wires have become taller in order to ameliorate their resistance without degrading the routing pitch, their coupling capacitance has started becoming more prominent. Furthermore, with the increase in the number of routing layers, the topmost layers are often reserved for tall and wide wires that can be used to route global signal, clock and power nets. Since these long wires have low resistance, their electrical characteristics are often dominated by coupling capacitance. Given that the routing stage occurs close to the end of typical design implementation flows, it is difficult to recover from design surprises arising from poorly optimized crosstalk during the routing stage. Hence, modern routers cannot afford to ignore coupling capacitance and the crosstalk between wires routed in adjacent tracks.

Historically, signal integrity concerns had been studied in the contexts of board-level and multi-chip module (MCM) routing and of analog circuits. However, the routing models and electrical considerations (such as transmission line effects) applicable to board-level and MCM routing do not translate easily to on-chip routing in digital designs. On the other hand, signal integrity concerns in analog and mixed-signal design were traditionally handled manually with design techniques such as spacing and guard ring insertion that were not very scalable to large digital designs without being reformulated. Design automation approaches to such techniques usually relied on augmenting block compaction algorithms with spacing constraints [1].

In the context of interconnect crosstalk optimization, Liu's pioneering contribution was to develop the first on-chip routing algorithm that directly optimized interconnect crosstalk during the process of generating a routing. This was presented as a mathematical programming formulation for track assignment during channel routing [10], and was soon extended to the more generalized switchbox routing model [11]. Remarkably enough, this work had been carried out even before there was a widespread appreciation of the importance of controlling on-chip crosstalk within the circuit design community [9]. Over the next decade, Liu's work spurred much academic research in the area of interconnect crosstalk optimization during routing. (See, for instance, [16][35][33][38][23][28][15][24]). By the end of the nineties, industrial circuit designers were beginning to get very concerned about this new emerging challenge of interconnect crosstalk. As a result, industrial routers also started supporting interconnect shielding and spacing constraints in signal integrity optimization modes by the early 2000s. Similar automated crosstalk-aware routing techniques were also demonstrated to have a significant impact on the convergence of challenging industrial designs, such as a high-volume 180 nm microprocessor operating at >2 GHz [26].

4. PERFORMANCE-DRIVEN LAYER ASSIGNMENT

Although semiconductor process scaling has enabled active devices to speed up by a factor of 0.7x in each generation, it has also resulted in the doubling of wire resistance per micron and therefore a wire delay degradation per scaled micron of 1.4x every generation (albeit ameliorated to some extent by improvements in manufacturing materials and processes). As a result, critical timing paths within modern electronic circuits are often dominated by wire delays. Consequently, modern routers need to address wire delays as a first-order objective in addition to their more traditional objectives. Furthermore, modern manufacturing processes allow 8-10 or even more routing layers. In response, the over-the-cell extensions of the channel routing models of the early nineties have evolved into generalized area routing models.

In the mid-1990s, most process technologies tended to use somewhat homogeneous interconnect stacks. However, this has been changing rapidly in recent years. Heterogeneous interconnect architectures with individually customized routing layers, which were earlier used almost exclusively by high-end designs such as microprocessors, are now being adopted by process technologies used for mainstream design, since such architectures can provide significantly improved wire delays [34]. (For instance, the Interconnect chapter of ITRS [14] states that the RC delay for a 1 mm copper wire at the 68 nm node can vary from 209 ps to 767 ps depending on its layer; indeed, this variation is even greater by almost an order of magnitude at modern process nodes such as the 22 nm node. Furthermore, the resistance of the vias between the different pairs of routing layers also varies tremendously). As a result, performance-driven layer assignment has become a critical sub-problem for routers dealing with modern routing architectures.

This problem of delay optimization through layer assignment and routing was first addressed in the context of the congestion-related tradeoffs inside a router by Liu's group in 1998 [27]. This work built upon an earlier technique for the performance-driven layer assignment of a single net [2], by extending that heuristic and embedding it within a fast iterative algorithm to enable efficient dynamic sharing of the desirable routing layers between different nets. The iteratively adjusted "dynamic area quotas" introduced in this work in order to prevent the nets routed early from greedily consuming the desirable routing resources, to the detriment of the nets routed later, are conceptually similar to the negotiation-based routing paradigm which iteratively incorporates the historical congestion of a given global routing cell into its current congestion cost; negotiation-based routing was introduced in the context of on-chip routing for standard cell designs in the mid-2000s and is used widely in leading academic and industrial routers. This work by Liu anticipated the increased importance of congestion-aware performance-driven layer assignment in modern routing practice. Indeed, today's industrial routers often provide support for performance-driven layer assignment during global routing that selects layers for different nets based on their timing criticality and the expected delay benefits accruing from a critical net being routed on a given desirable layer as opposed to some other layer, in a manner similar to that proposed in [27].

5. CONCLUDING REMARKS

The two problems discussed above, namely, interconnect crosstalk optimization and performance-driven layer assignment, merely serve as two illustrations of Liu's perspicacity in defining important sub-problems in the field of routing, thus opening up new research avenues and enabling the development of a mature corpus of viable solutions by the time these problems become critical to industry. He has demonstrated similar vision and impact in many other areas of physical design and synthesis. Several researchers have acknowledged Liu's influence in transforming the nature of EDA from *ad hoc* heuristics to algorithmic research, as well as his prominent role as an outstanding mentor and educator. In this paper, we have attempted to highlight another one of Liu's strengths, namely, his remarkable ability to peer into the future of the evolution of digital design, EDA, and process technology, and identify problems whose relevance would only be vindicated by industrial practitioners several years later.

6. REFERENCES

[1] Chaudhary, K., Onazawa, A., and Kuh, E. S. 1993. A spacing algorithm for performance enhancement and crosstalk reduction. In *Proc. Int. Conf. Computer-Aided Design* (Nov. 1993). ICCAD'93. 697-702.

[2] Cong, J., and Leung, K. S., 1995. Optimal wiresizing under Elmore delay model. *IEEE Trans. Computer-Aided Design* 14 (Mar. 1995), 321-336.

[3] Cong, J., and Liu, C. L. 1988. Over-the-cell channel routing. In *Proc. Int. Conf. Computer-Aided Design* (Nov. 1988). ICCAD'88. 80-83.

[4] Cong, J., and Liu, C. L. 1990. On the k-layer planar subset and via minimization problems. In *Proc. Euro. Design Automation Conf.* (Mar. 1990). EDAC'90. 459-463.

[5] Cong, J., Preas, B., and Liu, C. L. 1990. General models and algorithms for over-the-cell routing in standard cell design. In *Proc. Design Automation Conf.* (June 1990). DAC'90. 709-715.

[6] Cong, J., Wong, D. F., and Liu, C. L. 1988. A new approach to three- or four-layer channel routing. *IEEE Trans. Computer-Aided Design* 7 (Oct. 1988), 1094-1104.

[7] Deutsch, D. 1976. A dogleg channel router. In *Proc. Design Automation Conf.* (June 1976). DAC'76. 425-433.

[8] Dong, S. K., Sun, Y., Sato, S., and Liu, C. L. 1993. Two channel routing algorithms for quickly customized logic. In *Proc. Euro. Conf. Design Automation* (Feb. 1993). ECDA'93. 122-126.

[9] Gal, L. 1995. On-chip crosstalk – the new signal integrity challenge. In *Proc. Custom Integrated Circuits Conf.* (May 1995). CICC'95. 12.1.1-12.1.4.

[10] Gao, T., and Liu, C. L. 1993. Minimum crosstalk channel routing. In *Proc. Int. Conf. Computer-Aided Design* (Nov. 1993). ICCAD'93. 692-696.

[11] Gao, T., and Liu, C. L. 1994. Minimum crosstalk switchbox routing. In *Proc. Int. Conf. Computer-Aided Design* (Nov. 1994). ICCAD'94. 610-615.

[12] Hasan, N., and Liu, C. L. 1987. In *Proc. Stanford Conf. Adv. Res. VLSI*, 135-150.

[13] Hashimoto, A., and Stevens, J. 1971. Wire routing by optimizing channel assignment. In *Proc. Design Automation Conf.* (June 1971). DAC'71. 214-224.

[14] Interconnect chapter. 2006. In *Int. Technology Roadmap for Semiconductors*. ITRS. 2006 update. http://www.itrs.net/Links/2006Update/FinalToPost/09_Interconnect2006Update.pdf

[15] Kim, K. W., Jung, S. O., Narayanan, U., Kang, S. M., and Liu, C. L. 2000. Noise-aware power optimization for on-chip interconnect. In *Proc. Int. Symp. Low Power Electronics and Design* (July 2000). ISLPED'00.

[16] Kirkpatrick, D. A., and Sangiovanni-Vincentelli, A. L. 1994. Techniques for crosstalk avoidance in the physical design of high performance digital systems. In *Proc. Int. Conf. Computer-Aided Design* (Nov. 1994). ICCAD'94. 616-619.

[17] Leong, H. W., and Liu, C. L. 1983. A new channel routing problem. In *Proc. Design Automation Conf.* (June 1983). DAC'83. 584-590.

[18] Leong, H. W., and Liu, C. L. 1985. Permutation channel routing. In *Proc. Int. Conf. Computer Design* (Oct. 1985). ICCD'85. 579-584.

[19] Leong, H. W., Wong, D. F., and Liu, C. L. 1985. A simulated annealing channel routing. In *Proc. Int. Conf. Computer-Aided Design* (Nov. 1985). ICCAD'85. 226-228.

[20] Liu, C. L. 1963. A property of partially specified automata. *Information and Control* 6 (Sep. 1963), 169-176.

[21] Liu, C. L. 1963. K-th order finite automaton. *IEEE Trans. Electronic Computers* EC-12 (Oct. 1963), 470-475.

[22] Liu, C. L. 1963. Determination of the final state of an automaton whose initial state is unknown. *IEEE Trans. Electronic Computers* EC-12 (Dec. 1963).

[23] Morton, P. B., and Dai, W. W.-M. 1999. An efficient sequential quadratic programming formulation of optimal wire spacing for crosstalk noise avoidance routing. In *Int. Symp. Physical Design* (Apr. 1999). ISPD'99. 22-28.

[24] Pan, S.-R., and Chang, Y.-W. 2000. Crosstalk-constrained performance optimization by using wire sizing and perturbation. In *Proc. Int. Conf. Computer Design* (Sep. 2000). ICCD'00. 581-584.

[25] Raman, S., Jones, L. G., and Liu, C. L. 1996. A timing-constrained incremental routing algorithm for symmetric FPGAs. In *Euro. Design Test Conf.* (Mar. 1996). EDTC'96. 170-174.

[26] Saxena, P., and Gupta, S. 2003. On integrating power and signal routing for shield count minimization in congested regions. *IEEE Trans. Computer-Aided Design* 22 (Apr. 2003), 437-445.

[27] Saxena, P., and Liu, C. L. 1998. A performance-driven layer assignment algorithm for multiple interconnect trees. In *Proc. Int. Conf. Computer-Aided Design* (Nov. 1998). ICCAD'98. 124-127.

[28] Saxena, P., and Liu, C. L. 2000. A postprocessing algorithm for crosstalk-driven wire perturbation. *IEEE Trans. Computer-Aided Design* 19 (June 2000), 691-702.

[29] Sun, Y., Dong, S. K., Sato, S., and Liu, C. L. 1991. A channel router for single layer customization technology. In

Proc. Int. Conf. Computer-Aided Design (Nov. 1991). ICCAD'91. 436-439.

[30] Sun, Y., and Liu, C. L. 1994. Routing in a new 2-dimensional FPGA/FPIC routing architecture. In *Proc. Design Automation Conf.* (June 1994). DAC'94. 171-176.

[31] Sun, Y., Wang, T. C., Wong, C. K., and Liu, C. L. 1993. Routing on orthogonal segments. In *Proc. 1993 VLSI Design/CAD Workshop* (Aug. 1993, Nantou, Taiwan). 35-38.

[32] Sun, Y., Wang, T. C., Wong, C. K., and Liu, C. L. 1993. Routing for symmetric FPGAs and FPICs. In *Int. Conf. Computer-Aided Design* (Nov. 1993). ICCAD'93. 486-490.

[33] Tseng, H.-P., Scheffer, L., and Sechen, C., 1998. Timing and crosstalk-driven area routing. In *Proc. Design Automation Conf.* (June 1998). DAC'98. 378-381.

[34] Venkatesan, R., Davis, J. A., Bowman, K. A., and Meindl, J. D. 2001. Optimal n-tier multilevel interconnect architectures for gigascale integration (GSI). *IEEE Trans. VLSI Sys.* 9 (Dec. 2001). 899-912.

[35] Xue, T., Kuh, E. S., and Wang, D. 1997. Post global routing crosstalk systems. *IEEE Trans. Computer-Aided Design* 16 (Dec. 1997), 1418-1430.

[36] Yoshimura, T., and Kuh, E. S. 1982. Efficient algorithms for channel routing. *IEEE Trans. Computer-Aided Design* CAD-1 (Jan. 1982), 25-35.

[37] Wong, D. F., and Liu, C. L., 1986. Compact channel routing with via placement restrictions. *Integration – the VLSI J.* 4 (Dec. 1986), 287-307.

[38] Zhou, H., and Wong, D. F. 1998. Global routing with crosstalk constraints. In *Proc. Design Automation Conf.* (June 1998). DAC'98. 374-377.

I Attended the Nineteenth Design Automation Conference

C. L. Liu
Department of
Computer Science
National Tsing Hwa
University

liucl@mx.nthu.edu.tw

ABSTRACT

I presented my first technical paper in the EDA area at the Nineteenth Design Automation Conference in 1982. The thirty years since then was such a short, long, and wonderful period of time!

Categories and Subject Descriptors

B.7.2 [Integrated Circuits]: Design Aids

General Terms

Algorithm, Design.

Keywords

Design Automation Conference.

ISPD'12, March 25–28, 2012, Napa, California, USA.
ACM 978-1-4503-1167-0/12/03.

Routability-driven Placement Algorithm for Analog Integrated Circuits

Cheng-Wu Lin, Cheng-Chung Lu, Jai-Ming Lin, and Soon-Jyh Chang

Department of Electrical Engineering, National Cheng Kung University, Tainan, Taiwan, R.O.C.

lcw@sscas.ee.ncku.edu.tw; louislu2@hotmail.com; jmlin@ee.ncku.edu.tw; soon@mail.ncku.edu.tw

ABSTRACT

To obtain good layout quality and reliability, placement is a very important stage during the physical design of analog circuits. Many works have been proposed to consider topological constraints for analog placement, and they devote to generate compact placements to minimize area and wirelength. However, a compact placement may induce unwanted routing issues. In order to reduce parasitics and cross-talk effects during the routing phase, wires are preferred not to pass above the active area of analog devices. Therefore, it is required to preserve enough routing spaces between devices for successful routing. Currently, there exists limited works studying routability for analog placement, but none of these works consider that symmetry property must be maintained during placement expansion. In this paper, we present a two-stage routability-driven analog placer based on ASF-B*-tree and HB*-tree representations. To reduce running time, our placement algorithm first generates a compact placement to minimize wirelength and area without considering congestion problem. Then, routing congestion regions are expanded locally to resolve the routability problem. Most importantly, the symmetry property of analog placement is always satisfied during the expansion process. Experimental results show that our analog placer can effectively minimize routing congestion without violating the symmetry property after placement expansion.

Categories and Subject Descriptors: B.7.2 [Integrated Circuits]: Design Aids – Layout, Placement and routing

General Terms: Algorithms, Design

Keywords: Analog placement, routability

1. INTRODUCTION

In analog design flow, layout synthesis is eminently critical and time-consuming [1] because it needs complicated considerations, such as parasitic interconnect capacitance, mismatch effects, and thermal gradients. To enhance the quality of analog layout, many topological constraints based on designer's expertise have been introduced into analog placement. The major topological constraints include device matching, device symmetry, and device proximity [2]. The matching constraint limits devices to a common-gate or interdigital structure. The symmetry constraint forces the devices of a differential circuit into a mirrored placement. For the same kind

of devices, the proximity constraint restricts them together to a common substrate region. These constraints help to reduce the parasitic mismatches and coupling effects.

With the continuous scaling of process technologies, more and more analog circuits can be integrated into a single chip. This event enhances the functionality of a chip, but complicates the interconnect among circuits. The increasing complexity of routing not only raises the chance of signal coupling but also makes routing congestion severer. This condition may induce unwanted parasitic effects and degrade layout quality. Since routing is greatly affected by placement results, the quality of routing can be improved if routability is considered in advance during the placement phase. Therefore, in addition to the topological constraints, routing congestion is also a desirable consideration for analog placement [3].

1.1 Previous Work

The problem of analog placement considering topological constraints has been extensively studied for decades. Most of these works used topological representations with simulated annealing algorithm [4] to tackle the placement problem. To deal with the symmetry constraint, symmetric-feasible conditions have been explored for several representations, such as sequence-pairs [5], O-trees [6], binary trees [7], TCG [8], CBL [9], and B*-trees [10]. The placement algorithms based on CBL [9] and HB*-trees [11] further considered thermal effect with the symmetry constraint. Ma et al. [12] and Xiao et al. [13] handled the common-centroid constraint by C-CBL and sequence-pairs respectively. Based on sequence-pairs [14], Tam et al. [15] used dummy nodes and additional constraint edges for symmetry groups with other placement constraints. Lin et al. [16] employed HB*-trees to deal with the matching, symmetry, and proximity constraints simultaneously. Strasser et al. [17] also considered several placement constraints and proposed a deterministic placement algorithm based on B*-trees using hierarchically bounded enumeration and enhanced shape functions. Lin et al. [18] further proposed symmetry-island boundary constraint for performance consideration and they extended ASF-B*-trees [10] to handle this constraint.

The placement problem considering routing congestion has been widely studied for digital designs. However, there exists limited works discussing that problem for analog circuits. In analog design, a compact placement is not practical. In order to reduce parasitics and cross-talk effects, the routing spaces above the active area of the analog devices are often avoided (i.e., they are considered as obstacles) in the routing phase [3]. Since wires are preferred not to pass above the active area of devices, it is required to preserve enough routing spaces between devices for successful routing. To ensure that there exists enough spaces between devices for laying out wires, it is necessary to perform blockage aware congestion estimation during the analog placement phase. According to this requirement, Xiao et al. [3]

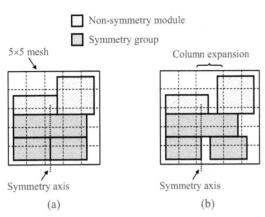

□ Non-symmetry module

▨ Symmetry group

Figure 1. (a) A compact placement which contains a symmetry group. (b) The third column is expanded according to the adjustment algorithm [3]. Note that the symmetry group becomes an asymmetric placement after the column expansion.

proposed the first work on implementing a blockage aware congestion-driven placer for analog designs. In their method, a routability-driven adjustment algorithm is performed after a compact placement is obtained. The algorithm first divides the whole placement region into an $n \times n$ mesh and estimates the vertical and horizontal congestion for each room. Then, rooms are expanded column-by-column and row-by-row based on the congestion map. Finally, The center of each device is moved to the same relative position of the same room. This algorithm is able to adjust the space between devices for successful routing. However, the symmetry constraint may be violated after the placement expansion. For example, Figure 1(a) shows a compact placement which contains a symmetry group. According to the adjustment algorithm [3], rooms in the same column will be expanded horizontally by an identical space. The placement after column expansion is shown in Figure 1(b). Note that the expanded placement becomes infeasible because the symmetry group is not in a symmetric placement.

1.2 Our Contributions

Analog placement problem has been extensively studied for decades. Most of these works used topological representations to minimize wirelength and area without considering congestion problem. However, compact placements are very impractical for analog designs. To the best of authors' knowledge, there exists only one work Xiao et al. [3] considering routability issue during analog placement. Although Xiao et al. [3] proposed an adjustment algorithm to maintain enough space between devices for routing nets, a symmetry group may fall into an asymmetric placement after the adjustment process. To resolve this problem, we extend ASF-B*-trees [10] to consider routability issues while keeping symmetry property. In ASF-B*-tree representation, only the spaces of the representatives which represent a half of a symmetry group have to be considered because the routing spaces for the other half of the symmetry group just duplicate their representatives while packing. By selecting suitable routing spaces for those representatives, each symmetry pair of the symmetry group will still keep symmetric with respect to the symmetry axis after placement expansion. Based on this concept, we propose a routability-driven analog placer based on ASF-B*-trees and HB*-trees [10] to handle multiple topological constraints for analog placement. Experimental results show that

our analog placer is effective to minimize routing congestion without breaking the symmetry property of analog placement.

The remainder of this paper is organized as follows. Section II formulates the problem of analog placement considering routability. Section III gives an overview of our congestion-aware analog placer. Section IV presents the approach to placement expansion for congestion elimination. Section V describes the detailed flow of our placement algorithm. Section VI reports the experimental results. Finally, section VII concludes this paper.

2. PROBLEM FORMULATION

Given a set of device modules and topologically constrained groups in which devices should be placed symmetrically, analog placement problem considering routability is to obtain a legal placement P for all devices such that routing congestion in the resulting placement can be minimized and all topological constraints are satisfied.

Since our placement result is generated by simulated annealing with ASF-B*-tree and HB*-tree representations, the cost function $\Phi(P)$ for evaluating the quality of placements is defined in the following:

$$\Phi(P) = \alpha \times A_P + \beta \times W_P + \gamma \times C_P, \qquad (1)$$

where α, β, and γ are user-specified parameters, A_P is the bounding-rectangle area of the placement, W_P is the total wire length measured by half-perimeter estimation, and C_P is the estimation of routing congestion.

3. OVERVIEW OF THE PROPOSED ANALOG PLACER

In this section, we first introduce the design flow of our analog placer. Then, we give a short review of the ASF-B*-trees [10] and the Steiner tree construction algorithm FLUTE [19], which are employed in our placer.

3.1 Design Flow

Figure 2 shows our design flow. Given a set of device modules, the information of interconnect, and topological placement constraints, we first generate a compact placement which satisfies all topological constraints using ASF-B*-tree and HB*-tree representation without considering routability. Next, to obtain routability information, a fast Steiner tree construction algorithm is employed in our design flow to generate a congestion map. According to the congestion map, we can eliminate routing overflow by expanding the over-congested regions. Based on the simulated annealing [4], these steps iterate until the placement area, total wirelength, and total overflow are minimized.

We employ ASF-B*-tree and HB*-tree representations in our placer because of their superior performance [10]. More importantly, ASF-B*-tree representation guarantees that each symmetry pair remains symmetric with respect to the symmetry axis after the placement is expanded. In order to estimate routing congestion, we perform Steiner tree construction to generate a congestion map. According to the exploration by Pan et al. [20], the way to predict congestion accurately is to use the same technique and parameters in both congestion estimation and global routing. We apply FLUTE for the estimation of routing congestion because FLUTE is efficient and effective in Steiner tree construction [19] and it has been imported into some excellent global routers, such as FastRoute [20] and NTHU-Route

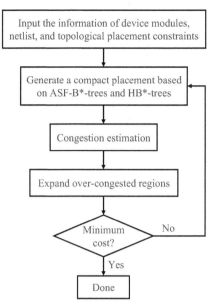

Figure 2. Overview of the proposed routability-driven analog placement flow.

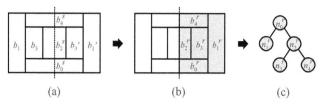

Figure 3. (a) Placement example of a symmetry group in vertical symmetry. (b) The right-half part of each self-symmetry module and symmetry pair is selected as a representative. (c) The ASF-B*-tree which represents the placement of the symmetry group.

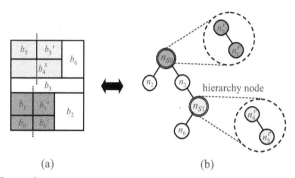

Figure 4. (a) A placement which contains two symmetry groups. (b) The corresponding HB*-tree.

[21]. In the following subsections, we review ASF-B*-trees, HB*-tree and FLUTE respectively.

3.2 Review of ASF-B*-tree

For a symmetry group, each module should abut at least one of other modules in the same group to ensure the electrical properties, such as parasitic matching and thermal gradient, can be satisfied. Thus, the modules of a symmetry group would form a connected placement. Based on this placement consideration, Lin et al. [10] introduced the concept of symmetry islands for symmetry groups and they proposed ASF-B*-trees to represent the symmetry islands.

In analog layout, the placement of a symmetry group can be in vertical symmetry or horizontal symmetry. The placement in vertical (horizontal) symmetry has the symmetry axis in the vertical (horizontal) direction. Without loss of the generality, the following description is addressed on vertical symmetry. The case of horizontal symmetry is just the one of vertical symmetry rotated by 90 degree. Figure 3(a) illustrates a placement example of a symmetry group in vertical symmetry. As defined by [10], the representative b_i^r is the right half of a self-symmetry module b_i^s, or it can be the b_i' of a symmetry pair (b_i, b_i'). For example, the b_1' of the symmetry pair (b_1, b_1') in Figure 3(a) is selected as the representative b_1^r in Figure 3(b). An ASF-B*-tree is a B*-tree in which each node n_i^r associates with a representative b_i^r. Figure 3(c) shows the ASF-B*-tree corresponding to the placement in Figure 3(b). The packing of an ASF-B*-tree can automatically form a symmetry island.

3.3 Review of HB*-tree

Figure 4(b) shows an HB*-tree for the placement in Figure 4(a). Two symmetry groups, $S_0 = \{(b_0, b_0'), (b_1, b_1')\}$ and $S_1 = \{b_4^s, (b_5, b_5')\}$, are represented by two hierarchy nodes, n_{S0} and n_{S1}, and each hierarchy node contains an ASF-B*-tree that corresponds to a symmetry island. Similar to the B*-tree packing, the HB*-tree packing runs in the DFS order. Before packing an HB*-tree, the ASF-B*-tree in each hierarchy node should be packed first to obtain the outline of the symmetry island. Besides the symmetry groups, hierarchy nodes can also handle the groups which require matching or proximity constraint. More details about HB*-tree can be found in [10].

3.4 Review of FLUTE

A rectilinear Steiner minimal tree (RSMT) is a tree with minimum total edge length in Manhattan distance to connect a given set of nodes possibly through some extra (i.e., Steiner) nodes. RSMT construction is a fundamental problem that has many applications in very large scale integration (VLSI) design. For global and detailed routing stages, it is used to generate the routing topology of each net.

FLUTE [19] is based on a pre-computed lookup table to make RSMT construction very fast and very accurate for low-degree nets. For high-degree nets, a net-breaking technique is employed to reduce the net size until the table can be used. FLUTE is optimal for low-degree nets (up to degree 9) and is still very accurate for nets up to degree 100. Therefore, it is particularly suitable for VLSI applications in which most nets have a degree of 30 or less.

4. PLACEMENT EXPANSION

In our routability-driven analog placer, we slightly expand the over-congested regions of a compact placement to eliminate routing overflow. To expand an over-congested region, our method first trace the key node which associates with the congested region from the corresponding ASF-B*-tree or HB*-tree, and then insert a dummy node into a suitable position near the key node according to different conditions. These conditions are classified into two types: the over-congested region is outside a symmetry group; or it is inside a symmetry group. In the following subsections, we first introduce the algorithm of congestion estimation and placement expansion. Then, we will illustrate how to insert dummy nodes to expand the placement.

4.1 Congestion Estimation and Placement Expansion

Since HB*-tree induces a bottom-left compact placement, it may not have enough routing spaces in some locations. To eliminate the routing overflow, the space between modules in these regions should be expanded slightly. In order to get sufficient routing spaces as well as minimum chip area usage, accurate congestion estimation is required to determine a suitable expansion rate. Thus, a fast Steiner tree construction algorithm is used in our design flow to generate accurate congestion map. In the following, we give the complete procedure for congestion estimation and placement expansion (please see Algorithm 1).

To estimate routing congestion, the whole placement region is divided into $m \times n$ global bins (i.e., a m-row and n-column routing grid), where m and n are computed by dividing chip's height and width respectively to make bin's height and width less than minimum module's height and width (see Line 1). Figure 5 shows a placement example with 6×6 global bins. Then, for each vertical and horizontal bin edge, we calculate its wire capacity, which defines how many wires can pass through it. Since the active area of all modules are avoided routing (i.e., all modules are considered as blockages), we compute the wire capacity for each edge according to its routable space not overlapped with the active area of modules (see Line 2). After the calculation of wire capacities, we use FLUTE to construct Steiner trees for all nets and generate a congestion map (see Line 3). Based on the congestion map, we evaluate the routing overflow and estimate a suitable expansion rate for each edge (see Lines 4-6 and 14-16).

To expand a placement, we first deal with the routing overflow of all horizontal edges and then handle that of all vertical edges. If a horizontal (vertical) edge has routing overflow, we trace which module is on that edge and push away its right (top) adjacent module according to the expansion rate (see Lines 7-10 and 17-20). This way can increase the wire capacity of the edge and thus decrease the overflow. To push a module in the corresponding ASF-B*-tree or HB*-tree, a dummy node can be inserted between the traced module and its adjacent module (see Lines 10 and 20). The suitable position for dummy node insertion is based on different conditions. We will illustrate these conditions in the following two subsections.

4.2 Dummy Node Insertion outside a Symmetry Group

In this subsection, we show how to deal with the condition that over-congested region is outside a symmetry group. Assume we have traced one non-symmetry module which is on the congestion edge. Another module which is adjacent to this module may be a non-symmetry module or a module belonging to a symmetry group. Figure 6 and Figure 7 illustrate the examples. For brief presentation, these examples deal with the condition that horizontal and vertical congestions are occurred simultaneously. The modules which are adjacent to the traced module's right and top sides will be pushed away.

Figure 6(a) and 6(b) show three non-symmetry modules and the corresponding HB*-tree respectively. If the module b_0 is on a horizontal congestion edge and two modules b_1 and b_2 are at its right side, we can insert a horizontal white space between them to push modules b_1 and b_2 rightward and thus increase the wire capacity of the horizontal edge. The width of the white space is determined by the expansion rate. If the module b_1 is on a vertical congestion edge and another module b_2 is on its top side, we can insert a vertical white space between them to push module b_2

1: Divide the placement into $m \times n$ global bins. Assume the bin width, bin height, and minimum wire width are w, h, and z respectively.

2: Calculate the routable space percentage R for each edge. The wire capacity of a horizontal (vertical) edge (i, j) is $C_h[i][j] = R_h[i][j] \times w / z$ ($C_v[i][j] = R_v[i][j] \times h / z$).

3: Construct Steiner trees for all nets and generate a congestion map. Assume the wire demand of a horizontal (vertical) edge (i, j) is $D_h[i][j]$ ($D_v[i][j]$).

4: **for** each horizontal edge (i, j) **do**

5: $\quad overflow_h[i][j] = \begin{cases} D_h[i][j] - C_h[i][j], & \text{if } D_h[i][j] > C_h[i][j] \\ 0, & \text{otherwise} \end{cases}$

6: $\quad expansion_h[i][j] = overflow_h[i][j] \times z$

7: $\quad$ **if** $overflow_h[i][j] > 0$ **then**

8: $\quad\quad$ Trace which module is on this edge. Assume the height of this module is h_m.

9: $\quad\quad$ **if** another module is adjacent to this module's right side **then**

10: $\quad\quad\quad$ Insert a dummy node between them. Dummy node width and height are $expansion_h[i][j]$ and h_m respectively.

11: $\quad\quad$ **end if**

12: $\quad$ **end if**

13: **end for**

14: **for** each vertical edge (i, j) **do**

15: $\quad overflow_v[i][j] = \begin{cases} D_v[i][j] - C_v[i][j], & \text{if } D_v[i][j] > C_v[i][j] \\ 0, & \text{otherwise} \end{cases}$

16: $\quad expansion_v[i][j] = overflow_v[i][j] \times z$

17: $\quad$ **if** $overflow_v[i][j] > 0$ **then**

18: $\quad\quad$ Trace which module is on this edge. Assume the width of this module is w_m.

19: $\quad\quad$ **if** another module is adjacent to this module's top side **then**

20: $\quad\quad\quad$ Insert a dummy node between them. Dummy node width and height are w_m and $expansion_v[i][j]$ respectively.

21: $\quad\quad$ **end if**

22: $\quad$ **end if**

23: **end for**

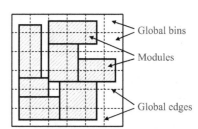

Figure 5. A placement example with 6×6 global bins. Routing congestion will be estimated for each bin edge.

---

upward and thus increase the wire capacity of the vertical edge. The height of the white space is determined by the expansion rate. Figure 6(c) shows the placement after expansion. Here we assume the three modules are involved in horizontal and vertical congestions simultaneously. To achieve this expanded placement, we should insert dummy nodes into the corresponding HB*-tree. Figure 6(d) shows the HB*-tree after the horizontal dummy node (colored in dark gray) and vertical dummy node (colored in gray) insertion.

Figure 7(a) shows another placement which contains a symmetry group $S_0 = \{(b_1, b_1'), (b_2, b_2')\}$. The corresponding HB*-tree is shown in Figure 7(b). Since the packing of a HB*-tree is hierarchical, if the module on a congestion edge belongs to a symmetry group, the expansion for this module should be

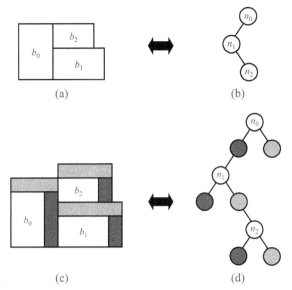

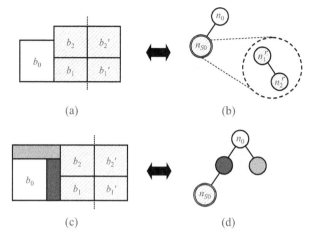

(a)   (b)

Figure 6. (a) The placement of three non-symmetry modules b_0, b_1, and b_2. (b) The corresponding HB*-tree. (c) The placement after expansion. Here the three modules are involved in horizontal and vertical congestions simultaneously. (d) The HB*-tree after the horizontal dummy node (colored in dark gray) and vertical dummy node (colored in gray) insertion.

handled in the associated ASF-B*-tree. We will detail how to tackle this condition in the next subsection. Here only the expansion for the top-level HB*-tree is discussed. Assume the module b_0 is involved in horizontal and vertical congestions simultaneously. The placement after expansion is shown in Figure 7(c). Figure 7(d) shows the corresponding HB*-tree after dummy node insertion.

4.3 Dummy Node Insertion in a Symmetry Group

In this subsection, we deal with the condition that over-congested region is inside a symmetry island. Under this condition, since the placement region which requires expansion is related to a symmetry group, we need to insert dummy node into the associated ASF-B*-tree. However, an ASF-B*-tree only corresponds to half the modules (i.e., the representatives) of a symmetry group. For a symmetry module in the congested region, we need to trace its representative so that we can insert dummy node into a suitable position of the ASF-B*-tree. The dummy node insertion is tackled according to the following conditions.

- **The symmetry module is a representative (i.e., it is on the right-half plane of a symmetry island):** For horizontal (vertical) expansion, a dummy node is inserted to be the left (right) child of the representative in the corresponding ASF-B*-tree.

- **The symmetry module is not a representative (i.e., it is on the left-half plane of a symmetry island):** For horizontal expansion, we trace which module is on its right side and find out the representative of the traced module. Then, a dummy node is inserted to be the left child of the representative. For vertical expansion, we trace the representative of this symmetry module, and then insert a dummy node to be the right child of the representative.

Figure 7. (a) The placement of a non-symmetry module b_0 and a symmetry group $S_0 = \{(b_1, b_1'), (b_2, b_2')\}$. (b) The corresponding HB*-tree. (c) The placement after expansion. Here only the expansion for the top-level HB*-tree is considered. (d) The HB*-tree after the horizontal dummy node (colored in dark gray) and vertical dummy node (colored in gray) insertion.

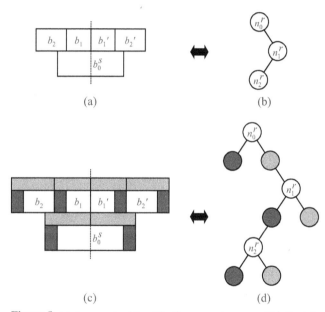

Figure 8. (a) A symmetry island for the symmetry group $\{b_0^s, (b_1, b_1'), (b_2, b_2')\}$. (b) The corresponding ASF-B*-tree. (c) The placement after expansion. Here all the representatives are involved in horizontal and vertical congestions simultaneously. (d) The ASF-B*-tree after the horizontal dummy node (colored in dark gray) and vertical dummy node (colored in gray) insertion.

Figure 8(a) shows a symmetry island for the symmetry group $\{b_0^s, (b_1, b_1'), (b_2, b_2')\}$. Modules b_0^s, b_1', and b_2' are the representatives of the symmetry island, which corresponds to the nodes n_0^r, n_1^r, and n_2^r of the ASF-B*-tree shown in Figure 8(b). If the module b_1' is on a horizontal congestion edge, a white space will be added to its right side to expand the placement. Since b_1' is a representative, we can directly insert a dummy node to be its left child in the ASF-B*-tree. If the module b_2 is on a horizontal congestion edge, a white space should be added to its right side for placement expansion. However, module b_2 is not a

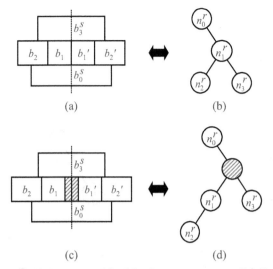

(a)   (b)

(c)   (d)

Figure 9. (a) A symmetry island for the symmetry group $\{b_0{}^s, (b_1, b_1'),$ $(b_2, b_2'), b_3{}^s\}$. (b) The corresponding ASF-B*-tree. (c) The placement after expansion. Here only the horizontal expansion between the symmetry pair (b_1, b_1') is considered. (d) The ASF-B*-tree after the horizontal dummy node (colored in gray pattern) insertion.

representative. Since b_2 does not exist in the ASF-B*-tree, we can not directly insert a dummy node to be its left child. Instead, we trace which module is on b_2's right side, and thus find module b_1. Then, we find out the representative of b_1: module b_1'. Finally, a dummy node is inserted to be the left child of b_1'. This way can add a white space to the module b_2's right side. Figure 8(c) shows the placement after expansion. For brief presentation, this example deals with the condition that horizontal and vertical congestions are occurred simultaneously. The corresponding ASF-B*-tree after dummy node insertion is shown in Figure 8(d).

Besides above general conditions, Figure 9 illustrates another special consideration. Figure 9(a) and 9(b) show a symmetry island for the symmetry group $\{b_0{}^s, (b_1, b_1'), (b_2, b_2'), b_3{}^s\}$ and the corresponding ASF-B*-tree respectively. In this example, the symmetry pair (b_1, b_1') abuts on the symmetry axis. If the module b_1, which is not a representative, is on a horizontal congestion edge, a white space should be added to its right side to expand the placement. However, this condition can not be well handled by the aforementioned process. To correctly expand the space between modules b_1 and b_1', a white space whose width is half the expansion rate can be added to the left side of b_1'. The placement after expansion is shown in Figure 9(c). Thus, a dummy node should be inserted to be the parent node of $n_1{}^r$ in the ASF-B*-tree. Figure 9(d) shows the corresponding ASF-B*-tree after dummy node insertion. Based on this scheme, we give a systematic process in the following:

● **Horizontal dummy node insertion for a symmetry pair** (b_i, b_i') **which abuts on the symmetry axis:** Let $n_i{}^r$ and n_d denote the representative node and the dummy node respectively in the corresponding ASF-B*-tree. First, the position of $n_i{}^r$ is replaced by n_d, and $n_i{}^r$ becomes the left child of n_d. Then, the right sub-tree of $n_i{}^r$ becomes that of n_d, and the left sub-tree of $n_i{}^r$ is still remained.

5. PLACEMENT ALGORITHM

In this section, we present the detailed flow of our routability-driven placement algorithm. Our algorithm is based on simulated

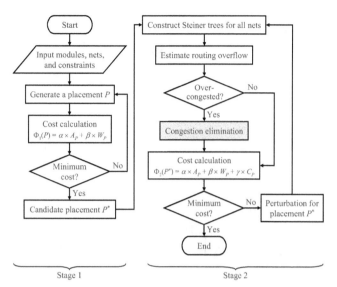

Figure 10. The proposed two-stage routability-driven placement flow.

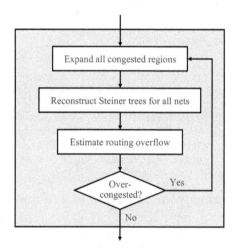

Figure 11. The flow of congestion elimination.

annealing [4]. The whole placement flow is divided into two stages. In the first stage, we focus on obtaining a placement with minimum chip area and wirelength without considering routing congestion. Thus, running time for unnecessary congestion estimation can be saved. Although congestion estimation and placement expansion are not executed in this stage, routing congestion can still be alleviated by wirelength minimization because a longer wire tends to cross more global bins and occupy the wire capacities of bin edges. Based on the placement solution from the first stage, the second stage starts to perform congestion estimation and placement expansion for routability. In addition to deal with routing congestion, we still need to consider the placement area and total wirelength in the second stage.

Figure 10 illustrates the flow of our routability-driven placement algorithm. Given the information of device modules, netlist, and topological placement constraints, Stage 1 first generates an initial placement P, and then tunes P based on the cost function $\Phi_1(P)$ iteratively until an optimized placement P^* is achieved. Next, Stage 2 tunes P^* with congestion estimation and placement expansion iteratively to eliminate routing congestion.

Table 1. Information of benchmark circuits

Circuit Name	# of Mod.	# of Sym. Mod.	# of Nets	Mod. Area (*mm²*)
ami33	33	7 (2+2+2+1)	49	1.16 = 100%
ami49	49	5 (2+2+1)	152	35.45 = 100%

Table 2. Comparison of area, wirelength, overflow, and runtime for benchmark circuits with and without placement expansion

Circuit Name	Without Placement Expansion				With Placement Expansion			
	Area (%)	HPWL (*mm*)	# of Overflow	Time (sec.)	Area (%)	HPWL (*mm*)	# of Overflow	Time (sec.)
ami33	107.11	37.38	154	103	113.25	47.73	0	556
ami49	108.44	569.25	253	206	115.32	677.24	0	1337

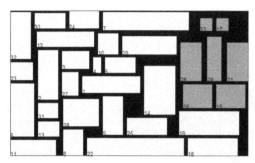

Figure 12. The placement result of ami33 with placement expansion. The symmetry group is colored in purple.

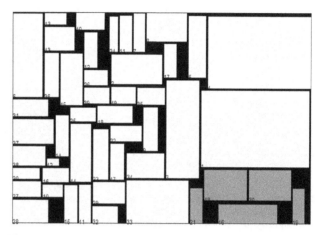

Figure 13. The placement result of ami49 with placement expansion. The symmetry group is colored in purple.

The objective of Stage 2 is to achieve a routable placement with minimum area and wirelength. In the beginning of the flow, we apply FLUTE to construct Steiner trees for all nets and generate a congestion map. Then, routing overflow is estimated according to the congestion map. If any bin edge is over-congested, the placement in the congested region will be expanded. After placement expansion, we reconstruct Steiner trees and deal with new congestion iteratively until congestion is resolved. Figure 11 shows the flow of congestion elimination. Then, the chip area and total wirelength of the expanded placement will be used to evaluate the cost for current candidate solution, which is more accurate than the original one. Finally, we can obtain the optimal placement based on the cost function $\Phi_2(P^*)$.

6. EXPERIMENTAL RESULTS

Our placement algorithm was implemented in C++ programming language and run on a 2.5 GHz SUN Fire-X4250 workstation with 16 GB memory. We performed the experiments based on two circuits, ami33 and ami49, from MCNC benchmark circuits. Since the symmetry groups of the original benchmark circuits only contain symmetry pairs, we randomly choose a non-symmetry module as a self-symmetry module into each symmetry group. Table 1 lists the information of benchmark circuits, including total number of modules, the number of modules in each symmetry group, the number of nets, and total module area. In these experiments, we randomly set the minimum wire width as 18 μm (i.e., $z = 18$ in Algorithm 1) for calculating the wire capacities of bin edges.

Table 2 compares the results of the benchmark circuits with and without placement expansion. The column "Without Placement Expansion" reports the results by only performing the Stage 1 in Figure 10, which does not consider routing congestion. The column "With Placement Expansion" shows the results by

our routability-driven placement flow (Stage 1 + Stage 2). For each approach, we list the placement area, the wirelength estimated by HPWL, the estimated number of routing overflow, and the runtime. The experimental results show that the placements expanded by our method achieve zero overflow with some additional overhead of area and wirelength. Figure 12 and Figure 13 show the placement results of ami33 and ami49 with placement expansion. In the figures, symmetry groups are colored in purple, and the empty space between modules is the expansion result for routability. Since the packing of ASF-B*-trees is always a symmetric placement, our method guarantees the symmetry property for each symmetry group after placement expansion.

7. CONCLUSION

In this paper, we have introduced the issue of analog placement considering routability. Based on ASF-B*-trees, we have proposed a congestion-aware placement algorithm to enhance routability without breaking the symmetry property of analog placement. The experimental results have shown the effectiveness of our approach. Since the placement problem considering routing congestion has been extensively studied for digital designs, we believe those approaches could be modified for analog placement in the future.

8. ACKNOWLEDGMENTS

This work was supported in part by National Science Council of Taiwan under Grant No's NSC-98-2221-E-006-156-MY3.

9. REFERENCES

[1] H. Graeb, F. Balasa, R. Castro-Lopez, Y.-W. Chang, F.V. Fernandez, P.-H. Lin, and M. Strasser, "Analog layout synthesis - Recent advances in topological approaches," in *Proc. DATE*, 2009, pp. 274-279.

[2] J. M. Cohn, D. J. Garrod, R. A. Rutenbar, and L. R. Charley, *Analog Device-Level Layout Automation*. Kluwer Academic Publishers, 1994.

[3] L. Xiao, E. F.Y. Young, X. He, and K.P. Pun, "Practical placement and routing techniques for analog circuit designs," in *Proc. ICCAD*, 2010, pp. 675-679.

[4] S. Kirkpatrick, C. D. Gelatt, and M. P. Vecchi, "Optimization by simulated annealing," *Science*, vol. 220, no. 4598, pp. 671-680, May 1983.

[5] F. Balasa and K. Lampaert, "Module placement for analog layout using the sequence-pair representation," in *Proc. DAC*, 1999, pp. 274-279.

[6] Y. Pang, F. Balasa , K. Lampaert, and C.-K. Cheng, "Block placement with symmetry constraints based on the O-tree non-slicing representation," in *Proc. DAC*, 2000, pp. 464-467.

[7] F. Balasa, "Modeling non-slicing floorplans with binary trees," in *Proc. ICCAD*, 2000, pp. 13-16.

[8] J.-M. Lin, G.-M. Wu, Y.-W. Chang, and J.-H. Chuang, "Placement with symmetry constraints for analog layout design using TCG-S," in *Proc. ASP-DAC*, 2005, pp. 1135-1138.

[9] J. Liu, S. Dong, Y. Ma, D. Long, and X. Hong, "Thermal-driven symmetry constraint for analog layout with CBL representation," in *Proc. ASP-DAC*, 2007, pp. 191-196.

[10] P.-H. Lin and S.-C. Lin, "Analog placement based on novel symmetry-island formulation," in *Proc. DAC*, 2007, pp. 465-470.

[11] P.-H. Lin, H. Zhang, M. D. F. Wong, and Y.-W. Chang, "Thermal-driven analog placement considering device matching," in *Proc. DAC*, 2009, pp. 593-598.

[12] Q. Ma, E. F. Y. Young, and K. P. Pun, "Analog placement with common centroid constraints," in *Proc. ICCAD*, 2007, pp. 579-585.

[13] L. Xiao and E. F. Y. Young, "Analog placement with common centroid and 1-D symmetry constraints," in *Proc. ASP-DAC*, 2009, pp. 353-360.

[14] H. Murata, K. Fujiyoshi, S. Nakatake, and Y. Kajitani, "Rectangle-packing-based module placement," in *Proc. ICCAD*, 1995, pp. 472-479.

[15] Y.-C. Tam, E. F. Y. Young, and C. Chu, "Analog placement with symmetry and other placement constraints," in *Proc. ICCAD*, 2006, pp. 349-354.

[16] P.-H. Lin and S.-C. Lin, "Analog placement based on hierarchical module clustering," in *Proc. DAC*, 2008, pp. 50-55.

[17] M. Strasser, M. Eick, H. Graeb, U. Schlichtmann, and F. M. Johannes, "Deterministic analog circuit placement using hierarchically bounded enumeration and enhanced shape functions," in *Proc. ICCAD*, 2008, pp. 306-313.

[18] C.-W. Lin, J.-M. Lin, C.-P. Huang, and S.-J. Chang, "Performance-driven analog placement considering boundary constraint," in *Proc. DAC*, 2010, pp. 292-297.

[19] C. Chu and Y.-C. Wong, "FLUTE: Fast lookup table based rectilinear Steiner minimal tree algorithm for VLSI design," *TCAD*, vol. 27, no. 1, pp. 70-83, Jan. 2008.

[20] M. Pan and C. Chu, "FastRoute: A step to integrate global routing into placement," in *Proc. ICCAD*, 2006, pp. 464-471.

[21] Y.-J. Chang, Y.-T. Lee, and T.-C. Wang, "NTHU-Route 2.0: A fast and stable global router," in *Proc. ICCAD*, 2008, pp. 338-343.

Keep it Straight: Teaching Placement how to Better Handle Designs with Datapaths

Samuel I. Ward[†], Myung-Chul Kim[‡], Natarajan Viswanathan[‡],
Zhuo Li[‡], Charles Alpert[‡], Earl E. Swartzlander, Jr.[†], David Z. Pan[†]
[†] ECE Dept. The University of Texas at Austin, Austin, TX 78712
[‡] IBM Austin Research Laboratory, 11501 Burnet Road, Austin, TX, 78758
{wardsi}@utexas.edu, {mckima}@umich.edu, {nviswan, lizhuo, alpert}@us.ibm.com,
{eswartzla}@aol.com, {dpan}@cerc.utexas.edu

ABSTRACT

As technology scales and frequency increases, a new design style is emerging, referred to as hybrid designs, which contain a mixture of random logic and datapath standard cell components. This work begins by demonstrating that conventional Half-Perimeter Wire Length (HPWL)-driven placers under-perform in terms of regularity and Steiner Wire Length (StWL) for such hybrid designs, and the quality gap between manual placement and automatic placers is more pronounced as the designs become more datapath-oriented. Then, a new unified placement flow that simultaneously handles random logic and datapath standard cells is proposed that significantly improves the placement quality of the datapath while leveraging the speed of modern state-of-the-art placement algorithms. The placement flow is built on top of a leading academic force-directed placer. It consists of a series of novel global and detailed placement techniques, collectively called Structure Aware Placement Techniques (SAPT). The techniques effectively integrate alignment constraints into placement, overcoming the deficiencies of the HPWL objective. Experimental results comparing our placement flow with six state-of-the-art placers on the ISPD 2011 Datapath Benchmark Suite show at least a 32% improvement in total StWL with over a 6× improvement in total routing overflow. In addition, the flow demonstrates an 8.25% improvement in total StWL on industrial hybrid designs.

Categories and Subject Descriptors

B.7.2 [**Hardware, Integrated Circuits**]: Design Aids—*Placement and Routing*

General Terms

Design

Keywords

Datapath, Placement, Physical Design

1. INTRODUCTION

As ASIC frequency exceeds $1GHz$ and shrinking schedules drive increased automation for microprocessor designs, the boundary be-tween manually designed datapath logic and random logic macros is blurring. A new design style, referred to as hybrid designs, is emerging that contains both random logic and datapath logic. The datapath logic generally refers to circuit structures containing highly parallel bit operations [1], (often called the bit-stack) and careful design is important for high frequency designs. Prior work [2] has shown that, with separate placement engines, a dedicated datapath placer may overly constrain the random logic placement solution causing overall degradation in congestion and wire length. A single placement flow handling both structures is extremely valuable, improving design time, quality, and saving development and maintenance costs. However, [3,4] demonstrate that most state-of-the-art placers are incapable of handling designs with regular structure. This is because, in part, current wirelength-driven placement algorithms are unaware of the structure of the datapath. However, with a bit of minimal design guidance, this work shows that an HPWL-driven placer can be taught to handle these situations much better than they do today. The clue is high fanout nets disrupting the structure of the datapath. Most placers will create a clique with a very low weight to model these nets, or treat them just like any other. To minimize the HPWL of a high fanout net, a placer naturally clumps it into a ball, but that is the exact opposite of what is required by a regular datapath structure, as shown in [5].

In this paper, effective techniques are proposed that can be incorporated within existing random-logic targeted placers to better handle designs with embedded datapaths. A novel placement flow is proposed that leverages the speed and flexibility of state-of-the-art HPWL-driven placers while imposing alignment constraints, to achieve better regularity of the datapaths and better StWL results.

The key contributions of this work are as follows:

1. A study of obstacles to current academic placers: the inadequacies and specifically the lack of fidelity of the HPWL model versus StWL model when evaluating datapath logic.

2. A key insight to bit-stack alignment: alignment of the bit-stack guides indirect StWL optimization, and significantly improves total StWL and routing congestion.

3. A novel placement flow: Structure Aware Placement Techniques (SAPT) that can be incorporated within existing HPWL-driven placers to enable better alignment of the embedded datapaths during both global and detailed placement.

Section 2 outlines the problem faced by current random logic placers when placing datapath logic. Section 3 describes the placement flow consisting of two global placement techniques and two detailed placement techniques, which provide alignment constraints to the datapath. Experimental results are presented in Section 4 and finally, conclusions and future work are presented in Section 5.

2. MOTIVATION

Datapath logic can refer to a wide variety of logic circuits including adders, multipliers, rotators, and other logics implemented within pipeline stages. This section discusses: (a) some of the benefits derived from a unified placement framework able to align cells during placement, (b) problems with existing random logic placers, and (c) datapath HPWL accuracy metrics; for designs containing datapath logic circuits.

2.1 The need for a Unified Placement Framework

Datapath logic circuits are traditionally placed by a separate datapath placer such as [1, 6, 7]. These separate datapath placement techniques generate highly efficient and tightly packed placements. After the datapath is placed, these methods generate a larger macro block or small individual bit-slice macro blocks, that are then placed similar to a movable macro block by the main random logic mixed-size placer. The primary drawback of these approaches is that even though a datapath placer may minimize the local wire length through cell ordering [8] or optimizing specific bit-stacks [9], the global connectivity of the placement problem with the embedded datapath is not taken into account. As shown by [2–4], the added constraints from this shortcoming, in general, produce suboptimal results. Additionally, a significant benefit from a unified placement framework comes in the form of reduced development and support costs derived from a single placement framework versus multiple.

2.2 StWL and HPWL Comparisons for Datapath Circuits

Most placers will create a clique with a very low weight to model high fanout nets, or treat them just like any other. To minimize the HPWL of a high fanout net, a placer naturally clumps it into a ball, but that is the exact opposite of what is required by a regular datapath structure, as shown in [5]. The clue to overcoming this problem is to apply alignment constraints during placement.

The way to measure alignment is to use StWL, rather than HPWL to measure the quality of placement, because it more accurately represents routability. To support this point, the following placers mPL6 v6 [11], CAPO v10.2 [12], FastPlace v3.0 [13], NTUPlace3 v7.10.19 [14], Dragon v3.01 [15], and SimPL [16] are compared on both, total Half-Perimeter Wire Length (HPWL) and total Steiner Wire Length (StWL) on the modified ISPD 2011 Datapath Benchmark Suite [10] [1] [3] shown in Table 1. For improved experimental control, all StWL measurements were performed using coalesC-grip [17], and all reported numbers are total wire length results for each design. The HPWL column in Table 1 is sorted from smallest to largest for each benchmark. The table reports both the actual measured HPWL and StWL for the benchmark circuits as well as the wire length ratio compared to the manually placed solution.

Careful examination of this table yields the following surprising results:

1. While HPWL for both benchmarks is very close to the manually placed solution, the StWL results degrade significantly from the manual solution, with the best automated solution at 82% increase in StWL for benchmark A and 227% increase for benchmark B.

2. While fidelity of the HPWL model is expected, for datapath logic it does not hold true. As Table 1 shows, the HPWL

[1]The MISPD 2011 Datapath Benchmark Suite was modified to contain unfixed latch rows compared to the original fixed latch placement reported in ISPD 2011. Benchmarks can be downloaded at: http://www.cerc.utexas.edu/utda/download/DP/

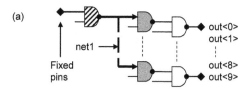

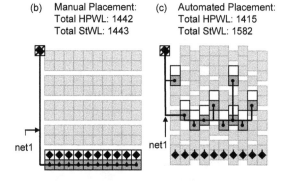

Figure 1: An example circuit where StWL of the manually placed design is better than that of the automated placement, but HPWL of the automated placement solution is better than that of the manual placement. Net1 has fanout of 10.

column is sorted by increasing value and it is generally expected that StWL would maintain that order. But in fact that does not happen. In both cases, the placer with the best HPWL does not generate the best StWL. For benchmark A, though CAPO generates the best HPWL, 5% larger than the manually placed solution, SimPL generates the best StWL, outperforming CAPO with a 82% increase over the manual solution. The same holds true for benchmark B. Again, CAPO generates the best HPWL, but SimPL generates the best StWL, outperforming CAPO with a 2.43% increase compared to the manual solution.

As Section 4.3 will show, the significant improvement in StWL also corresponds to vastly improved congestion metrics.

2.3 Implicit StWL Optimization through Bit-Stack Alignment

There has been prior work in optimizing StWL directly. As reported in [18], StWL generally correlates with routed wire length (rWL) much better than HPWL. However, optimizing StWL directly during global placement is a hard problem, and iteratively computing StWL can be time consuming. Alternately, this work shows that if a HPWL-driven placer can obtain better alignment for regular structures, it will have better StWL. An example is shown in Figure 1. In Figure 1(a), a partial logic netlist with one NAND gate, shown as hashed, drives net $net1$ with a fanout of 10. All the input and output pins are fixed objects placed on top of the gate. Figure 1(b) shows a manually placed solution for this partial circuit and Figure 1(c) shows a solution from an existing placer. The dark shaded cells match the same dark shaded NAND gates in Figure 1(a). The light shaded grey cells are other logic placed within the design.

For both solutions, we measure the total HPWL and StWL, and the numbers are shown in the figure. As Section 2.2 pointed out, even though the HPWL of the manual solution (1442) is greater than the HPWL of the automated placer (1415), the StWL shows the reverse trend. While it is impractical to list the HPWL and StWL of every single net, clearly for net $net1$, the StWL in Fig-

Table 1: Legalized HPWL and StWL comparison on the ISPD 2011 Datapath Benchmark Suite [10] between manually placed and automated placement solutions. Placement results are sorted by increasing HPWL value. To note: (1) Best HPWL solution does not indicate the best StWL solution. (2) Bold numbers are the best automated placement wire length.

| | ISPD Datapath Benchmark A | | | | | ISPD Datapath Benchmark B | | | |
	Total HPWL		Total StWL			Total HPWL		Total StWL	
Manually Placed	11000365	1.00	11066683	1.00	Manually Placed	8642097	1.00	9823680	1.00
CAPO	11535525	**1.05**	21516128	1.94	CAPO	10338805	**1.20**	23881606	2.43
SimPL	11837307	1.08	20180311	**1.82**	NTUPlace3	10433894	1.21	26110039	2.66
mPL6	12919955	1.17	23950663	2.16	SimPL	10631304	1.23	22319594	**2.27**
NTUPlace3	13447753	1.22	24673151	2.23	Dragon	12229019	1.42	28577316	2.91
FastPlace3	15672727	1.42	27115750	2.45	FastPlace3	14537026	1.68	36642434	3.73
Dragon	16424739	1.49	26182449	2.37	mPL6	16263018	1.88	28846387	2.94

ure 1(b) is better than the StWL in Figure 1(c). This is due to the better alignment of the whole net in one horizontal row, which produces much better StWL. Also the solution of Figure 1(c) shows the existing placer trying to clump the net into smaller HPWL but causing the StWL to be worse. This paper presents techniques to teach the existing placer to generate a placement solution similar to Figure 1(b) with better StWL than the one in Figure 1(c).

By providing alignment constraints to small portions of the datapath, it is observed that during the iterative placement process, other surrounding cells become aligned as well. Previous works, like post ECO datapath placement, or placing the datapath as a macro block, tend to ignore the connection between random logic and datapath cells since they place every datapath cell in priori. The alignment constraints presented in this work however are providing hints to the placer directing it toward a more globally optimal solution. Thus, as results will show, with relatively few manually defined bit-stacks, the placer generates significantly reduced overall wire length and congestion. The next section outlines the details of supplying these alignment constraints during placement.

3. UNIFIED PLACEMENT WITH ALIGNMENT CONSTRAINTS

Given a netlist $N = (V, E)$ with nodes V and nets E, placement obtains locations (x_i, y_i) for all the movable nodes, such that the area of nodes within any rectangular region does not exceed the area of cell sites in that region.

With $\vec{x}, \vec{y} = \{x_i, y_i\}$, HPWL is defined as:

$$HPWL(\vec{x}, \vec{y}) = HPWL(\vec{x}) + HPWL(\vec{y}) \qquad (1)$$

$$HPWL(\vec{x}) = \sum_{e \subset E} [MAXx_i - MINx_i] \qquad (2)$$

Modern placers often approximate HPWL by a differentiable function using the quadratic objective, defined as:

$$\Phi_G(\vec{x}, \vec{y}) = \sum_{i,j} w_{i,j} [(x_i - x_j)^2 + (y_i - y_j)^2] \qquad (3)$$

From Equation 3, (x_i, y_i) represents the coordinates of cell i, and $w_{i,j}$ represents the weight between cells i and j. In this work, a force-directed global placer in the spirit of SimPL [16], where $w_{i,j}$ is given by the Bound2Bound net model [19], is used along with a detailed placer similar to FastPlace-DP [20]. Briefly, SimPL is a flat, force-directed global placer. It maintains a lower-bound and an upper-bound placement and iteratively narrows the displacement between the two to yield a final placement solution. The upper-bound placement is generated by applying lookahead legalization, which is based on top-down geometric partitioning and non-linear scaling. The coordinates obtained from the upper bound placement are used to generate the fixed-points and pseudo nets for force-directed placement. The lower-bound placement is then generated by minimizing the quadratic objective in Equation 3.

In this paper, it is assumed a set of T datapath groups and their directions are given. Each datapath group $g_k \in G$, $0 < k < T$, is an unordered nonoverlaping subset of cells from V. Generally, each g_k corresponds to the bit-stack in the datapath, but can be other elements such as cells connected to a single high fanout net that improves through alignment, buffers that need careful placement to facilitate routing of large buses, or for structured latch placement. The direction of datapath g_k is defined as $\vec{d_k}$. In this paper, only horizontal and vertical directions are considered, which means $\vec{d_k} \in (0, 90)$. In the example shown in Figure 1, $\vec{d_k} = 0$.

The above assumptions that the datapath groups and directions are given are valid and practical. One may use datapath extractors such as [21, 22], based on circuit properties, to generate the datapath. This information could also be provided by designers, which may come directly from the logic description of the nelist, or designer experience. As an example, if a designer is trying to structure the latch placement to be vertical, it is trivial for him to provide the vectored latch name and the direction (horizontal or vertical).

3.1 Alignment Nets

Adding pseudo nets is a common method used by modern placers to provide spreading forces. In this work, similar to SimPL, during every iteration of global placement, pseudo nets are added after lookahead legalization to enforce spreading during the subsequent linear system solver.

Definition 1. A pseudo net $c(f, i)$ is a weighted two-pin connection between a fixed-point f and a cell i in the circuit netlist. The pseudo net has a weight equal to $\alpha \cdot w_{i,j}$, defined in [16], and does not exist in the circuit netlist.

It shall be noted that, these nets are added for every movable cell in the design. In addition, existing pseudo nets are discarded at the end of the current iteration, and a new set is added to enforce spreading during the subsequent placement iteration. The pseudo net weighting technique with increasing parameter α is described in [16], and controls the rate of overlap removal during global placement. During early iterations, greater significance is given to interconnect minimization while the relative cell ordering stabilizes. This is accomplished by starting with a small α value and gradually increasing through each iteration. This scheme provides flexibility to the placer during the early stages, while tightening the constraints for no overlap towards the end of global placement.

To generate better datapath alignment, one approach is direct manipulation of existing nets between the datapath cells. But this approach interferes with other prior placement enhancements. Specifically, direct weighting manipulation of current nets disrupts timing aware placement and net weighting for those cells. Due to the above problem, a new method is instead proposed. A new category of nets, refereed as alignment nets is defined.

Definition 2. An alignment net s_k, where $0 < k < T$ and $T = |G|$, is a weighted multi-pin connection between all cells in the datapath group g_k. For placement, the alignment net is modeled using the Bound2Bound net model [19].

These nets are created at the beginning of global placement and remain persistent during the entire global and detailed placement stages. A skewed net-weight schedule (Section 3.3), helps these nets align the cells within the corresponding datapath group g_k inside the placement region. By applying the alignment constraints to a new net s_k, prior techniques continue to function as before.

3.2 Unified Placement Flow Overview

The proposed new placement flow is presented in Figure 2, where the shaded boxes highlight the enhancements applied to each g_k. During global placement, after pseudo net insertion for all cells, the modified flow applies the *skewed weighting with step size scheduling* to each datapath group. Then after the linear solver and fixed-point generation, the second global placement modification is applied called *fixed-point and pseudo net alignment constraint*. Once the global placement solution has converged as defined in [16], two detailed placement steps that act only on the datapath logic are presented. The first is *bit-stack aligned cell swapping* and the second is *datapath group repartitioning*. Detailed placement for the random logic cells is implemented using the techniques presented in [20]. At each step, the modifications apply only to the defined placement groups g_k leaving all other random logic cells to be placed as they would before. Though in this work SimPL and FastPlace-DP are used as an example, the techniques can be adapted for other force directed global placement and detail placement methods as well.

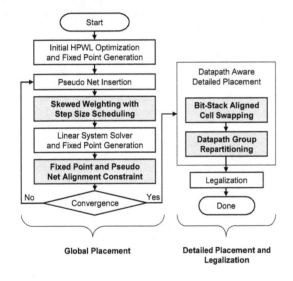

Figure 2: Proposed unified datapath-aware placement flow. The baseline components are shown in transparent boxes and the added datapath-aware components are shaded.

3.3 Skewed Weighting with Step Size Scheduling

In this section, a skewed weighting process applied to each alignment net s_k is described that improves alignment along the datapath. The high level idea is to add a skewed weight for each datapath group, with the weight gradually increasing during each iteration. The rate of change of the weighting value increases slowly during the initial stages of global placement, increases rapidly during the middle stages, and slows again near the end of global placement.

Applying hard constraints (forced alignment) in the early stage of wire length optimization can disrupt the original optimization

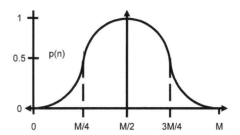

Figure 3: Bell-shaped step size scheduling function.

and often can lead to a solution that suffers from sub-optimality in terms of overall wire length.

Thus, let n be the global placement iteration number and M be its upper bound[2] and define $p(n)$ as the alignment weight schedule function for each iteration n. The following equation for $p(n)$ is proposed:

$$p(n) = \begin{cases} \frac{8n^2}{M^2} & 0 \le n < \frac{M}{4} \\ 1 - \frac{8(n-\frac{M}{2})^2}{M^2} & \frac{M}{4} \le n \le \frac{3M}{4} \\ \frac{8(n-M)^2}{M^2} & \frac{3M}{4} < n \le M \end{cases} \quad (4)$$

To minimize hard constraints during the initial stages of global placement, $p(n)$ gradually increases during the initial iterations and to minimize large constraint changes during the final stages, the function decreases toward zero at the last iteration. This function is also used in [23] as a penalty function, but it serves a *completely different purpose* here as a scheduling function. Using $p(n)$, Equation 5 displays the skewed monotonically increasing weighting parameters γ^n and δ^n for alignment net s_k. Using $p(n)$ directly generates very large weighting steps therefore a constant scaling factor β is added. This parameter is left default throughout all placement runs. Let $\hat{x}, \hat{y}$ be the directional unit vectors and $\sigma_{x,y}^2$ the nth iteration's variance in either the x or y direction. Finally, the modified placement equation is shown in Equation (6). For non-alignment nets, $\delta_{i,j} = 0$ and $\gamma_{i,j} = 0$.

$$\gamma^n = \gamma^{n-1} + \hat{y} \cdot \overrightarrow{\mathbf{d_k}} * \beta * p(n) * \sigma_x^2(n) \quad \text{where } \gamma^0 = 1$$
$$\delta^n = \delta^{n-1} + \hat{x} \cdot \overrightarrow{\mathbf{d_k}} * \beta * p(n) * \sigma_y^2(n) \quad \text{where } \delta^0 = 1 \quad (5)$$

$$\Phi_G(\vec{x}, \vec{y})^n = \sum_{i,j} [(\gamma_{i,j}^n + w_{i,j})(x_i - x_j)^2 + (\delta_{i,j}^n + w_{i,j})(y_i - y_j)^2]$$
$$(6)$$

3.4 Fixed-Point Alignment Constraint

Modern force-directed global placement frameworks use fixed-points and pseudo nets to increase spreading. By gradually perturbing the unconstrained linear system solver, consecutive placement solutions with less overlap are generated In SimPL [16], after each global placement iteration, lookahead legalization generates a fixed zero-area anchor with two-pin pseudo nets. During the following global placement iteration, these pulling forces reduce the amount of cell overlap. For datapath logic, the lookahead legalization step and subsequent pseudo net insertion step cause misalignment within the bit-stack requiring a constraint forcing alignment which minimizes wrong-way perturbations in the bit-stack.

The proposed fixed-point alignment constraint is applied in two steps. First, lookahead legalization generates a fixed-point location for all cells. Second, for all cells in datapath group g_k, a modified fixed-point is added. The defined location of this fixed-point for

---

[2] M is typically upper-bounded by 50 [16]

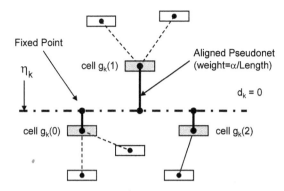

Figure 4: Example of a fixed-point alignment constraint for a horizontal bit-stack. Lookahead legalization generates new zero area fixed-points and the locations of these points are modified to be in alignment with η_k^n.

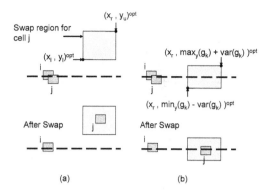

Figure 5: Swap region shift for cell j when the datapath direction is parallel to the x-axis. The upper y coordinate location is defined by cell i plus the variance between cell i and cell j.

cell i as $\eta_{k,i}^n$ for the nth iteration. In this paper, $\eta_{k,i}^n$ is computed as follows:

$$\eta_{k,i}^n = (x_i, \, _{|g_k|}\!\sqrt{\prod_{j=1,\ldots,|g_k|} y_j}), \quad \text{if } d_k = 0$$
$$\eta_{k,i}^n = (\, _{|g_k|}\!\sqrt{\prod_{j=1,\ldots,|g_k|} x_j}, y_i), \quad \text{if } d_k = 90 \tag{7}$$

An example of the modified fixed-point locations and corresponding pseudo nets for a horizontal datapath is shown in Figure 4. In this example, the three grey cells, $g_k(0:2)$ are in one datapath group g_k. The other cell connections are shown with the dashed line connected to the hollow cells. For random logic cells, the fixed-point will be determined based on the lookahead legalization step alone. For the datapath cells shown, after lookahead legalization generates a new fixed-point location, those locations are modified based on the geometric mean parallel to the datapath direction $\vec{d_k}$.

Modifying the fixed-point locations enables the global placer to progressively reduce cell overlap while maintaining bit-stack alignment. Two items should be noted about this process. First, this work *only modifies the fixed-point location* for datapath logic, not the weighting of the pseudo net. The pseudo net weighting, disparate from the alignment net weighting proposed in Section 3.3, acts on datapath and random logics the same. Second, though this technique violates the overlap constraint during global placement, the overlap reduces with consecutive global placement iterations [16] and small overlaps can be easily removed during legalization without undermining the overall wire length.

3.5 Bit-Stack Aligned Cell Swapping

This detailed placement technique modifies global cell swapping from [20] for nodes within each g_k by modifying the "swap region" while keeping the overlap penalty the same. Assuming all cells in the placeable region are fixed except for cell j, the "swap region", based on the median idea from [8], is the location where the wire length for cell j is improved if it is swapped with a cell k located in the swap region. This technique looks for cells to swap between the current location of cell j and all cells within the swap region. If a swap produces improved HPWL, the cell locations are updated.

This work, unlike [20], bounds the swap region perpendicular to $\vec{d_k}$. More specifically, for each net $p \in E$, the left, right, lower and upper edges of the bounding box are: $(x_l[p], x_r[p], y_l[p], y_u[p])$ and the x^{opt} and y^{opt} from [8] is the median of the x series $(x_l[1], x_r[1], x_l[2], x_r[2], \ldots)$ and y series $(y_l[1], y_u[1], y_l[2], y_u[2], \ldots)$.

Because the number of elements is generally even, the x^{opt} and y^{opt} becomes a region with bounding box $(x_l^{opt}, y_l^{opt}, x_r^{opt}, y_u^{opt})$. The modified swap region assuming the alignment net s_k is parallel to the x-axis is shown in Equation 8, and assuming the alignment net s_k is parallel to the y-axis is shown in Equation 9.

$$x_l^{opt}, min_y(g_k) - var_y(g_k)$$
$$x_r^{opt}, max_y(g_k) + var_y(g_k)$$
$$when \qquad (\vec{\mathbf{d_k}} = 0) \tag{8}$$

$$min_x(g_k) - var_x(g_k), y_l^{opt},$$
$$max_x(g_k) + var_x(g_k), y_u^{opt}$$
$$when \qquad (\vec{\mathbf{d_k}} = 90) \tag{9}$$

Figure 5 Illustrates the difference between the original potential swap region and the datapath aware swap region. In the original example from Figure 5(a) on the left, the swap region for cell j, based on the HPWL would cause j to move out of line with $\vec{d_k}$ for that group, thus disrupting the alignment. In the proposed method shown in Figure 5(b) on the right, the swap region is shifted down to maintain alignment for that group.

3.6 Datapath Group Repartitioning

The second detailed placement technique is a top down recursive repartitioning for each g_k along alignment net s_k, referred to as datapath partitioning. With datapaths, traditional HPWL metrics can at times fail to detect alignment improvements. This technique minimizes internal net cut values potentially improving both HPWL and StWL metrics for all nodes in g_k along s_k. The partitioning iterates through each node in g_k and swaps among other cells in g_k that minimize the total net cut of that partition. By minimizing cut value, improved alignment and routability is possible. The base cut algorithm is from [24], but there are a couple of key differences to the repartitioning method. First, the swap is only accepted when the HPWL after the swap is less than or equal to the HPWL before the swap. Second, the initial median is the midpoint between the nodes. All nodes in g_k with values less than the median go in one partition, the other nodes in the other partition.

For each defined datapath in the design, the algorithm calculates the median or middle point of s_k. Once a median point has been identified, the algorithm partitions all nodes connected to the datapath alignment group using KL partitioning. The partitioning solution is made HPWL aware by evaluating the KL solution for HPWL changes. If the HPWL increases, the solution is discarded and KL

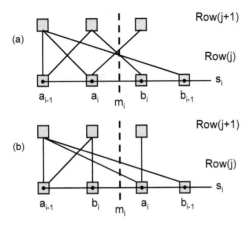

Figure 6: Group repartitioning example swapping the positions of cell a_i and b_i for an improved net cut.

evaluates a new solution for a higher net cut value that does not cause an increase in HPWL or total net cut. Once a partition has been selected, the design is legalized and the loop at that partition level is complete. The algorithm continues to hierarchically break, using recursive repartitioning, each datapath group into smaller partitions until a predefined minimum partition size is met. As an example, consider Figure 6(a) with median location m_i, datapath alignment group s_i in $Row(j)$. In this placement, the initial cut value across m_i is three. After swapping nodes b_i and a_i, shown in Figure 6(b), the total net cut value along m_i is reduced to one.

4. EXPERIMENTAL RESULTS

The placer in this work is compatible with the Bookshelf format and requires an additional datapath definition file as input. This file is loaded prior to global placement, and includes each manually identified datapath groups to be aligned and provides a direction for each datapath group. For improved experimental control, all HPWL numbers and StWL estimates were generated using coalesCgrip [17]. The ISPD 2011 Datapath Benchmark Suite for the manual placement and spacing variations were modified to make all latches movable compared to the fixed latch placement in the original work. All logical connections and input and output pin locations remained the same. Briefly, the ISPD 2011 Datapath Benchmark Suite [10] contains two common datapath circuits each with a series of eight different utilizations to examine the ability of automatic placers to generate placement solutions at different densities.

All placement numbers are the ratio of the *total wire length* of the placed solution vs. the *total wire length* of the manually designed placement as described in [3]. All wire length numbers are from legalized placements and in cases where overlaps were generated, post placement legalizers were used to generate a legal placement. This work focuses on the placement solution, so each datapath was manually defined for improved experimental control.

The proposed approach is referred to as Structure Aware Placement Techniques (SAPT). In the tables that follow, *SAPTgp* refers to the proposed structure aware global placement techniques with the base FastPlace-DP [20] detailed placer, and *SAPTdp* refers to the wire length results when running both the proposed global and detailed placement techniques. All placers were supplied a target density requirement of 1 as defined in the ISPD placement contests [25]. The placer ROOSTER [18], a variant of Capo that optimizes StWL in global and detailed placement, was also run and we observed slightly improved HPWL and StWL results with little impact on overflow numbers when compared to CAPO.

Table 2 provides benchmark characteristics. Of note is the number of datapath groups g_k in each design and the datapath ratio. The datapath ratio is defined as the ratio of datapath cells to random logic cells in each design. Though the hybrid designs are on the smaller side, they are state-of-the-art industrial circuits containing both datapath and random logic cells. They are included to demonstrate the impact of the proposed techniques on modern designs. Due to page limitations, run times are reported for only the hybrid industrial designs.

4.1 Results on the Modified ISPD2011 Datapath Benchmark Suite

Table 3 shows the total HPWL ratio (including both random and datapath logic nets in the design) for the ISPD2011 Datapath Benchmark Suite comparing prior placers to the manually generated layout solutions. In these runs for benchmark A, CAPO generated the best total HPWL results for all placers coming within 2% of the manually placed benchmark at 82% design utilization. The SimPL placer also generated very competitive HPWL results at only 4% increase at 77% utilization. NTUPlace3 failed to run on benchmark A. For benchmark B, both NTUPlace3 and CAPO generated the best overall total HPWL result among all placers at 12% worse than the manual solution. For all proposed techniques, the HPWL numbers oscillate within a few percentage points of the original placement solution from SimPL and FastPlace-DP detailed placement. As will be shown, though the total HPWL numbers are approximately the same as the manually placed solution, the total StWL of the automated placers is significantly worse. This result reinforces that for datapath designs, total HPWL is a bad indicator of placement quality.

Total steiner wire length (StWL) results (including all nets in the design) for each datapath benchmark A and B variant are shown in Table 4. As previously shown, total StWL results of prior placement algorithms were abysmal compared to the total StWL of the manually placed benchmark. However, the proposed global placement (SAPTgp) solutions improved the StWL from 1.78 to 1.35 compared to the manual solution and the detailed placement methods (SAPTdp) further improves the ratio from 1.78 to 1.30. The proposed placer on benchmark B also significantly outperformed prior placers with SAPTgp achieving 1.48 and SAPTdp achieving 1.46 compared to the manual designed solution. The results in bold represent the best published automated StWL placement solution.

These results show that the presented placer for benchmark A outperforms all other automated placers with SimPL coming closest at a 36% increase compared to SATPdp. For benchmark B, the proposed placer outperforms all other placers with SimPL again being the closest at a 48% increase over SAPTdp.

Figure 7 (a) displays the datapath placement solution from SimPL. In this figure, a random selection of bit-stack cells are plotted with a black line connecting them. In the manually placed solution, the bit stack is aligned, either vertically or horizontally depending on the group, which allows the placer to obtain a more compact placement solution. As shown in Figure 7 (a), clearly these cells are not placed in alignment. The modified placement solution generated using the proposed placer is shown in Figure 7 (b). The same set of datapath groups shown in Figure 7 (a) are displayed with a black line connecting each cell. Clearly there is significant straightening improvement in each bit-stack group.

4.2 Hybrid Placement Results

In addition to significantly improved StWL results on the datapath benchmarks, the proposed placer generates improved hybrid design StWL results as shown in Table 5. For each placer, the num-

Table 2: Circuit statistics. Datapath ratio is calculated as the total number of datapath cells divided by the total number of cells.

	ISPD2011 Datapath Benchmarks		Industrial Hybrid Designs			
	Benchmark A	Benchmark B	Hybrid C	Hybrid D	Hybrid E	Hybrid F
Total node count	160416	152668	17922	55387	83802	263906
Total pin count	637984	653116	64078	94682	130000	397652
Total net count	157849	148682	16874	14458	16422	53884
Datapath groups g_k	1425	1932	35	110	60	131
Datapath ratio	0.920	0.850	0.010	0.012	0.008	0.007

Table 3: Total HPWL ratio comparison on the modified ISPD 2011 Datapath Benchmark A and B variants with legalized placement. The ratios are computed with respect to the manually placed solution.

	ISPD 2011 Datapath Benchmark A: Total HPWL								ISPD 2011 Datapath Benchmark B: Total HPWL							
Utilization	94	91	89	86	84	82	79	77	95	93	91	89	86	84	81	79
CAPO	1.05	1.04	1.04	1.04	1.03	1.02	1.06	1.03	1.20	1.18	1.17	1.12	1.13	1.13	1.14	1.12
mPL6	1.17	1.19	1.22	1.14	1.16	1.20	1.17	1.16	1.64	1.86	1.72	1.64	1.65	1.65	1.78	1.78
NTUPlace3	1.22	1.19	1.16	1.19	1.15	1.19	1.23	1.26	1.25	1.19	1.17	1.15	1.16	1.15	1.12	1.15
Dragon	1.49	1.58	1.63	1.60	1.51	1.62	1.66	1.60	1.40	1.40	1.35	1.32	1.32	1.30	1.31	1.31
FastPlace3	1.42	1.50	1.53	1.54	1.53	1.67	1.70	1.75	1.69	1.66	1.73	1.71	1.77	1.86	1.77	1.87
SimPL	1.08	1.07	1.06	1.07	1.05	1.06	1.05	1.04	1.23	1.22	1.21	1.20	1.17	1.16	1.16	1.15
SAPTgp	1.10	1.12	1.07	1.05	1.06	1.05	1.04	1.04	1.21	1.20	1.17	1.16	1.16	1.16	1.16	1.15
SAPTdp	1.09	1.07	1.05	1.05	1.04	1.04	1.03	1.04	1.21	1.19	1.17	1.16	1.15	1.15	1.14	1.15

Table 4: Total StWL ratio comparison on the modified ISPD 2011 Datapath Benchmark A and B variants with *unfixed latches* after legalized placement. The ratios are computed with respect to the manually placed solution. Numbers in bold are the best automated placement results published for these benchmarks.

	ISPD 2011 Datapath Benchmark A: Total StWL								ISPD 2011 Datapath Benchmark B: Total StWL							
Utilization	94	91	89	86	84	82	79	77	95	93	91	89	86	84	81	79
CAPO	1.94	1.94	1.91	1.93	1.90	1.90	1.80	1.90	2.40	2.40	2.38	2.35	2.36	2.36	2.35	2.32
mPL6	2.16	2.14	2.16	2.08	2.10	2.12	2.11	2.09	2.94	3.29	3.06	3.01	2.97	2.95	3.20	3.21
NTUPlace3	2.23	2.18	2.15	2.15	2.11	2.16	2.19	2.09	2.66	2.48	2.47	2.44	2.44	2.44	2.32	2.44
Dragon	2.37	2.44	2.53	2.48	2.36	2.48	2.56	2.43	2.91	2.87	2.84	2.80	2.79	2.77	2.75	2.74
FastPlace3	2.45	2.53	2.56	2.59	2.56	2.71	2.75	2.79	3.73	3.58	3.78	3.79	3.97	4.13	3.96	4.14
SimPL	1.82	1.83	1.80	1.81	1.78	1.78	1.78	1.75	2.27	2.30	2.25	2.24	2.23	2.19	2.24	2.22
SAPTgp	1.43	1.46	1.39	1.36	1.37	1.36	1.35	1.35	1.59	1.56	1.54	1.50	1.51	1.50	1.50	1.48
SAPTdp	1.38	1.35	1.34	1.34	1.32	1.32	**1.30**	1.32	1.58	1.55	1.52	1.49	1.49	1.48	1.48	**1.46**

Table 5: The impact of datapath placement on hybrid designs. The wire length ratios are compared to the proposed placer.

	Hybrid C			Hybrid D			Hybrid E			Hybrid F		
	Total HPWL	Total StWL	Run Time(s)	Total HPWL	Total StWL	Run Time(s)	Total HPWL	Total StWL	Run Time(s)	Total HPWL	Total StWL	Run Time(s)
CAPO	1.13	1.26	94.6	1.17	1.32	74.0	1.12	1.27	83.4	1.19	1.17	480.3
mPL6	1.05	1.15	48.5	1.02	1.14	32.4	1.20	1.32	36.2	1.37	1.30	161.7
NTUPlace3	0.95	1.10	13.0	0.95	1.13	30.0	0.99	1.19	70.0	1.30	1.30	278.0
Dragon	1.10	1.20	425.9	2.11	2.04	193.0	1.32	1.38	283.9	1.29	1.24	927.4
FastPlace3	0.95	1.04	13.0	0.96	1.16	10.7	1.22	1.30	17.4	1.17	1.14	55.3
SimPL	1.02	1.10	9.2	0.97	1.16	12.6	1.03	1.12	27.1	1.04	1.04	59.2
SAPTdp	1.00	1.00	15.9	1.00	1.00	16.7	1.00	1.00	38.2	1.00	1.00	70.9

Table 6: The Total Overflow (x 1e+5) using the router and evaluation script from the ISPD 2011 routability-driven placement contest on the modified ISPD 2011 Datapath Benchmark A and B variants with unfixed latches after legalized placement. The "Total Overflow", a measure of the routing congestion of the placement solution, is reduced to zero on six of the benchmark A variants and reduced by at least 6.7x for all benchmark B variants.

	ISPD 2011 Datapath Benchmark A: Routing Overflow								ISPD 2011 Datapath Benchmark B: Routing Overflow							
Utilization	94	91	89	86	84	82	79	77	95	93	91	89	86	84	81	79
CAPO	2.29	2.17	1.72	1.83	1.84	1.68	1.10	2.18	9.16	7.28	7.05	6.68	7.17	7.01	7.13	6.98
mPL6	4.66	4.38	4.44	3.40	3.38	3.65	6.03	5.02	12.7	16.4	14.0	13.6	12.8	12.6	15.3	15.3
NTUPlace3	5.54	5.12	4.63	5.19	4.92	5.63	6.03	5.02	10.2	8.41	8.3	8.09	8.92	9.07	8.21	9.92
Dragon [3]	-	-	-	-	-	-	-	-	12.8	12.7	12.5	12.4	12.6	12.7	12.8	12.9
FastPlace3	7.23	8.10	8.72	9.08	8.80	10.4	11.8	12.1	20.8	19.3	21.6	21.7	23.7	25.5	23.5	25.6
SimPL	1.28	1.28	1.22	0.98	0.87	0.87	0.85	0.77	5.98	6.24	5.65	5.49	5.26	4.85	5.21	5.25
SAPTgp	0.0012	0.032	0.0	0.0	0.0	0.0	0.0	0.0	0.90	0.70	0.56	0.45	0.48	0.59	0.62	0.59
SAPTdp	0.0014	0.038	0.0	0.0	0.0	0.0	0.0	0.0	0.88	0.70	0.55	0.43	0.67	0.58	0.60	0.58

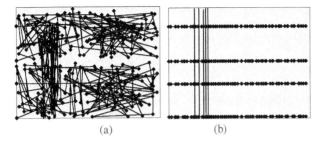

(a)   (b)

Figure 7: Forty structured bit-stacks are randomly chosen to show the impact of the proposed placer on structured nets in Benchmark A. (a) is generated by SimPL [16] whereas (b) is generated by the proposed placer. Movable cells are shown lightly shaded while the bit-stack connectivity is shown in black.

bers indicate the ratio of the total wire length obtained by the placer to that obtained by our techniques (given under SAPTdp). Though the HPWL results are similar, the proposed placer obtains an improvement in StWL between 4% and 13%. For these designs, the wire length improvement is significant considering the percentage of datapath logics within the designs is less than 1.5%.

By providing alignment constraints to portions of the datapath, it is observed that neighboring cells also get aligned during the iterative placement process. The alignment constraints provide hints, directing the placer in the correct gradient. These hints help to overcome local optima, driving placement towards a more globally optimal solution. Thus, with relatively few manually pre-defined bit-stacks, this work shows that a HPWL-driven placer can generate improved solutions for the other cells, resulting in significantly improved total wire length.

4.3 Routing Congestion Results

To empirically prove our claim that StWL accurately approximates routability, Table 6 displays the total overflow ($\times 1e + 5$), as defined in the ISPD 2011 routability-driven placement contest [26]. Reported overflow numbers are provided using the contest evaluation script on legal placements. As seen in Table 6, SAPT produces the smallest overflow across all test cases. For benchmark A, SAPT produced a routable placement solution with zero (0) overflow for all but two of the variations. For benchmark B, SAPT improves total overflow by 6.7×, 23.54×, 14.44×, 11.56×, 14.3×, and 10.36× versus SimPL, FastPlace3, Dragon, NTUPlace3, mPL6, and CAPO respectively. Though SAPT is not a congestion aware placer, the significant improvement in routing congestion indicates the strong correlation between alignment, congestion and the importance of StWL for datapath logics.

5. CONCLUSIONS

This work presents a unified framework to enhance current random logic placers to better handle designs containing datapath logics. A set of new global and detail placement techniques, including skewed weighting with step size scheduling, fixed-point and pseudo net alignment constraint, bit-stack aligned cell swapping and group recursive repartitioning, were presented that seamlessly integrate alignment constraints into a state-of-the-art placement engine to overcome the shortcomings of the HPWL model for datapaths. Experimental results show at least a 32% improvement in total StWL compared with six state-of-the-art academic placers for the ISPD 2011 Datapath Benchmark Suite and a 8.25% average improvement in total StWL over six state-of-the-art placers for industrial hybrid designs. Though comparisons do not report the timing impact because the current implementation is limited to reading the

Bookshelf format, significant improvements in wire length are generally attributed to improved timing. Future research will including quantifying this effect in addition to automatically extracting the datapath.

6. REFERENCES

[1] R. X. T. Nijssen and J. A. G. Jess, "Two-dimensional datapath regularity extraction," in *IFIP Workshop on Logic and Architecture Synthesis*, pp. 110–117, 1996.

[2] P. Ienne and A. GrieBing, "Practical experiences with standard-cell based datapath design tools," in *Proceedings of DAC*, pp. 396–401, 1998.

[3] S. I. Ward, D. A. Papa, Z. Li, C. N. Sze, C. J. Alpert, and E. Swartzlander, "Quantifying academic placer performance on custom designs," in *Proc. ISPD*, pp. 91–98, 2011.

[4] D. A. Papa, S. N. Adya, and I. L. Markov, "Constructive benchmarking for placement," in *ACM Great Lakes Symposium on VLSI*, pp. 113–118, 2004.

[5] T. Kutzschebauch and L. Stok, "Regularity driven logic synthesis," in *Proc. ICCAD*, pp. 439–446, 2000.

[6] T. Serdar and C. Sechen, "Automatic datapath tile placement and routing," in *Design, Automation and Test in Europe, 2001. Conference and Exhibition 2001. Proceedings*, pp. 552 –559, 2001.

[7] T. Ye, S. Chaudhuri, F. Huang, H. Savoj, and G. D. Micheli, "Physical synthesis for ASIC datapath circuits," in *Circuits and Systems, 2002. ISCAS 2002. IEEE International Symposium on*, vol. 3, pp. III–365 – III–368 vol.3, 2002.

[8] S. Goto, "An efficient algorithm for the two-dimensional placement problem in electrical circuit layout," *Circuits and Systems, IEEE Transactions on*, vol. 28, pp. 12 – 18, jan 1981.

[9] C. Yang, X. Hong, Y. Cai, W. Hou, T. Jing, and W. Wu, "Physical synthesis for ASIC datapath circuits," in *ASIC, 2003. Proceedings. 5th International Conference on*, vol. 1, pp. 97–100, 2003.

[10] S. I. Ward, D. Z. Pan, and E. Swartzlander, "2011 ISPD Datapath benchmark suite." http://www.cerc.utexas.edu/utda/download/DP/, Mar 2011.

[11] T. F. Chan, J. Cong, J. R. Shinnerl, K. Sze, and M. Xie, "mPL6: enhanced multilevel mixed-size placement," in *Proc. ISPD*, pp. 212–214, 2006.

[12] J. A. Roy, D. A. Papa, S. N. Adya, H. H. Chan, A. N. Ng, J. F. Lu, and I. L. Markov, "Capo: robust and scalable open-source min-cut floorplacer," in *Proc. ISPD*, pp. 224–226, 2005.

[13] N. Viswanathan, M. Pan, and C. Chu, "FastPlace 3.0: A fast multilevel quadratic placement algorithm with placement congestion control," in *Proceedings of ASPDAC*, pp. 135–140, 2007.

[14] T.-C. Chen, Z.-W. Jiang, T.-C. Hsu, H.-C. Chen, and Y.-W. Chang, "A high-quality mixed-size analytical placer considering preplaced blocks and density constraints," in *Proc. ICCAD*, pp. 187–192, 2006.

[15] M. Wang, X. Yang, and M. Sarrafzadeh, "Dragon2000: Standard-cell placement tool for large industry circuits," in *Proc. ICCAD*, pp. 260–263, 2000.

[16] M.-C. Kim, D.-J. Lee, and I. L. Markov, "simPL: an effective placement algorithm," in *Proc. ICCAD*, pp. 649–656, 2010.

[17] H. Shojaei, A. Davoodi, and J. Linderoth, "Congestion analysis for global routing via integer programming," in *Proc. ICCAD*, pp. 256–262, 2011.

[18] J. A. Roy and I. L. Markov, "Seeing the forest and the trees: Steiner wirelength optimization in placement," *CAD of Integrated Circuits and Systems, IEEE Transactions on*, vol. 26, no. 4, pp. 632–644, 2007.

[19] P. Spindler, U. Schlichtmann, and F. M. Johannes, "Kraftwerk2 - a fast force-directed quadratic placement approach using an accurate net model," *IEEE TCAD*, vol. 27, no. 8, pp. 1398–1411, 2008.

[20] M. Pan, N. Viswanathan, and C. Chu, "An efficient and effective detailed placement algorithm," in *Proc. ICCAD*, pp. 48–55, 2005.

[21] A. Chowdhary, S. Kale, N. Saripella, N. Sehgal, and R. Gupta, "Extraction of functional regularity in datapath circuits," *IEEE TCAD*, vol. 18, no. 9, pp. 1279–1296, 1999.

[22] A. Rosiello, F. Ferrandi, D. Pandini, and D. Sciuto, "A hash-based approach for functional regularity extraction during logic synthesis," in *Proc. ISVLSI*, pp. 92–97, 2007.

[23] A. B. Kahng, S. Reda, and Q. Wang, "Architecture and details of a high quality, large-scale analytical placer," in *Proc. ICCAD*, pp. 890–897, 2005.

[24] B. W. Kernighan and S. Lin, "An efficient heuristic procedure for partitioning graphs," *Bell Systems Technical Journal*, vol. 49, no. 1, pp. 291–307, 1970.

[25] G.-J. Nam, C. J. Alpert, P. Villarrubia, B. Winter, and M. Yildiz., "ISPD 2005 placement contest benchmark suite," in *Proc. ISPD*, pp. 216–220, 2005.

[26] N. Viswanathan, C. J. Alpert, C. Sze, Z. Li, G.-J. Nam, and J. A. Roy, "The ISPD-2011 routability-driven placement contest and benchmark suite," in *ACM International Symposium on Physical Design*, pp. 141–146, 2011.

Mixed Integer Programming Models for Detailed Placement

Shuai Li and Cheng-Kok Koh
School of Electrical and Computer Engineering, Purdue University
West Lafayette, IN, 47907-2035
{li263, chengkok}@purdue.edu

ABSTRACT

Existing detailed placement optimization methods typically involve the use of enumeration to determine the optimal location of a small number of cells. We propose two Mixed Integer Programming (MIP) models that can optimize the detailed placement of more cells efficiently. Compared with existing models, the first proposed model has fewer integer variables. The second proposed model, derived based on Dantzig-Wolfe decomposition principle, is with tighter bounds during its solution. Experimental results show that both models are capable of optimizing in reasonable time the detailed placement of much larger problem instances than existing models. Experiments on large-scale real benchmark circuits also show that detailed placer based on advanced MIP models can effectively reduce half-perimeter wirelengh (HPWL), as well as routed wirelength and vertical vias, of the original placement results generated by enumeration approach.

Categories and Subject Descriptors

B.7.2 [**Integrated Circuits**]: Design Aids—*Placement and Routing*

General Terms

Algorithms, Theory

Keywords

Detailed Placement, MIP

1. INTRODUCTION

The quality of a placement solution can greatly affect the performance, power consumption, and reliability of a VLSI circuit. Research in placement has flourished in the past several years after it has been demonstrated in [1] that there is still significant room left for the improvement in placement quality. Moreover, instead of only minimizing half-perimeter wirelength (HPWL), recent research in placement have also focused on the routability, delay, and power issues.

The placement for standard-cell circuits is typically a two-step approach: global placement followed by detailed placement. Most papers on placement deal with global placement (examples of such papers are [2, 3, 4]). However, recent studies in [5, 6, 7] also demonstrate that detailed placement is critical to the quality of placement solutions.

Dealing with the exact location of cells, detailed placement is a discrete optimization problem with huge solution space. Consider the placement of n cells in a row of sites that accommodate all the cells exactly. There are $n!$ permutations to place the cells. In a more general case when m sites would be left empty after all the n cells are placed, the number of all the possible permutations would be $(m+n)!/m!$. With a large placement area, the number of permutations may be much larger than $n!$.

The horizontal whitespace distribution technique in [8] shows that without changing the order of cells, the detailed placement of cells in one row can be optimized in polynomial time by means of dynamic programming. Using the concept of optimal region, the cell swap technique is effective in reducing HPWL [9]. In Domino [10], the whole chip is partitioned into clusters of cells and the placement of each cluster of cells is transformed into a transportation problem and solved with a network flow algorithm.

The sliding window technique is usually deployed in detailed placement [11, 12, 13]. The whole chip is partitioned into overlapping windows, each typically with 6 cells or fewer [6]. Enumeration is usually used to find the optimal placement of cells within a sliding window. With the branch and bound technique, the problem of placement of cells in a sliding window can be optimized more efficiently [12].

In [6], a mixed integer programming (MIP) model is built for the optimization of detailed placement of cells in sliding windows. Experiments on regular mesh grids show that the MIP-based placement algorithm is powerful in optimizing large windows that are far beyond the capability of enumeration. However, the model considers only uniform-width cells.

A more generalized MIP model for the detailed placement of cells with different widths is presented in [14], based on which a detailed placer is implemented and successfully parallelized. However, the MIP model in [14] is a compact model defined with binary decision variables determining whether a site is occupied by a cell. Such a model has a large number of integer variables, referred to as the site occupation variables.

In this paper, we propose two more efficient MIP models that are capable of handling cells of different widths. Instead of site occupation variables, the first model is defined with row occupation variables and column occupation variables (both to be defined later). The number of integer variables is reduced and the solution time is shorter. Derived by applying Dantzig-Wolfe decomposition, the second model uses single-cell-placement variables (to be defined later). With tighter bounds in the optimization process, its solution turns out to be even more efficient.

We have implemented all the models with commercial MIP solver CPLEX. Experimental results show that when applied to optimize the placement of the same sliding windows, our new models have shorter solution times compared with existing models. Furthermore, experiments on large-scale benchmark circuits also show that the detailed placer based on advanced MIP models can effectively generate placement results with better HPWL and routed wirelength than the enumeration approach.

The paper is organized as follows. Section 2 provides a brief review of MIP. Section 3 covers different MIP models for detailed placement. Section 4 presents experimental results comparing different models, as well as the results obtained on large-scale benchmark circuits. Section 5 concludes the paper.

2. MIXED INTEGER PROGRAMMING

MIP technique has been used to solve various NP-hard problems in the past decades, mainly because of the maturity of the branch-and-cut technique. By means of branch-and-cut, the size of solvable Traveling Salesman Problem has reached over 20,000 cities [15, 16]. Another technique for MIP solution, namely, the branch-and-price technique [17], has also proved to be efficient in solving Vehicle Routing problem [18], Airline Crew Scheduling problem [19], as well as global routing [20, 21].

Both branch-and-cut and branch-and-price are based on the branch-and-bound technique for MIP. A branch-and-bound tree, the size of which can be exponential in terms of the number of integer variables in the model, is built in the optimization process. At the root of the tree, the MIP is relaxed and "solved" as a Linear Programming (LP) problem. The objective value of the optimal relaxed LP solution provides a bound for the objective of MIP. If the optimal relaxed LP solution is non-integral, the root is branched into child nodes with a particular integer variable being restricted to different ranges. A binary variable, for example, will be selected and restricted to 0 in one node and 1 in the other node. Subsequently, at each child node (leaf node), the relaxed LP is solved, and branching takes place again except when the relaxed LP is infeasible, the optimal relaxed LP solution is integral, or the objective value of the optimal relaxed LP solution is worse than that of the best integer solution found so far. In the second case, the best integer solution may be updated. In the third case, the node can be pruned without leading to the loss of better integer solutions.

In branch-and-cut, constraints or *cuts* that are valid for MIP but are violated by relaxed LP solution are added at every node of the branch-and-bound tree, and the relaxed LP is solved again after the addition. The added cuts may lead to tighter bound at the node and pruning may take place as a result. Moreover, in branch-and-cut, the global

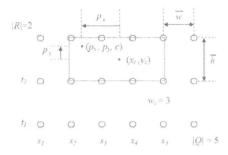

Figure 1: An illustration of cell c ($w_c = 3$) placed in a 2×5 window.

cuts added at one particular node are applicable to all the nodes throughout the tree.

Branch-and-price is usually applied in solving the Master Problem (MP) derived from Dantzig-Wolfe decomposition [22]. Applying Dantzig-Wolfe decomposition on Cutting Stock problem is discussed in [23]. Normally the derived MP is *tighter* than the original compact model, i.e., the objective value of its optimal relaxed LP solution is closer to that of the optimal MIP solution. With tighter bounds, more pruning will happen in branch-and-bound. However, it is much likely that the derived MP has too many integer variables and branch-and-price must be adopted for its solution.

Branch-and-price is a combination of branch-and-bound and column generation. The column generation technique is used to solve the relaxed LP at every node of the branch-and-bound tree. Only a small portion of integer variables or *columns* in MP are kept to form a Restricted Master Problem (RMP). The relaxed LP of RMP is solved first. Then, *pricing* subproblems are solved to identify new columns that can be introduced in the basis. After adding the new columns to RMP, the relaxed LP of RMP is solved again. This iterative process of alternating between solving RMP and subproblems continues until no such columns exist.

3. MIP MODELS FOR DETAILED PLACEMENT

In detailed placement for standard cell circuits, a large number of small logical elements called *cells* are to be placed in the placement region with rows of discrete locations, called *sites*, that are uniformly placed. (Consequently, we also have columns of sites.) Each standard cell is of uniform height, but of different widths, expressed in terms of the number of sites that it occupies. In a legal placement, each cell occupies a number of contiguous sites in one row. Each *net* connects multiple *pins* located in cells. The objective of detailed placement is to minimize the total wirelength of all the nets.

Three MIP models are discussed next. Section 3.1 presents the existing model with site occupation variables; Section 3.2 describes the model with row occupation and column occupation variables; Section 3.3 introduces the model with single-cell-placement variables.

In our models, the sliding window is considered to be a rectangular area composed of uniform-width, uniform-height sites. An illustration of cell c placed in a 2×5 window is given in Figure 1.

The definitions used in the models are:

Q set of all columns in the window;

R set of all rows in the window;

$\overline{w}$ separation between adjacent columns;

$\overline{h}$ separation between adjacent rows;

s_q horizontal coordinate of column $q \in Q$;

t_r vertical coordinate of row $r \in R$;

C set of all cells;

w_c width of cell $c \in C$ measured by the number of sites it occupies;

N set of all nets;

(p_x, p_y, c) a pin on cell $c \in C$ with (p_x, p_y) being the offset from c's centroid;

P_n set of all pins connected by net $n \in N$;

The continuous variables shared in the models are:

(x_c, y_c) coordinates of the centroid of cell $c \in C$;

(l_n^x, u_n^x) horizontal span of net $n \in N$;

(l_n^y, u_n^y) vertical span of net $n \in N$;

3.1 S Model

The MIP model with site occupation variables, called the S Model, is a variant of the model in [14]. In the S Model, the binary site occupation variables are defined as:

p_{crq} 1 if cell $c \in C$ occupies the site at row $r \in R$, column $q \in Q$; 0 otherwise

Then, the MIP model for wirelength-driven detailed placement is:

$$\text{Min} \quad \text{HPWL} = \sum_{n \in N} (u_n^x - l_n^x + u_n^y - l_n^y) \qquad (1)$$

subject to

$$l_n^x \leq x_c + p_x \leq u_n^x, \quad \forall (p_x, p_y, c) \in P_n, n \in N, \qquad (2)$$

$$l_n^y \leq y_c + p_y \leq u_n^y, \quad \forall (p_x, p_y, c) \in P_n, n \in N, \qquad (3)$$

$$\frac{1}{w_c} \sum_{r \in R} \sum_{q \in Q} (s_q + \frac{\overline{w}}{2}) p_{crq} = x_c, \quad \forall c \in C, \qquad (4)$$

$$\sum_{r \in R} \sum_{q \in Q} (t_r + \frac{\overline{h}}{2}) p_{crq} = y_c, \quad \forall c \in C, \qquad (5)$$

$$\sum_{r \in R} \sum_{q \in Q} p_{crq} = w_c, \quad \forall c \in C, \qquad (6)$$

$$p_{crq} + p_{cr'q'} \leq 1, \quad \forall c \in C, r \neq r' \text{ or } |q - q'| \geq w_c, \qquad (7)$$

$$\sum_{c \in C} p_{crq} \leq 1, \quad \forall q \in Q, r \in R, \qquad (8)$$

$$p_{crq} \in \{0, 1\}, \quad \forall c \in C, q \in Q, r \in R.$$

(2)–(5) define net bounding boxes and cell centroids. Constraint (6) makes sure that the number of sites a cell occupies is the same as the cell width. Constraint (8) ensures that no two cells overlap at one particular site.

Moreover, for cells occupying multiple sites, contiguity constraints are required to ensure cells occupy only contiguous sites at the same row. In [14], the following contiguity

constraints based on cell centroids are used:

$$s_q p_{crq} + \overline{w} \leq x_c + w_c \frac{\overline{w}}{2}, \qquad \forall c \in C, r \in R, q \in Q,$$

$$t_r p_{crq} + \overline{h} \leq y_c + \frac{\overline{h}}{2}, \qquad \forall c \in C, r \in R, q \in Q.$$

In this paper, we adopt a tighter contiguity constraint (7), i.e., two sites in different rows ($r \neq r'$) or separated horizontally with no less than w_c sites ($|q - q'| \geq w_c$) cannot be occupied by the same cell c.

The advantage of constraint (7) over centroid-based constraints can be seen in the following example. At one node of the branch-and-bound tree, $p_{cr^*q^*}$ may be restricted to be 1. Then with constraint (7), all the other p_{crq}'s are forced to be 0 automatically except for those of sites around (r^*, q^*). On the contrary, if the model is with centroid-based constraints, p_{crq}'s of sites far from (r^*, q^*) may still be nonzero because fractional p_{crq} may easily satisfy the centroid-based constraints. As a result, the solution of the relaxed LP at this node possibly will not provide as tight a bound as the solution of the relaxed LP of the proposed S Model with contiguity constraint (7).

Experimental results in Section 4.1 give a more convincing comparison of the two groups of contiguity constraints. The model with constraint (7) (S Model) and the model with centroid-based constraints (S-cen Model) [14] are applied on the same groups of sliding windows, and the results are shown in Table 1 and Table 2 respectively. We will elaborate more in Section 4.1.

The number of inequalities in (7) seems to be too large. However, as inequalities in (7) involve only binary variables, they can be merged as follows:

$$\begin{aligned} p_{cr_1q_1} + p_{cr_2q_2} &\leq 1 \\ p_{cr_1q_1} + p_{cr_3q_3} &\leq 1 \iff p_{cr_1q_1} + p_{cr_2q_2} + p_{cr_3q_3} \leq 1 \\ p_{cr_2q_2} + p_{cr_3q_3} &\leq 1 \end{aligned}$$

In fact, given m site occupation variables of the same cell, the original $\binom{m}{2}$ inequalities in the form of (7) can be merged in the same way shown above, as long as any two variables are of two sites in different rows or separated with a distance no less than w_c. This can be proved by mathematical induction. Examples are given below showing how inequalities in (7) can be merged, $\forall 1 \leq i < w_c$:

$$\sum_{r \in R} p_{cr1} + \sum_{r \in R} p_{cr(1+w_c)} + \sum_{r \in R} p_{cr(1+2w_c)} + \cdots = 1,$$

$$\sum_{r \in R} p_{cr1} + \sum_{r \in R} p_{cr(1+i+w_c)} + \sum_{r \in R} p_{cr(1+i+2w_c)} + \cdots \leq 1,$$

$$\sum_{r \neq 1} p_{cr(1+i)} + \sum_{r \neq 1} p_{cr(1+i+w_c)} + \sum_{r \neq 1} p_{cr(1+i+2w_c)} + \cdots$$
$$+ p_{c11} + p_{c1(1+w_c)} + p_{c1(1+2w_c)} + \cdots \leq 1.$$

Note that the $2w_c - 1$ merged inequalities cover all the contiguity constraints related to p_{c11}, as well as a large portion of contiguity constraints related to other variables, say p_{c21}. Inequalities related to other site occupation variables can also be merged in the same way, after which the number of contiguity constraints will be reduced to $O(|C||R||Q|)$.

3.2 RQ Model

The number of integer variables in MIP problem affects its solution time because the size of the branch-and-bound tree can be exponential in terms of the number of integer variables. In the S Model, the number of binary site occupation variables is $O(|C||R||Q|)$.

Here, instead of site occupation variables, we propose a different MIP model, called the RQ Model, based on row occupation variables and column occupation variables defined below:

$\overline{p}_{cq}$ 1 if cell $c \in C$ occupies column $q \in Q$; 0 otherwise
$\hat{p}_{cr}$ 1 if cell $c \in C$ occupies row $r \in R$; 0 otherwise

Only $O(|C|(|R|+|Q|))$ occupation variables are used. When the number of rows is large, the reduction in the number of binary variables is significant.

After replacing p_{crq} with new occupation variables, the RQ Model has the same objective function (1). The model also retains the constraints (2)–(3). But constraints (4)–(8) should be replaced by constraints (9)–(16) below:

$$\frac{1}{w_c} \sum_{q \in Q} (s_q + \frac{\overline{w}}{2}) \overline{p}_{cq} = x_c, \qquad \forall c \in C, \qquad (9)$$

$$\sum_{r \in R} (t_r + \frac{\overline{h}}{2}) \hat{p}_{cr} = y_c, \qquad \forall c \in C, \qquad (10)$$

$$\sum_{q \in Q} \overline{p}_{cq} = w_c, \qquad \forall c \in C, \qquad (11)$$

$$\sum_{r \in R} \hat{p}_{cr} = 1, \qquad \forall c \in C, \qquad (12)$$

$$\overline{p}_{cq} + \overline{p}_{cq'} \leqslant 1, \qquad \forall c \in C, |q - q'| \geqslant w_c, \quad (13)$$

$$\sum_{c \in C} \overline{p}_{cq} \leqslant |R|, \qquad \forall q \in Q, \qquad (14)$$

$$\sum_{c \in C} w_c \hat{p}_{cr} \leqslant |Q|, \qquad \forall r \in R, \qquad (15)$$

$$\overline{p}_{cq} + \hat{p}_{cr} + \overline{p}_{c'q} + \hat{p}_{c'r} \leqslant 3, \qquad \forall q \in Q, r \in R, c \neq c', \quad (16)$$

$$\overline{p}_{cq}, \hat{p}_{cr} \in \{0,1\}, \qquad \forall c \in C, q \in Q, r \in R.$$

The constraints for the definition of cell centroids (9)–(10), cell occupation limit (11)–(12), and cell contiguity (13) are similar to those in the S Model. Contiguity constraints can be merged in a similar way, too.

However, additional attention has to be paid to cell overlap. In the S Model, constraint (8) is sufficient to make sure no two cells overlap. In the new model, constraints (14)–(15) also help avoid cell overlap by preventing too many cells occupying the same row or the same column. But they are far from enough, as shown in the example in Figure 2. Four cells, each with width 2, are to be placed in a 2×4 placement area. Two cells have already been placed. Next, even if the remaining two cells are placed such that Cell 3 and Cell 4 overlap with Cell 1 and Cell 2, respectively, the resulted placement with cell overlap still can satisfy constraints (14)–(15).

As a result, constraints (16) must be included in the model. Given binary variables, the following relation holds:

$$\overline{p}_{cq} \hat{p}_{cr} + \overline{p}_{c'q} \hat{p}_{c'r} \leqslant 1 \iff \overline{p}_{cq} + \hat{p}_{cr} + \overline{p}_{c'q} + \hat{p}_{c'r} \leqslant 3$$

where $\overline{p}_{cq} \hat{p}_{cr}$ is equivalent with p_{crq}. Thus, constraints (16) make sure no two cells occupy the same site at the same time.

The necessity of including constraints (16) leads to an obvious disadvantage: $O(|C|^2|R||Q|)$ constraints must be

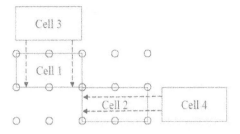

Figure 2: An illustration of the necessity of constraints (16).

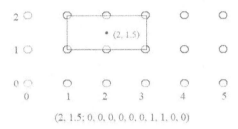

$$(2, 1.5; 0, 0, 0, 0, 0, 0, 1, 1, 0, 0)$$

Figure 3: An illustration of single-cell-placement pattern.

added to prevent cell overlap. This number increases quadratically with the number of cells. In windows with large number of cells, it can be a bottleneck.

However, even with the disadvantage of having larger number of constraints, the RQ model proves to be more efficient than the S Model, as shown in Section 4.

3.3 SCP Model

The SCP Model is derived from the S Model by applying Dantzig-Wolfe decomposition principle in Integer Programming [22]. In the S Model, constraints (4)–(7) with binary occupation variables are independent constraints for cells. For each cell, this group of independent constraints define a *pattern* how the single cell is legally placed in the window, occupying exact number of contiguous sites in one row of the window. Suppose the set of all the patterns for cell c is K_c, with the kth pattern being described as a vector $(x_c^k, y_c^k; p_{crq}^k), r \in R, q \in Q$. An illustration of a single-cell-placement pattern and the corresponding vector is shown in Figure 3. Then we have,

$$x_c = \sum_{k \in K_c} x_c^k \lambda_c^k, \quad y_c = \sum_{k \in K_c} y_c^k \lambda_c^k, \quad p_{crq} = \sum_{k \in K_c} p_{crq}^k \lambda_c^k,$$

$$\sum_{k \in K_c} \lambda_c^k = 1, \quad \lambda_c^k \in \{0,1\}, \quad \forall k \in K_c.$$

After replacing x_c, y_c, p_{crq} in the S Model with the preceding equations, we obtain the following MIP model, called the SCP model, with single-cell-placement variables:

$$\text{Min} \quad \text{HPWL} = \sum_{n \in N} (u_n^x - l_n^x + u_n^y - l_n^y) \qquad (17)$$

Table 1: Experiments applying MIP models in the optimization of 2-row or 8-row sliding windows with different number of cells; for each window size, over 300 windows are randomly extracted from ibm01e.

	#cell	S Model				RQ Model				SCP Mode			
		t(s)	red.	opt	o.t(s)	t(s)	red.	opt	o.t(s)	t(s)	red.	opt	o.t(s)
2-row	8	7.88	138.4	94.7%	5.28	6.27	138.5	96.3%	4.68	0.82	138.5	100%	0.82
	9	15.11	199.1	88.0%	10.12	13.46	194.1	86.0%	7.38	1.17	202.7	100%	1.17
	10	23.64	232.4	73.8%	13.17	22.83	237.7	78.6%	15.44	1.93	250.4	99.7%	1.74
	12	42.39	222.6	31.1%	20.88	41.41	205.3	28.1%	19.66	5.86	321.2	96.4%	4.22
	14	–	–	–	–	–	–	–	–	14.90	453.6	88.7%	9.34
8-row	10	4.74	133.0	98.0%	3.89	0.46	133.7	100%	0.46	0.34	133.7	100%	0.34
	11	11.46	159.9	91.8%	7.93	0.91	168.8	100%	0.91	0.65	168.8	100%	0.65
	12	15.04	192.6	87.5%	10.33	2.50	216.7	99.0%	2.03	1.22	217.1	99.7%	1.09
	14	43.18	184.0	37.9%	23.58	13.94	293.8	87.0%	8.17	3.40	305.8	98.3%	2.61
	16	–	–	–	–	28.94	250.1	59.1%	13.29	7.37	313.4	94.7%	5.03

subject to

$$l_n^x \leqslant \sum_{k \in K_c} x_c^k \lambda_c^k + p_x \leqslant u_n^x, \quad \forall (p_x, p_y, c) \in P_n, n \in N, \quad (18)$$

$$l_n^y \leqslant \sum_{k \in K_c} y_c^k \lambda_c^k + p_y \leqslant u_n^y, \quad \forall (p_x, p_y, c) \in P_n, n \in N, \quad (19)$$

$$\sum_{c \in C} \sum_{k \in K_c} p_{crq}^k \lambda_c^k \leqslant 1, \quad \forall q \in Q, \forall r \in R, \quad (20)$$

$$\sum_{k \in K_c} \lambda_c^k = 1, \quad \forall c \in C, \quad (21)$$

$$\lambda_c^k \in \{0, 1\}, \quad \forall c \in C, k \in K_c.$$

As it is derived from Dantzig-Wolfe decomposition, the SCP Model is *tighter* than the original model. Moreover, for cell c, the exact number of single-cell-placement variables is only $|R|(|Q| - w_c + 1)$, smaller than the number of site occupation variables. Since the size of the SCP Model is not large, we need not resort to branch-and-price for its solution.

As a comparison, the model in [6] is also derived from a model with site occupation variables by applying Dantzig-Wolfe decomposition. The difference is that in the derived master problem (MP) in [6], every binary variable corresponds to a single-net-placement pattern denoting how cells interconnected by a single net are placed in the window. With a large number of such patterns, branch-and-price must be adopted for the solution of MP. Column generation is used to solve the relaxed LP. Only a small portion of columns in MP are kept to form a Restricted Master Problem (RMP). A pricing subproblem is defined for each net, whose solution space is the set of all the patterns for the net. For each subproblem, with cost coefficients of objective function coming from dual values in RMP, an integer solution with negative objective value will provide RMP with a new column to be introduced in the basis. The iterative process of alternating between solving RMP and subproblems continues until no such columns exist.

The branch-and-price model in [6] considers only uniform-width cells. However, with the addition of contiguity constraints, it can be updated to handle cells with different widths, too.

4. EXPERIMENTAL RESULTS

We consider IBM version 2 easy and hard benchmark circuits [24] in our experiments. For each benchmark, the orig-

inal placement result is generated by the routability-driven placer proposed in [25]. All experiments are performed on 2.66GHz CPU with 2GB memory.

Sliding windows in experiments are all extracted from IBM version 2 benchmark circuits. We derive MIP models for each sliding window. In a extracted rectangle window, if originally some cells are not completely located inside the window but are on the edge of the window, those cells are considered fixed and not included in C. Moreover, if some nets in N have pins outside the window, each of such external pins is projected onto the nearest point in the window to form a fixed pseudo pin. The MIP models can be easily updated according to the two changes. The objective function of the updated MIP models is the HPWL inside the window.

4.1 Comparison of MIP Models

Different MIP models are implemented and performed on 2-row windows and 8-row windows with different number of cells. For each window size, experiments are performed on over 300 sliding windows randomly extracted from benchmark circuit ibm01e.

In the experiments, a tolerance time of 40 seconds is imposed as follows: The optimization for a window will terminate if no better placement results (i.e., no better integer solutions) are found in 40 seconds, and every time a better placement result is obtained, the timer for the tolerance time restarts. Hence, the run-time for some windows may be larger than 40 seconds. Also, with the tolerance time constraint, some windows cannot be optimized fully, i.e., solved completely. Even then, it is possible that the solver has obtained better placement results before termination.

First, a comparison of the three models discussed in Section 3 is given in Table 1. In the table, "t(s)" means the average run-time, in seconds, for all windows. "red." denotes the average HPWL reduction, in micrometers, for all windows. "opt" gives the percentage of windows that are fully optimized, and "o.t(s)" is the average run-time, in seconds, for windows that are fully optimized.

The advantage of the SCP Model over the S Model is quite obvious. Most smaller windows can be optimized with both models. But the average run-time with the SCP Model is generally over 9 times shorter than that of the S Model. For larger windows, over one half of windows cannot be optimized with the S Model because of the 40 seconds con-

Table 2: Experiments applying the S-cen Model [14] on the same windows used to generate Table 1.

	#cell	t(s)	red.	opt	o.t(s)
2-row	8	34.06	91.2	26.9%	8.52
	9	38.75	112.0	14.0%	11.77
8-row	10	35.03	83.8	24.3%	9.14
	11	38.43	106.3	15.8%	12.95

Table 3: Experiments applying the branch-and-price model with net subproblems on the same windows used to generate Table 1.

	#cell	t(s)	red.	opt	o.t(s)	#site
2-row	8	40.12	49.6	23.9%	28.81	2×41.3
	9	44.36	41.1	6.3%	35.61	2×44.1
8-row	10	35.64	91.5	38.5%	21.11	8×30.0
	11	43.44	93.7	22.7%	21.25	8×29.7

straint. With the SCP model, however, nearly 90% 2-row windows with 14 cells and 95% 8-row windows with 16 cells can be optimized, and the average run-time for optimized windows is around 10s and 5s, respectively. This makes it possible to make use of the SCP Model to optimize windows that are much larger than the enumeration approach. Larger windows usually lead to more wirelength reduction in the optimization. In the table, with the SCP Model, the average HPWL reduction of 2-row windows with 14 cells is over 3 times larger than that of 2-row windows with 8 cells.

The RQ Model also outperforms the S Model in the optimization of 8-row windows. It achieves this by having fewer integer variables. For 8-row windows with no more than 12 cells, the average run-time of the RQ Model is also over 6 times shorter than that of the S Model. However, the scalability of the RQ Model is not as good as the SCP Model because the RQ Model for larger windows has too many constraints.

Next, for the same randomly-extracted sliding windows, the performance of the model with centoid-based contiguity constraints (S-cen Model) [14] and the branch-and-price model with net subproblems are shown in Table 2 and Table 3, respectively.

As introduced in Section 3.1, the S-cen Model is equivalent to the S Model except for the contiguity constraints. With centoid-based contiguity constraints in the S-cen Model, only less than 30% 2-row windows with 8-9 cells and 8-row windows with 10-11 cells can be optimized. In comparison, with contiguity constraints (7) in the S Model, over 88% windows of the same window sizes can be optimized. The advantage of contiguity constraints (7) is obvious.

The branch-and-price model with net subproblems is a generalized version of the model in [6]. With the addition of contiguity constraints, it is capable of handling cells with different widths. We solved the branch-and-price model with our implementation of the solver proposed in [6]. When optimizing the placement of 4×4, 5×5, 6×6 regular mesh grids with uniform-width cells, our implementation of the branch-and-price solver is as efficient as that in [6].

However, with the branch-and-price model, the placement of cells with different widths cannot be optimized as efficiently as the S Model, the RQ Model and the SCP Model.

With the branch-and-price model, the placement of as many as 16 uniform-width cells in 4×4 regular mesh grids can be optimized within 40 seconds, but over 70% windows with 8 cells of different widths cannot be optimized. Two possible reasons for its poor performance are the larger window size and the presence of high degree nets. Table 3 provides the average number of sites in extracted windows of every window size, "#site", which are all larger than that in 4×4 regular mesh grids. Although the number of cells is smaller, with more sites, the number of single-net-placement patterns may be larger. The size of restricted master problem and subproblems may be larger, too, leading to longer solution time. Besides, unlike regular mesh grids, high degree nets may appear in extracted windows, and solving subproblems of high-degree nets usually is time-consuming. Therefore, compared with the RQ Model and the SCP Model, it is not efficient to apply the branch-and-price model with net subproblems to optimize the placement of cells with different widths.

4.2 Experiments on Large-Scale Benchmarks

With the detailed placer based on the SCP Model, we also optimize the detailed placement across the whole chip for IBM version 2 benchmark circuits. The original placement results are generated by the routability-driven placer proposed in [25]. The placement results have been optimized for routability and routed wirelengths. Under the column labeled "Original", Table 4 lists HPWL, "HPWL(m)", routed wirelength, "r-WL(m)", and the count of vertical vias, "#vias", of the original placement result for each benchmark. The routed wirelengths and via counts are obtained from the routed solutions generated by Cadence WROUTE with all settings on default. As a reference, we also obtain the placement results generated by other placement tools: RUDY [26], CHKS+WSA [27], mPL-R+WSA [13], APlace 2.0 [28], and mPL6 [29]. All the placement results are routed by Cadence WROUTE with all settings on default, too, and Table 5 lists the routed wirelengths of the routed solutions. For each placement tool, the sum of routed wirelengths is normalized against the sum of routed wirelengths under the "Original" column in Table 4.

Note that in the original placement results generated by the placer in [25], detailed placement has already been optimized. With the sliding window technique and the enumeration approach proposed in [13], each benchmark circuit has already been scanned for 6 times. Our experiments demonstrate how detailed placement can be further optimized with MIP-based detailed placer.

In the MIP-based detailed placer, the sliding window technique in [14] is adopted. The whole chip is partitioned into rectangle windows with no more than 13 cells, with each window overlapping with neighbor windows in four directions. Each benchmark is scanned for 5 times and all the windows across the chip are optimized in each scan. The tolerance time for optimizing each window is 40 seconds. Moreover, the optimization of windows is parallelized with Message Passing Interface as in [14]. The parallelization is realized in a Master-Slave structure, with one master allocating independent windows of placement problems to multiple slaves, each of which optimizes the allocated window.

Under the column labeled "MIP", Table 4 lists the HPWL and routability metrics of the placement results generated by the MIP-based detailed placer. Compared with the orig-

Table 4: Comparison of MIP approach and enumeration approach to further optimize the original placement results generated by the routability-driven placer in [25] for IBM version 2 benchmark circuits.

	Original			MIP			Enumeration		
	HPWL(m)	r-WL(m)	#vias	HPWL(m)	r-WL(m)	#vias	HPWL(m)	r-WL(m)	#vias
ibm01e	47.58	0.6245	85848	46.19	0.6163	83873	47.20	0.6243	85398
ibm01h	47.45	0.6127	85266	45.94	0.6065	83638	47.06	0.6131	85379
ibm02e	132.00	1.6823	194294	129.49	1.7047	192181	130.71	1.7144	194071
ibm02h	128.94	1.7301	198177	127.15	1.7175	196303	128.33	1.7554	197124
ibm07e	296.46	3.3743	351669	291.72	3.3489	346840	294.67	3.3781	352105
ibm07h	289.52	3.3754	359261	285.43	3.3533	353902	288.38	3.3761	358464
ibm08e	310.64	3.7528	430905	306.91	3.7315	424543	309.45	3.7456	429679
ibm08h	306.18	3.7289	432689	302.89	3.7055	427173	305.33	3.7234	430502
ibm09e	267.25	2.9201	368283	262.68	2.8864	361312	266.14	2.9137	366252
ibm09h	260.98	2.8690	368610	256.79	2.8331	361360	260.08	2.8609	366780
ibm10e	505.19	5.6628	584716	498.05	5.6051	574144	503.06	5.6515	582472
ibm10h	498.20	5.6252	584079	492.21	5.5770	574712	496.86	5.6199	582660
ibm11e	408.05	4.3640	474085	401.23	4.3109	464933	405.76	4.3520	471026
ibm11h	395.66	4.2695	479599	390.17	4.2283	469384	394.30	4.2622	476849
ibm12e	712.10	8.1989	707431	699.91	8.0897	692005	705.21	8.1555	702892
ibm12h	697.47	8.0553	703139	686.61	7.9658	689523	692.71	8.0201	698443
Avg Red	–	–	–	1.663%	0.827%	1.707%	0.546%	-0.062%	0.383%
Norm	1.0000	1.0000	1.0000	0.9849	0.9907	0.9825	0.9946	0.9987	0.9956

inal placement results, on the average, HPWL is reduced by 1.663%. Moreover, routed by Cadence WROUTE with all settings on default, all the generated placement results are routable with no violations, and the routed wirelength is reduced for all benchmarks except ibm02e. On the average, the routed wirelength is reduced by 0.827% and the count of vertical vias is reduced by 1.707%. The normalization of total HPWL, total routed wirelength, and total count of vertical vias for all benchmarks also show the advantage of the generated placement results, over the original placement results as well as the placement results of other placement tools in Table 5. The results of the MIP-based detailed placer are obtained on a computing system with 33 cores. It takes 110 minutes to 480 minutes for benchmark circuits of different size.

In comparison, the enumeration approach [13] is also applied to further optimize the original placement results. Each chip is scanned for at least 60 times with the enumeration approach until we obtain little difference in the solutions in successive scans. The final results of the enumeration approach, as illustrated under the "Enumeration" column in Table 4, show that even applying the enumeration approach for many times cannot generate placement results as good as those generated by the MIP approach.

5. CONCLUSION

Two MIP models for the detailed placement of cells with different widths are presented in the paper. The first model features fewer integer variables. The second model derived from Dantzig-Wolfe decomposition is with tighter bounds during its solution. When applied to optimize the detailed placement of sliding windows extracted from real benchmarks, both models are shown to be more efficient than the existing model with site occupation variables, as well as the branch-and-price model with net subproblems. Further-

more, experiments on large-scale benchmarks show that as an alternative of the enumeration approach, detailed placer based on the proposed MIP models manages to generate better placement results with reduced HPWL and routed wirelength.

6. ACKNOWLEDGEMENTS

This work was supported in part by SRC (Task ID 1822.001). We also thank the reviewers for their constructive comments and suggestions.

7. REFERENCES

[1] C.-C. Chang and J. Cong. Optimality and scalability study of existing placement algorithms. *IEEE Trans. on Computer-Aided Design of Integrated Circuits and Systems*, 23(4):537–549, April 2004.

[2] T. Chan, J. Cong, T. Kong, and J. Shinnerl. Multilevel generalized force-directed method for circuit placement. In *Proc. IEEE/ACM Intl. Conf. on Comput.-Aided Des.*, pages 171–176, 2000.

[3] T. Chan, J. Cong, and K. Sze. Multilevel generalized force-directed method for circuit placement. In *Proc. Intl. Symp. on Physical Design*, pages 185–192, 2005.

[4] A. B. Kahng and Q. Wang. Implementation and extensibility of an analytical placer. In *Proc. Intl. Symp. on Physical Design*, pages 18–25, 2004.

[5] S. Ono and P. H. Madden. On structure and suboptimality in placement. In *Proc. Asia South Pacific Design Automation Conf.*, 2005.

[6] P. Ramachandaran, A. R. Agnihotri, S. Ono, P. Damodaran, K. Srihari, and P. H. Madden. Optimal placement by branch-and-price. In *Proc. Asia South Pacific Design Automation Conf.*, pages 858–871, 2005.

Table 5: Routed wirelength, r-WL(m), of the placement results generated by other placement tools for IBM v2 benchmark circuits.

	RUDY	CHKS +WSA	mPL-R +WSA	APlace2.0	mPL6
ibm01e	0.6493	0.6985	0.7615	0.8015	0.7612
ibm01h	0.6412	0.7468	0.7358	0.7988	0.7591
ibm02e	1.8432	1.8738	2.0790	2.0802	2.0498
ibm02h	1.8926	1.8898	2.2887	2.0513	2.0361
ibm07e	3.5777	3.8694	3.8487	4.1224	3.9750
ibm07h	3.5773	3.6095	3.9059	4.2366	4.0425
ibm08e	4.0298	3.8918	4.4643	4.1605	4.2369
ibm08h	3.9513	3.8910	4.6828	4.1468	4.2028
ibm09e	2.9241	3.0222	3.3698	3.1617	3.0660
ibm09h	2.9392	3.0723	3.3357	3.1429	3.0642
ibm10e	5.7190	5.9044	6.5663	5.9690	6.0082
ibm10h	5.7505	6.2941	6.4881	6.0162	6.0140
ibm11e	4.3633	4.4004	5.3466	4.8273	4.6216
ibm11h	4.3212	4.3399	4.8630	4.8209	4.5633
ibm12e	8.3565	8.4578	9.5176	9.0592	8.6683
ibm12h	8.3325	8.4579	10.2315	8.9418	8.9605
Norm	1.0333	1.0587	1.1913	1.1231	1.1016

[7] A. R. Agnihotri, S. Ono, and P. H. Madden. Recursive bisection placement: feng shui 5.0 implementation details. In *Proc. International Symposium on Physical Design*, 2005.

[8] A. B. Kahng, P. Tucker, and A. Zelikovsky. Optimization of linear placements for wirelength minimization with free sites. In *Proc. Asia and South Pacific Design Autom. Conf.*, pages 241–244, 1999.

[9] M. Pan, N. Viswanathan, and C. Chu. An efficient and effective detailed placement algorithm. In *Proc. Int. Conf. Comput.-Aided Des.*, pages 48–55, 2005.

[10] K. Doll, F. M. Johannes, and K. J. Antreich. Iterative placement improvement by network flow methods. *IEEE Trans. Computer-Aided Design of Integrated Circuits and Systems*, 13(10):1189–1200, 1994.

[11] A. Agnihotri, M. C. Yildiz, A. Khatkhate, A. Mathur, S. Ono, and P. H. Madden. Fractional cut: Improved recursive bisection placement. In *Proc. Int. Conf. Comput.-Aided Des.*, pages 307–310, 2003.

[12] A. E. Caldwell, A. B. Kahng, and I. L. Markov. Optimal partitioners and end-case placers for standard-cell layout. *IEEE Trans. on Computer-Aided Design*, 19:1304–1313, 2007.

[13] C. Li, M. Xie, C.-K. Koh, J. Cong, and P. H. Madden. Routability-driven placement and white space allocation. *IEEE Trans. on Computer-Aided Design of Integrated Circuits and Systems*, 26(5):858–871, May 2007.

[14] S. Cauley, V. Balakrishnan, Y. C. Hu, and C.-K. Koh. A parallel branch-and-cut approach for detailed placement. *ACM Transactions on Design Automation of Electronic Systems*, 16(2), March 2011.

[15] M. Padberg and G. Rinaldi. Optimization of a 532-city symmetric traveling salesman problem by branch and cut. *Operations Research Letters*, 6(1):1–7, Mar. 1987.

[16] http://www.tsp.gatech.edu/.

[17] C. Barnhart, E. L. Johnson, G. L. Nemhauser, M. W. P. Savelsbergh, and P. H. Vance. Branch-and-price: Column generation for solving huge integer programs. *Operations Research*, 46(3):316–329, May - Jun. 1998.

[18] M. Desrochers, J. Desrosiers, and M. M. Solomon. A new optimization algorithm for the vehicle routing problem with time windows. *Operations Research*, 40(2):342–354, 1992.

[19] P. H. Vance, C. Barnhart, E. L. Johnson, and G. L. Nemhauser. Airline crew scheduling: A new formulation and decomposition algorithm. *Operations Research*, 45:188–200, 1997.

[20] T.-H. Wu, A. Davoodi, and J. T. Linderoth. GRIP: Scalable 3D global routing using integer programming. In *Proc. Proc. Design Automation Conf.*, pages 320–325, 2009.

[21] T.-H. Wu, A. Davoodi, and A. Zelikovsky. A parallel integer programming approach to global routing. In *Proc. Design Automation Conf.*, 2010.

[22] F. Vanderbeck. On Dantzig-Wolfe decomposition in integer programming and ways to perform branching in a branch-and-price algorithm. *Operations Research*, 48(1):111–128, Jan. - Feb. 2000.

[23] W. E. Wilhelm. A technical review of column generation in integer programming. *Optimization and Engineering*, 2:159–200, 2001.

[24] X. Yang, B.-K. Choi, and M. Sarrafzadeh. Routability-driven white space allocation for fixed-die standard-cell placement. *IEEE Trans. on Computer-Aided Design of Integrated Circuits and Systems*, 22(4):410–419, April 2003.

[25] K. Tsota. Algorithms and methodologies for routability-driven VLSI placement. *Ph.D. Dissertation, Purdue University*, 2011.

[26] P. Spindler and F. Johannes. Fast and accurate routing demand estimation for efficient routability-driven placement. In *Proc. Design Automation and Test in Europe Conference*, pages 1–6, 2007.

[27] C. Li and C.-K. Koh. Recursive function smoothing of half-perimeter wirelength for analytical placement. In *Proc. International Symposium on Quality Electronic Design*, pages 829–834, 2007.

[28] A. B. Kahng and Q. Wang. Implementation and extensibility of an analytic placer. *IEEE Trans. Computer-Aided Design*, 24(5):734–747, May 2005.

[29] T. Chan, J. Cong, M. Romesis, J. R. Shinnerl, K. Sze, and M. Xie. mPL6: A robust multilevel mixed-size placement engine. pages 227–229, Apr. 2005.

Power-Grid (PG) Analysis Challenges for Large Microprocessor Designs

(Our Experience with Oracle Sparc Processor Designs)

Alexander Korobkov

Oracle Corp.

4190 Network Circle

Santa Clara, CA 95054

alexander.korobkov@oracle.com

ABSTRACT

True to Moore's law, chip sizes have been growing exponentially. For signal nets, this implies an increase in the number of nets, for power-grids, this implies and increase in size/complexity[1].

Monolithic power-grids on large composed flat blocks, which we will call clusters in this talk, present a capacity challenge for all stages of analysis from extraction to simulation. It is not uncommon to see larger than 100G parasitics files (with more than half a billion nodes and a billion elements) for a single power net.

Advances in automated place&route technology makes it advantageous[2] to compose large designs flat up from library cells (without intermediate hierarchy). This reduces the scope/potential for hierarchical analysis as a solution for the capacity challenges. Luckily, most of the grid density is on lower metal layers within library cells so gate level analysis allows some reduction of data size but this is not a scalable advantage like true hierarchical analysis. Artificial hierarchy injection into power-grid is one approach but it has its own limitations.[3]

Another challenge of large designs such as clusters in SPARC processors is the aggregation of ECO fixes into fewer designs. To avoid tapeout bottlenecks it is important to fix as many of these violations as early as possible in design cycle, preferably during composition where the flexibility for width increase is greatest. At Oracle, we developed a set of in-house power-grid analysis solutions that analyze various stages of design with appropriate performance-accuracy tradeoff.[4]

This range of power-grid analysis solutions uses a variety of simulation techniques for accuracy-performance tradeoff as listed:

- Solvers: The power-grid being composed of a sparse network of passive elements is efficiently modeled as a sparse positive definite matrix. For smaller testcases with a large number of input vectors Direct solvers are the most efficient while iterative solvers are useful for larger designs with few vectors (static/pseudo-dynamic vectors)
- Dynamic vs Pseudo Dynamic methods: The transistor currents used for power-grid analysis can be simulated by targeted vectors (ideal for simple designs like SRAMs whose functionality can be comprehensively covered with few vectors) or by static (non-vector based) estimates. Pseudo-dynamic methods involve circuit modifications to simulate switching of most of the design without long vectors and yet avoid the inaccuracies of most static estimators. These have as such completely replaced static estimators. There is an ongoing research to further improve the accuracy of pseudo-dynamic methods by sophisticated accounting of activity and mutual exclusivity.
- Multi-threading vs multi-processing: To contain run time and memory explosion with increasing design size, various algorithms for partitioning and concurrent processing have been tried out. A key tradeoff in concurrent processing is multi-threading[5] vs multi-processing[6]. Multi-threading avoids multiple copies of a source data if all processes may need to access any part of the data while multiprocessing allows breaking of one large memory requirement to n-machines of smaller memory footprint. The right choice depends on the modeling of the problem.

---

1 Except for the trend towards independent controlled voltage subgrids.
2 Larger designs allow for greater optimization of area by moving cells independent of bounding box restrictions.
3 Significant part of cross-border current for artificial hierarchies in horizontal at lower metal levels.
4 Gross violations fixed as early as possible with quick and dirty analysis while borderline violations are fixed

with highest accuracy analysis close to tapeout.
5 Multiple processes running concurrently on the same machine often sharing the same memory space.
6 Multiple processes running independently on multiple machines with limited data exchange during run.

ISPD'12, March 25–28, 2012, Napa, California, USA.
ACM 978-1-4503-1167-0/12/03.

Besides the standard power-grids (vdd/vss), the bulk grids (vnw/vsb) are also large (grid) nets sharing many of the problems and peculiarities of power-grids. As such, they are also analyzed for IR/EM problems by the same set of analysis tools but some of their finer differences require different interpretation of results. The first difference of bulk grids (w.r.t. regular power-grids) is that the tap-point for transistor current is not localized[7]. The solution is to extract the substrate resistances on the bulk grid too. While these can be the dominant part of the bulk-grid, the overall size of the bulk-grid still does not exceed that of regular power-grids. The second peculiarity of the bulk grids is that there are 2 components of current we need to factor

• Leakage current, unidirectional and small in magnitude
• Charging/Discharging current (from capacitance Cgb, Csb, Cdb) which swing large in both (+ve, -ve) directions
Depending on the current direction, we risk:
• Transistor slowing down
• Latch
So, results interpretation is dependent on design budget and methodology.

The EM violations on Power-grid analysis are located very easily by coordinates of extracted nodes/resistors and can be debugged on GUI displays (overlaying violations to layout) for even large designs. These localized violations are also compatible with automated fixing tools[8]. IR violations on the other hand are flagged far from the bottleneck[9]. Automated tracing of monotonic voltage change paths help trace the critical path (and hence bottleneck) between IR violation and its source. Such automated traces help GUI debugging as well as automated fixes (similar to EM).

While the above described challenges get carried over or magnified in all upcoming process nodes, some new challenges may threaten 14 nm and beyond.

Inductance effects have been the boogeyman for power-grid analysis for a long time. This may change in 14nm. For now, a well-designed power-grid[10] has enabled inductive effects (and decaps to address them) to be

analyzed independently of IR. However, newer processes involve:

• Reduced supply voltage (cant afford separate IR and LdI/dt budgets)
• Gated power-grids which switch during operation generating huge current swings.
• Faster gates with high current consumption.
• Longer metal lines running parallel to each other

At some point, we may need to involve on-chip inductance extraction and simulation into power-grid analysis. This will dramatically increase the capacity/ performance challenges.

FinFets are the new wonder devices that enable more transistors on the same die area but the interconnects that support them (on power-grid as well as signals) are not likely to get more cross-section in the 3rd dimension. This will exacerbate the EM problems. Further, the device modeling of FinFets is still ongoing and may present newer challenges for fast tap current estimation.

Categories and Subject Descriptors
B.7.2 [Integrated Circuits]: Design Aids – Verification; I.6.4 [Simulation and Modeling]: Model Validation and Analysis

General Terms
Performance, Design, Reliability, Verification.

Keywords
Power-grid, VLSI, Microprocessor designs.

---

7   The bulk terminal of transistors is not uniquely contacted to the bulk grid, rather it is loosely defined in the area/well of the transistor.
8   Involving metal widening, via increase, cell replacement, double routing etc.
9   At the transistor rather than the bottleneck in path from voltage source to transistor.
10 Ideally thin power lines spaced evenly throughout the chip and preferably aligned well on alternate layers (that run parallel) minimize, IR drops, LdI/dt drops and provide noise shielded channels for signals.

Efficient On-line Module-Level Wake-Up Scheduling for High Performance Multi-Module Designs

Ming-Chao Lee[*], Yiyu Shi[**], Yu-Guan Chen[*], Diana Marculescu[***], Shih-Chieh Chang[*]

[*]Dept. of CS, National Tsing Hua University, HsinChu, Taiwan

[**]Dept. of ECE, Missouri University of Science and Technology, Rolla, MO 65409, US

[***]Dept. of ECE, Carnegie Mellon University, Pittsburgh, PA15213, US

chao@nthucad.cs.nthu.edu.tw, yshi@mst.edu, scchang@cs.nthu.edu.tw

ABSTRACT

Power consumption has become the major bottleneck for modern high-performance architectures, which typically contain large numbers of modules. To suppress leakage power, sleep transistors have been extensively used, and wake-up scheduling is needed to determine the wake-up times and order of these sleep transistors. Most existing works on wake-up scheduling are based on sleep transistors and delay buffers in daisy-chains; they work well for the gate-level scheduling within a module when all the gates need to be turned on. Yet, for state-of-the-art designs, the number of modules that need to be turned on and their locations may vary depending on the task to be performed at runtime. Accordingly, we cannot extend the existing gate-level scheduling algorithms to decide the module-level wake-up order. To address the problem, we propose to first off-line construct a multi-conflict graph (MCG) based on the noise constraints; based on the graph, we then develop an on-line algorithm to decide the wake-up order. Experimental results show that on average, the wake-up latency from our approach is not only 46.01% shorter compared with the existing work but also conservatively only 0.45% longer than that from a Monte Carlo search-based evaluation, which is orders of magnitude slower. To the best of our knowledge, this is the first in-depth study on on-line module-level wake-up scheduling for high-performance architectures.

Categories and Subject Descriptors

B.8.2 [PERFORMANCE AND RELIABILITY]

General Terms

Algorithms

Keywords

Power gating, module-level, wakeup scheduling, system-level.

1. INTRODUCTION

Accompanied by the drastically increasing number of modules in high performance computing architectures [6], the power consumption has risen to be the bottleneck for design reliability. Figure 1 shows that the leakage power dissipation takes over 30% of total power consumption among state-of-the-art graphic processing units (GPUs), a typical representative of multi-module

designs. To suppress the leakage power efficiently, power gating [8] is commonly used, which inserts sleep transistors between logic gates and ground rail as shown in Figure 2(b). The sleep transistors can be turned off when its corresponding logic gates are idle.

One well-discussed issue related to the power gating is the large wake-up surge current: simultaneously turning all the sleep transistors from the "off" state to the "active" or "on" state will induce a large and sudden current, degrading the signal and power integrities in the nearby active gates. One effective method to regulate the surge current during wake-up process is the wake-up scheduling for sleep transistors, which decides the turn-on time and order of the sleep transistors in each module. This problem has been well-addressed in [1][10][11][12][13][14].

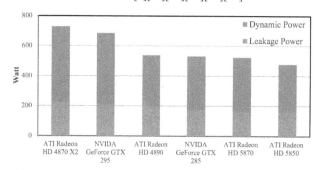

Figure 1. Power consumption of state-of-the-art GPUs [17].

Traditionally, the modules are turned on one at a time [4]. With the rapid increase in the number of modules, however, such a practice is over-pessimistic and causes large wake-up latencies. Therefore, it becomes important to determine the wake-up order at module-level to minimize latency. Noise is again the major concern here: turning on multiple modules simultaneously without caution can induce large noise on the global power/ground rail coupling to active modules. To avoid confusion, in this paper we denote the wake-up scheduling for the sleep transistors in a module as gate-level scheduling, and that for the modules as module-level scheduling.

There are two fundamental differences between the gate-level and the module-level wake-up scheduling: First, while in gate-level wake-up scheduling all sleep transistors are turned on after the power mode transition, in module-level wake-up scheduling the number of modules to be turned on and their locations are highly application-dependent [7], and can only be determined at runtime. Second, due to the physical complexity, the gate-level wake-up scheduling is typically implemented by sleep transistors and delay buffers in daisy-chains, which turns on serially. However, more flexibility is available in the module-level scheduling through separate wake-up signals to each module. Accordingly, gate-level scheduling algorithms cannot be directly applied to module-level.

There are a few related works on the module-level wake-up scheduling. The research [9] used genetic algorithm to minimize the noise on the power rail, which ignores the difference discussed above and does not work if only some of the modules need to be turned on. The researchers in [15][16] proposed to use some greedy heuristics for the wake-up scheduling. However, as shown by the details in Section II.B, it is not clear how they directly relate to the voltage drop (or noise) bound specified by the designers.

It is obvious that an effective module-level wake-up scheduling should be best decided at runtime, based on the number of modules to be turned on and their locations. Yet this imposes high demand on the algorithm: it cannot be too complicated to use on-line, and it should also be effective to produce a minimal wake-up latency. Unfortunately, due to the inherent difficulty of the problem, no such algorithm exists yet in literature.

In this paper, we propose an on-line module-level wake-up scheduling framework for high performance multi-module designs. We first use a heuristic algorithm to construct a multi-conflict graph (MCG) from all modules of the design off-line. Then, based on the graph, we develop a fast O(nlogn) on-line scheduling algorithm to determine the wake-up order for a given set of n modules attached to the same power management unit. Experimental results show that our approach can achieve 46.01% speedup on average over the industrial approach [4] in the wake-up latency without violating the noise constraint. To the best of our knowledge, this is the first in-depth study on on-line module-level wake-up scheduling for high-performance architectures.

The remainder of the paper is organized as follows. The preliminaries are shown in Section 2. We propose our algorithm in Section 3. Experimental results are given in Section 4 and concluding remarks are given in Section 5.

2. PRELIMINARIES

In this section, we first show the problem definition of the module-level wake-up scheduling. Related works are then briefly reviewed in Section 2.2.

2.1. Problem Definition

Similar to [9], in this paper we assume a multi-module architecture as shown in Figure 2(a), which shows a power-gated design with six modules. Each of these modules can be ALUs, memories, processor cores, etc. We assume that all modules are connected to a global power supply network. Meanwhile, each module has its own sleep transistors and wake-up signal (WS), as shown in Figure 2(b). We further assume that each module has its own gate-level wake-up scheduling, i.e., the wake-up order and timing for its sleep transistors, which can be decided by any of the existing algorithms [1][10][11][12][13][14]. As such, turning on each module will not cause any noise violation inside the same module.

With the structure in Figure 2(a), multiple modules may need to go to sleep or to wake up, depending on the tasks to be performed. If not properly scheduled, the surge current of multiple modules can potentially introduce noise and generate signal and power integrity issues on other active modules through the global power supply network. In view of this, it is important to determine the *module-level* wake-up scheduling for those modules to be turned on, such that the noise is regulated within a user-specified bound. Such a bound can be determined in a variety of ways, such as from the noise margin of the modules when they are in the active mode.

Take Figure 2(a) as an example. We would like to find the *module-level* wake-up scheduling for three modules M3, M4, and M5, instead of the gate-level wake-up scheduling for the sleep transistors in the modules, such as Figure 2(b).

We formally state the module-level wake-up scheduling problem as follows.

Problem Definition: Find the optimal order for turning on a set of modules at runtime, such that the total wake-up latency is minimized under the given noise bound.

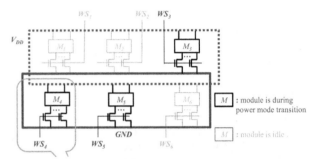

Figure 2(a). A system with six power-gated modules.

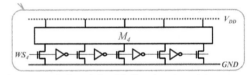

Figure 2(b). A power-gated module with five sleep transistors.

Figure 2. An example for power-gated system with six modules.

2.2. Related Works

Most existing works [1][10][11][12][13][14] focus on the *gate-level* wake-up scheduling, which is for sleep transistors in one module as in Figure 2(b). They cannot be applied to the *module-level* wake-up scheduling problem for high performance multi-module design, due to two major differences: First, while in the *gate-level* wake-up scheduling all sleep transistors should be turned on after the power mode transition, the number of modules to turn on and their locations are highly application-dependent and can only be determined at runtime. Second, due to the physical complexity, the *gate-level* wake-up scheduling is typically implemented as a daisy-chain including sleep transistors and delay buffers, which turns on serially. In Figure 2(b), the gate-level wake-up scheduling decides timings of delay buffers, which are placed between sleep transistors, to turn on sleep transistors from left to right in serial order. On the other hand, for the module-level wake-up scheduling, there is no such constraint. The modules can be turned on in an arbitrary sequence through separate wake-up signals (*WS*).

As an initial attempt to tackle the module-level wake-up scheduling, [9] used genetic algorithm to decide the wake-up order of a given set of power-gated modules to minimize the noise on the power rail. However, the method still turns on all modules. As we have discussed earlier, this is not the case for the module-level wake-up scheduling.

As a conservative approach, the researchers in [4] proposed to use delay buffers and daisy chains, which turns on one module at a time. This is obviously a sub-optimal solution in terms of wake-up latency.

The research [15] proposed a method to shape the current according to some current constraints and adopted a multi-thread activation technique to turn on the modules. However, it is not clear how to obtain the target current shape based on the power grid noise constraint.

Most recently, in [16], an on-line task-scheduling algorithm was proposed for a power-gated multi-module design. For each module, an impact range is characterized, which is defined as an area where the voltage drop is over the specified bound during the wake-up process. Then a greedy heuristic algorithm is adopted to decide the sequence to turn on modules by checking whether the impact ranges of a few modules overlapped. If not, these modules can be turned on together. We use Figure 3 as an example to illustrate the idea. The impact range of each module is identified with a solid circle. According to [16], since they do not overlap, both modules M_1 and M_2 can be turned on simultaneously. However, it is possible that the superposed noise can exceed the bound on module M_3. It is unclear how to properly select the noise level for the impact range given a noise constraint.

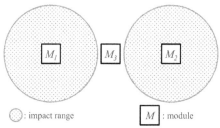

: impact range M : module

Figure 3. An example for [16].

3. EFFICIENT MODULE-LEVEL WAKE-UP SCHEDULING ALGORITHM

In this section, we propose our efficient module-level wake-up scheduling algorithm. We first discuss our circuit model for noise evaluation in Section 3.1. And then, we model the wake-up scheduling problem as a *multi-conflict graph* (*MCG*) in Section 3.2. In Section 3.3, we show how to construct the MCG. After that, in Section 3.4, using the MCG, we adopt a vertex-coloring-based heuristic to on-line determine the optimal order to turn on the modules such that the wake-up latency is minimized without noise violation. The overall algorithm is shown in Figure 7 (next page).

3.1. Modeling

Obtaining the noise on the power rail via transistor-level transient simulation is an expensive task due to its extremely high complexity. To alleviate the problem, we first assume that the *gate-level* wake-up scheduling of each module is given so that we have surge current waveforms of modules1. As such, we can

---

[1] This assumption has already been considered and extensively studied in literature (e.g. [9][16]).

model each module as a current source with the corresponding surge current waveform. We then model the global power supply network as an RLC network, which includes the on-chip and in-package capacitance, as well as the package inductance, as shown in Figure 4. We connect each current source to the power supply network in turn, and obtain the corresponding noise waveform using SPICE. As such, to evaluate the maximum noise when a few modules are to be turned on simultaneously, we can simply superpose the corresponding noise waveforms.

While the RLC model can simplify the problem significantly, determining the optimal module-level wake-up scheduling at runtime is still not an easy task. It is possible to use simulated annealing or genetic algorithms, but their high complexity prevents them from being applied at runtime. In the remainder of this paper, we will propose an efficient framework for the module-level wake-up scheduling.

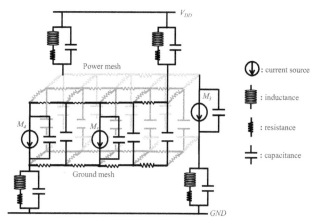

Figure 4. Equivalent models for RLC evaluation.

3.2. The Concept of Multi-Conflict Graph (MCG)

In the module-level wake-up scheduling problem, the objective is to turn on several modules simultaneously while minimizing the wake-up latency. There is no existing work addressing this problem, mainly due to the difficulty in determining which of the modules can be turned on simultaneously. A brute-force approach would enumerate all the possible combinations. However, such an approach would result in exponential complexity. In addition, it is not clear how to perform the optimal scheduling based on enumeration results.

In graph theory, it is common knowledge that a set of mutually exclusive constraints can be represented using a conflict graph (CG), where an edge represents a conflict between the two vertices. Such an approach is poised to provide some insight and a possible solution to this challenging problem. We may simply construct a CG by adding an edge between a pair of vertices/modules that should not be turned on simultaneously. While the complexity is only quadratic, unfortunately such a graph does not represent all possible conflict scenarios. For example, for a six-module system with the CG shown in Figure 5, since there is no edge between modules M_1 and M_2, they can be turned on simultaneously. For the same reason, M_2 and M_6, M_1 and M_6 can be turned on simultaneously as well. However, even though there is no edge between M_1, M_2 and M_6, it is not guaranteed that the three modules can be turned on simultaneously. In other words, the graph is only good for determining any two modules can be turned on simultaneously, but not more.

To alleviate the problem, we propose to extend the concept of the conflict graph into a multi-conflict graph (MCG), which is defined as follows:

Figure 5. The conflict graph (CG) for a six-module system.

Definition 1: *A multi-conflict graph (MCG) is an undirected graph satisfying the following two constraints:*

a. Each vertex represents a module;

b. A group of modules can be turned on simultaneously if their corresponding vertices form an independent set[2] (IS).

For example, for a six-module system with an MCG shown in Figure 7, modules M_1, M_2 and M_6 can be turned on at the same time while M_2, M_4 and M_5 cannot. The key idea of our algorithm is based on the following definition and theorem.

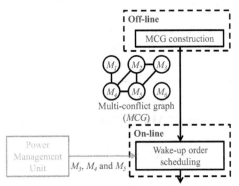

The optimal wakeup scheduling is (M_3, M_4), M_5.

Figure 7. The flow of our proposed algorithm.

Definition 2: *If all corresponding modules in an IS of a CG can be turned on simultaneously, we say that the IS is coherent.*

Theorem 1: *A graph is a multi-conflict graph (MCG) if any maximal independent set[3] (MIS) in the CG is coherent.*

The proof is straightforward and is thus omitted here in the interest of space.

3.3. Multi-Conflict Graph (MCG) Construction

This section describes a heuristic algorithm for constructing an MCG in an off-line manner. Then, based on the derived MCG, Section 3.4 proposes a O($n\log n$)-time scheduling algorithm to on-line determine the wake-up order.

---

[2]An independent set (IS) is a set of vertices without any edge between them.

[3]A maximal independent set (MIS) is an IS not contained by other IS's.

The above procedure has three essential issues. First, in the above algorithm, we need to frequently check the noise of a given set of vertices (modules). For the sake of efficiency, rather than running SPICE simulation each time, we pre-calculate and store the noise at each node of the power supply network caused by each current source (with the others removed). Then, we can use the superposition theorem to get noise values for any set of modules.

To construct an MCG, our basic procedure is as follows. We first build a CG by checking whether every two modules can be turned on simultaneously. From the initial CG, we iterate through all MIS's. During each iteration, if an MIS, MIS_i, is not coherent, we add an edge to MIS_i so that vertices in MIS_i no longer form an MIS in the modified CG. If any of the two IS's from split MIS_i is not coherent, we further split the IS as well. The process continues until all MIS's and the IS's split from them are coherent. As such, the final CG is an MCG. The pseudo code of our MCG construction algorithm is shown in Figure 6.

Second, while adding an edge between any two vertices of an incoherent MIS can split it into two. It is natural to choose the edge which will result in the smallest maximum noise in the two IS's. To achieve this, we need to enumerate the noise for the two IS's resulted from an added edge between any pair of vertices. If the MIS has m vertices, then the total number of edges that can be added is $m(m-1)/2$. A naive implementation will calculate the noise for $m(m-1)$ IS's, resulting in O(m^2) complexity. However, a lot of these calculations are indeed redundant: if only one edge is to be added, there can only be m unique IS's. Accordingly, we can

MCG_construction(noise constraint *bound*)
1: **Function** *MCG_construction*(*bound*)
//Construct the conflict graph *cg* by any pair of vertices
2: **for** any two vertices *x* and *y* **do**
3:    **if** (*noise*(*x*) + *noise*(*y*) > *bound*) **then**
4:        Add an edge between *x* and *y* in *cg*;
5:    **endif**
6: **endfor**
//Check noises of each MIS
7: *mcg* = *cg*; //initialize *mcg*
8: Find all MIS's in *mcg* and save them in φ[];
9: **while** (φ[] != NULL) **do** //until φ[] is empty
10:    choose an MIS *i* for φ[];
11:    **if** (*noise*(*i*) > *bound*) **then**
12:       *smallest_noise* = ∞; //Initialize
13:       **for** any two vertices *x* and *y* in *i* **do**
14:          *j* = *i* excludes *x*;
15:          *k* = *i* excludes *y*;
//Split *i* into two IS's *j* and *k*
16:          **if** (*noise*(*j*) and *noise*(*k*) < *smallest_noise*) **then**
17:             *temp_x* = *x*;
18:             *temp_y* = *y*;
19:             update *smallest_noise*;
20:          **endif**
21:       **endfor**
22:    **endif**
23:    Add an edge between *temp_x* and *temp_y* in *mcg* and update φ[];
24: **endwhile**
25: **return** *mcg*; //Multi-conflict graph
Figure 6. The pseudo code of MCG construction.

simply calculate the noise for all the possible IS's and then use the aforementioned look-up-table method to get noises for the two IS's resulting from any added edge. Such an implementation can bring the complexity down to O(m).

Third, the number of MIS's is exponentially proportional to the number of nodes in a graph [3]. Accordingly, the above algorithm cannot be directly applied to large number of modules. To resolve this issue, we adopt a divide-and-conquer heuristic. We first partition all vertices into M small groups, and then construct the MCG for each small group using the proposed algorithm. When combining the MCG's of two groups, we iterate through all pairs of vertices (p, q) with vertex p and vertex q in different groups. An edge needs to be added between (p, q) unless any MIS containing p and any MIS containing q are coherent when put together. Towards this, during the course of construction, in each vertex we also store the maximum noise in every node of the power network caused by all the MIS's containing that vertex. Then we can simply superpose the noise data stored in p and q, and if such superposed noise exceeds the bound in some nodes of the power supply network, it means that at least one MIS containing p and one MIS containing q cannot be turned on consistently. Accordingly, an edge should be added between p and q. For a fixed group size, it is easy to verify that the complexity of such a divide-and-conquer approach is $O(N^3)$.

To conclude this section, we use a six-module example in Figure 8 to illustrate our MCG construction algorithm. Initially, the CG has three MIS's, among which only the MIS $\{M_1, M_3, M_4, M_6\}$ has a noise violation. We add an edge between M_1 and M_4, and thus the MIS $\{M_1, M_3, M_4, M_6\}$ becomes two IS's $\{M_1, M_3, M_6\}$ and $\{M_3, M_4, M_6\}$. Then we verify that the maximum noise of both IS's do not exceed the bound, and thus the MCG is finalized.

3.4. On-line Scheduling

Given the multi-conflict graph (MCG) and the modules to be turned on specified by the power management unit, we can extract a sub-graph only consisting of these modules. The complexity of such a step is $O(n^2)$ where n is the number of modules to be turned-on.

Now the scheduling problem simply reduces to clustering these vertices into as few independent sets as possible. Since the total number of clusters is proportional to the total wake-up latency, accordingly, we would like to minimize the total number of clusters. In addition, since the modules in each independent set can be turned on simultaneously, the wake-up latency minimization problem can be cast as finding the minimum number of independent sets in an MCG.

We notice that there is a synergy between this problem and the well-known vertex coloring problem, which can be stated as follows: Given a graph, label all the vertices with the minimum number of colors, such that any two vertices connected by an edge are labeled with different colors. As such, we can simply cluster the vertices with the same color into one cluster, and therefore, minimizing the number of colors is equivalent to minimizing the number of clusters.

We choose to apply a heuristic algorithm [2] to solve the coloring problem, which is described as follows. For simplicity of the presentation, we label the colors with positive integers {1, 2, 3, ...} and the pseudo code of the algorithm is shown in Figure 9.

Back to the above example in Figure 7, assume the power management unit decides to turn on modules M_3, M_4, and M_5 this time. In Figure 10, we construct the sub-MCG from the MCG in Figure 8. Then, we choose the vertex M_5 which is the vertex with the maximum degree in the sub-MCG, and color M_5. After that, we update saturation degrees of those uncolored vertices, and then choose the vertex with the maximum saturation degree, M_3, for coloring. Finally, we update the saturation degree of M_4 and select a possible color for M_4. As a result, the order of turning on M_3, M_4 and M_5 are (M_3, M_4), M_5, which stands for turning on M_3 and M_4 simultaneously and then turning on M_5. Compared with the previous work [4] which is to turn on modules one by one, we have a 33% improvement on the wake-up latency for turning on three modules M_3, M_4 and M_5.

The well-known vertex coloring problem is NP-complete and the complexity of the heuristic algorithm is $O(n\log n)$ where n is the number of vertices [2]. Hence, our module-level wake-up scheduling is efficient enough so that it can decide the wake-up order of modules at runtime.

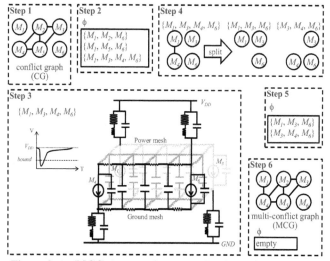

Figure 8. An MCG construction example with a six-module design.

Find the minimum number of colors in the sub-*mcg* H (modules *w_modules*[] to be turned on, multi-conflict graph *mcg*)

1: **Function** *Vertices_Coloring*(*w_modules*[], *mcg*)
2: Construct the sub-MCG H from *mcg* by *w_modules*[];
3: Find the vertex i with the maximum degree in H;
4: *Color*[i] = 1;   //Color the vertex i with color 1
5: **while** (!all vertices in H are colored) **do**
6:   Update *saturation_degree* of the uncolored vertices in H;
   //*saturation_degree* of a vertex is the number of different colors which is connected.
7:   Find the vertex j with the maximum *saturation_degree* in H;
8:   *Color*[j] = the least possible color; //lowest number of color
9: **endwhile**
10: **return** *Color*[];

Figure 9. The pseudo code of the vertex coloring algorithm.

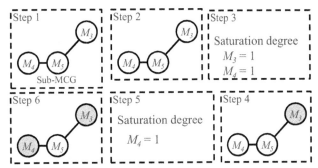

Figure 10. A vertex coloring example with only three modules in a six-module design.

Before we end the section, we would like to point out that the MCG construction is a one-time effort and can include a trade-off between runtime and optimality. Furthermore, extracting the sub-MCG at runtime based on the modules to turn on may lose some optimality: our MCG construction algorithm may force to sequentially turn on some modules which can in fact be turned-on simultaneously. However, experimental results indicate that on average our heuristic can achieve a significant improvement over existing approaches.

4. EXPERIMENTAL RESULTS

We have implemented the module-level wake-up scheduling described in Section 3. In our experiments, we use TSMC 90nm CMOS technology and perform experiments on three homogeneous architectures containing 16, 64, and 256 modules. The module is an industrial design with 55,794 gate count. The sizes of the sleep transistors and the gate-level wake-up scheduling for the module are designed using the algorithms in [5] and [10] respectively. The maximum surge current during the wake-up process is 240mA, and the wake-up latency from the gate-level wake-up scheduling is 18ns. The placement of and interconnections between the modules, the number and the locations of power/ground pads on the designs are properly designed. The DIP-40 package [11] is adopted as the power/ground pad in the SPICE simulation.

Based on the SPICE simulation results, we first implement the MCG construction as mentioned in Section 3.3 in MATLAB. The degree distribution of vertices on the MCG for the 16-module system is shown on the left of Figure 11 while that for the

Degree distribution

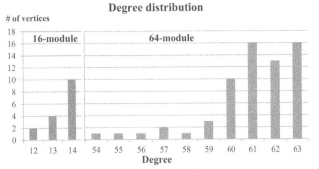

Figure 11. The degree distribution of vertices in the MCG's for 16-module and 64-module designs.

64-module system is on the right. Moreover, the MCG for the 16-module system is shown in Figure 12.

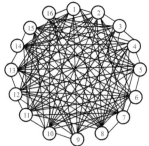

Figure 12. The MCG for the 16-module design.

We assume that we are provided with the set of modules to be turned on from the power management unit at runtime. We can then decide the module-level wake-up schedule by our on-line scheduling algorithm as discussed in Section 3.4. To see how the wake-up latency from our method compare with the minimum possible latency, we implement a Monte Carlo search algorithm, which chooses the best result among one million random wake-up orders without noise violations on the same set of modules. We anticipate that this will provide a sufficiently close-to-optimal solution. As a comparison, we also re-implement the industrial approach [4], which turns on one module at a time, and a heuristic that is simple yet can guarantee the noise bound: we first group all the modules off-line using the best result from the Monte Carlo search, and at runtime the wake-up order simply follows the grouping results[4].

Table 1 shows wake-up latencies for turning on different number of modules for the four methods mentioned above, the Monte Carlo search (**mc**), our work (**ours**), [4], and off-line grouping (**og**), on various designs. On average, the wake-up latency of our approach is 14.98% and 46.01% less than off-line grouping and [4], respectively, and has only 0.45% overhead compared with the minimum wake-up latency from the Monte Carlo search.

Next, we randomly select 10 sets (each containing 64 modules) to be turned on in the 256-module design, and compare the wake-up latency for each set obtained from the four methods. The result is depicted in Figure 13. For each set, we first use the boxplot to show the wake-up latency distribution from one million Monte Carlo search, with the two terminals of the vertical line indicating the minimum and maximum values and the box indicating the first and third quartiles. The medium is also shown in the box. Obviously, the bottom terminal and the top terminal correspond to wake-up latencies of the Monte Carlo search (the smallest latency) and [4] (the largest latency), and therefore we do not explicitly label them in the figure. The diamond and the triangle stand for wake-up latencies for ours and off-line grouping, respectively. The wake-up latency of our approach is only 2% longer than the minimum latency from the Monte Carlo search while those from off-line grouping and [4] are 16.23% and 18.08% longer, respectively.

Finally, we would like to mention that the runtime of our off-line MCG construction algorithm for the 256-module design is 1.64 seconds and the on-line scheduling algorithm is 0.12 μs while that of the Monte Carlo search algorithm is over 3 hours. Meanwhile,

---

[4] Any modules in the same group will be turned on simultaneously, and vice versa. Use Figure 5 as an example. Initially, we off-line group the modules, such as: (M_1, M_2), (M_3, M_4), (M_5, M_6). If we need to turn on M_3, M_4 and M_5 on-line, then we would turn on (M_3, M_4) simultaneously as a group and then M_5.

Table 1. Wake-up latencies (ns) to turn on different number of modules in different designs from the four methods.

# of modules in multi-module design	# of modules to turn on								(**mc**: Monte Carlo search, **og**: off-line grouping)			
	8				16				32			
	mc	ours	og	[4]	mc	ours	og	[4]	mc	ours	og	[4]
16	72	72	72	144	144	144	144	288	-	-	-	-
64	72	72	72	144	144	144	162	288	306	306	360	576
256	108	108	144	144	216	216	288	288	450	450	576	576
Average	**1.00**	**1.00**	**1.11**	**1.78**	**1.00**	**1.00**	**1.15**	**1.78**	**1.00**	**1.00**	**1.23**	**1.58**
# of modules in multi-module design	# of modules to turn on								(**mc**: Monte Carlo search, **og**: off-line grouping)			
	64				128				256			
	mc	ours	og	[4]	mc	ours	og	[4]	mc	ours	og	[4]
16	-	-	-	-	-	-	-	-	-	-	-	-
64	720	720	864	1152	-	-	-	-	-	-	-	-
256	954	972	1134	1152	1998	2016	2268	2304	4140	4176	4608	4608
Average	**1.00**	**1.01**	**1.19**	**1.40**	**1.00**	**1.01**	**1.14**	**1.15**	**1.00**	**1.01**	**1.11**	**1.11**

the MCG construction for each multi-module design is an one-time off-line effort, while the Monte Carlo search must be executed repeatedly for different sets of modules to be turned on.

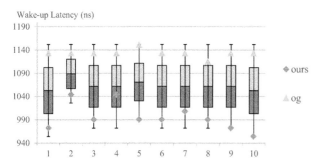

Figure 13. The wake-up latency of mc (entire range), ours (diamond), og (triangle), and [4] (max value) to turn on 64 modules on the 256-module design over ten runs.

5. CONCLUSIONS

In this paper, we discuss a novel on-line module-level wake-up scheduling for high performance computing architectures. We propose an off-line algorithm to characterize the multi-conflict graph (MCG) for the high performance multi-module design, and an on-line scheduling algorithm to efficiently decide the order to turn on modules. Experimental results show that for a 256-module system, the wake-up latency for our approach on average achieves 23.06% and 23.67% wake-up latency reduction compared with [16] and [4], respectively. And our approach only results in 0.6% latency overhead compared with the result from the Monte Carlo search.

REFERENCES

[1] A. Abdollahi, F. Fallah and M. Pedram, "A Robust Power Gating Structure and Power Mode Transition Strategy for MTCMOS Design," *IEEE Transactions on VLSI Systems*, vol. 15, no. 1, January 2007.

[2] D. Brelaz, "New Methods to Color the Vertices of a Graph," *Communications of the ACM*, Vol.22, Issue 4, Apr. 1979.

[3] C. Bron and J. Kerbosch, "Algorithm 457: Finding All Cliques of an Undirected Graph," *Communications of the ACM*, Vol. 16, Issue 9, 1973.

[4] S. H. Chen and J. Y. Lin, "Experiences of low power design implementation and verification," *Prof. of the ASPDAC*, pp. 742-747, 2008.

[5] D. S. Chiou, D.C. Juan, Y.T. Chen, and *S.C. Chang*, "Fine-Grained Sleep Transistor Sizing Algorithm for Leakage Power Minimization," *Proc. of the DAC*, pp. 81-86, 2007.

[6] A.H. Farrahi, D.J. Hathaway, M. Wang, M. Sarrafzadeh, "Quality of EDA CAD tools: definitions, metrics and directions," *Proc. of the ISQED*, pp. 395 – 405, 2000.

[7] S. Hong and H. Kim, "An Integrated GPU Power and Performance Model," *Proc. of ISCA*, pp. 280-289, 2010.

[8] H. Jiang, M. Marek-Sadowska and S. R. Nassif, "Benefits and Costs of Power-Gating Technique," *Proc. of the ICCD*, pp. 559-566, 2005.

[9] H. Jiang and M. Marek-Sadowska, "Power Gating Scheduling for Power/Ground Noise Reduction," *Proc. of the DAC*, pp.980-985, 2008.

[10] D. C. Juan, Y. T. Chen, M. C. Lee and S. C. Chang, "An Efficient Wake-Up Strategy Considering Spurious Glitches Phenomenon for Power Gating Designs," *IEEE Transactions on VLSI Systems*, Vol.18, Issue 2, pp. 246 – 255, 2010.

[11] S. Kim, S. V. Kosonocky and D. R. Knebel, "Understanding and Minimizing Ground Bounce During Mode Transition of Power Gating Structures," *Proc. of the ISLPED*, pp. 22-25, August 25-27, 2003.

[12] Y. Lee, D-K Jeong, and T. Kim, "Simultaneous Control of Power/Ground Current, Wakeup Time and Transistor Overhead in Power Gated Circuits," *Proc. of the ICCAD*, pp. 169-172, 2008.

[13] A. Ramalingam, A. Devgan and D. Z. Pan, "Wakeup Scheduling in MTCMOS Circuits Using Successive Relaxation to Minimize Ground Bounce," *Journal of Low Power Electronics*, vol.3, pp. 1-8, 2007.

[14] K. Shi, and D. Howard, "Challenges in Sleep Transistor Design and Implementation in Low-Power Designs," *Proc. of the DAC*, pp. 113-116, 2006.

[15] H. Xu, R. Vemuri, W.-B. Jone, "Current shaping and multi-thread activation for fast and reliable power mode transition in multicore designs," *Proc. of the ICCAD*, pp. 637–641, 2010.

[16] Y. Xu, W. Liu, Y. Wang, J. Xu, X. Chen, H. Yang, "On-line MPSoC Scheduling Considering Power Gating Induced Power/Ground Noise," *Proc. of the ISVLSI*, pp.109-114, 2009.

[17] http://www.legitreviews.com/article/1228/12/

Low-Power Gated Bus Synthesis for 3D IC via Rectilinear Shortest-path Steiner Graph

Chung-Kuan Cheng[1,2], Peng Du[1], Andrew B. Kahng[1,2] and Shih-Hung Weng[1]
[1]Dept. of Computer Science and Engineering, University of California San Diego, La Jolla, CA, 92093
[2]Dept. of Electrical and Computer Engineering, University of California San Diego, La Jolla, CA, 92093
ckcheng@ucsd.edu, pedu@ucsd.edu, abk@ucsd.edu, s2weng@ucsd.edu

ABSTRACT

In this paper, we propose a new approach for gated bus synthesis [16] with minimum wire capacitance per transaction in three-dimensional (3D) ICs. The 3D IC technology connects different device layers with through-silicon vias (TSV), which need to be considered differently from metal wire due to reliability issues and a larger footprint. Practically, the number of TSVs is bounded between layers; thus, we first devise dynamic programming and local search techniques to determine the optimal TSV locations. We then employ two approximation algorithms to generate a rectilinear shortest-path Steiner graph in each device layer. One algorithm extends the well-known greedy heuristic for the Rectilinear Steiner Arborescence problem and handles large cases with high efficiency. The other algorithm utilizes a linear programming relaxation and rounding technique which costs more time and generates a nearly-optimal Steiner graph. Experimental results show that our algorithms can construct shortest-path Steiner graphs with 22% less total wire length than the previous method of Wang et al. [16].

Categories and Subject Descriptors
B.7.2[**Integrated Circuits**]:Design Aids–placement and routing
General Terms
Algorithm
Keywords
3D IC, TSV, shortest-path Steiner graph, Gated Bus

1. INTRODUCTION

Three-dimensional (3D) IC technology offers the potential for improving performance and power consumption for bus architectures in SoC design [4]. 3D IC technology uses through-silicon vias (TSVs) to connect device layers. In SoC designs, TSVs can potentially reduce global interconnect among IPs, improving communication power efficiency and performance of bus-based architectures. For example, Pathak et al. [12] show that the timing of a 3D LEON3

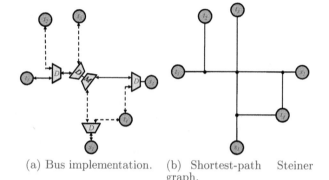

(a) Bus implementation.   (b) Shortest-path Steiner graph.

Figure 1: Gated bus architecture.

multi-core design with AMBA bus architecture connected by TSVs is better than that of a 2D design by about 79%.

A 3D IC design needs many TSVs for data bus, address bus and control wires to connect IPs that are on different layers. However, the number of TSVs must be limited since TSV is costly to fabricate, and a large number of TSVs may degrade manufacturing yield, test cost efficiencies and available layout area on-chip. Given this consideration, designers cannot generously use TSVs for communication between device layers. Therefore, selection of TSV positions becomes important when designers seek to increase performance and reduce power consumption of bus architectures while using a restricted number of TSVs.

Many researchers [5][11][16] have worked on low-power bus architectures, reducing unnecessary wire loading or signal switching when the bus transfers data. Bus segmentation [5] and bus splitting [11] methods reduce wire load by masking off certain bus segments, but only consider where to mask off within a given bus topology to achieve maximum power savings. In the gated bus architecture [16], Wang *et al.* synthesize a gated bus topology to reduce the wire capacitance by using distributed multiplexers and demultiplexers.

Figure 1(a) shows an implementation of the gated bus architecture with two masters s_1, s_2 and four slaves $t_1, \ldots, t_4$. The gray multiplexers and demultiplexers mask off the unused path when transferring data from s_2 to t_1 so that s_2 only needs to drive the path consisting of the three solid arrows. Therefore, driven wire capacitance is greatly reduced. Figure 1(b) shows the Steiner graph representation of Figure 1(a). The Steiner points of Figure 1(b) indicate where to place multiplexers/demultiplexers in the bus architecture. To have the minimum wire capacitance for every data transaction, the Steiner graph must contain a shortest path with length equal to the Manhattan distance between each master-slave pair. We call such a Steiner graph a *shortest-path Steiner graph*; our objective is to find a shortest-path Steiner graph with smallest total wire length. In [16], Wang *et al.* proposed a heuristic method that first constructs a minimum rectilinear Steiner arborescence [8][13] starting from a master, and then adds other masters one by one in each iteration to obtain a shortest-path Steiner graph.

In this paper, we investigate the problem of gated bus synthesis in 3D ICs to minimize total power consumption of the gated bus architecture. Since constraints of TSVs are different from these of on-die vias, e.g., larger footprint and keep-out zones, we must consider TSVs separately in this problem. We first develop dynamic programming and local search algorithms to determine TSV locations that minimize the sum of weighted shortest distances over all master-slave pairs in a 3D IC design. Then, given the TSV locations thus determined, we propose two approximation algorithms to synthesize a shortest-path Steiner graph with smallest total wire length on each layer of the 3D IC stack; this graph determines locations of multiplexers and demultiplexers for the gated bus architecture. The two algorithms include a greedy heuristic that can handle large instances and a linear programming relaxation and rounding method [15] that is more accurate and suitable for solving cases with relatively small scale. Overall, our method can reduce wire length by up to 22% when compared to [16].

The remainder of this paper is organized as follows. In Section 2, we introduce the problem of gated bus synthesis in 3D ICs and formally state two problem formulations on TSV location determination and construction of shortest-path Steiner graphs. In Section 3, we give two algorithms for determination of TSV locations; these address the cases of one TSV and multiple TSVs between adjacent layers, respectively. Section 4 describes our approximation algorithms to generate a shortest-path Steiner graph with minimum total wire length. Experimental results are shown in Section 5, and conclusions are given in Section 6.

2. PROBLEM FORMULATION

We assume that we are given locations of masters and slaves on device layers, and the communication frequency for each master-slave pair. Our goal is to minimize total power consumption over all master-slave pairs while keeping the total wire length as small as possible. Total power is estimated as the summation of frequency multiplied by capacitance, i.e., only dynamic power. Note that since introduced auxiliary gates in our bus architecture only take a small percentage of total gate count in a design, the increase of leakage power is negligible (ignoring possible effects of power-gating).

As noted above, we cannot place an arbitrary number of TSVs between adjacent device layers. Hence, we assume that there is a given constant B indicating the maximum number of TSVs available for connecting masters and slaves between two device layers. Our experimental results (below, Section 5) show that when $B > 2$, this trade-off achieves nearly the optimal power consumption when the number of TSVs between layers is unbounded ($B = \infty$). Besides, with so few TSVs, the power impact of large capacitance TSV is insignificant. The overhead of the control wires in the

gated bus architecture is small compared to the data bus and the address bus because the number of such wires is much less than the bus width. Moreover, since the control signals do not switch during data transaction, the dynamic power associated with control signal is small. We therefore ignore effects of control wiring in this study.

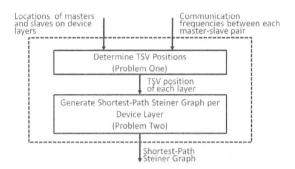

Figure 2: Overall flow.

Figure 2 shows the overall flow of our method. The locations of masters and slaves can be obtained after performing 3D floorplanning [2][6]. The communication frequency can be calculated from gate-level simulation. We address the following two sub-problems in Sections 3 and 4, respectively.

DEFINITION 1. *A rectilinear shortest path between a master-slave pair in the same layer is a shortest path with length equal to the Manhattan distance between them.*

- Problem One: Find locations of TSVs between adjacent layers so that the total length of weighted shortest paths between master-slave pairs is minimized. We take frequencies to be weights and only B TSV locations can be assigned between two adjacent layers.

- Problem Two: Given optimal (fixed, from Problem One) locations of TSVs, construct a rectilinear shortest-path Steiner Graph in each layer that minimizes the total wire length subject to the existence of a layer-wise rectilinear shortest path for each master-slave pair. This is the *minimal rectilinear shortest-path Steiner graph* (RSSG) problem, which was first considered in [16].

3. PROBLEM ONE: DETERMINATION OF TSV LOCATIONS

The input of our problem includes the locations of n masters $\{s_1, s_2, \ldots, s_n\}$ and m slaves $\{t_1, t_2, \ldots, t_m\}$ on L device

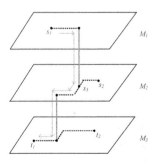

Figure 3: Two TSVs between three device layers.

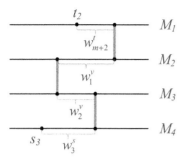

Figure 4: Contribution of an s-t path $s_3 - t_2$ to weights.

layers from M_1 to M_L, and the communication frequency $c_{i,j}$ for each master-slave pair (s_i, t_j). We assume that all device layers have the same size $W \times H$. Given the upper bound B of the number of TSVs between adjacent device layers, our first objective is to determine the best locations of TSVs so that the sum of weighted shortest-path lengths over all $s_i - t_j$ pairs, for $1 \le i \le n$ and $1 \le j \le m$, is as small as possible. The weight for the shortest path between s_i and t_j is chosen to be $c_{i,j}$. For example, Figure 3 shows a possible assignment of two sets of TSVs in blue between three device layers where the shortest path through TSVs from s_1 to t_1 is denoted by red arrows. The length of TSVs will be neglected in the following discussion since it does not affect the objective value. We denote the planar coordinates and layers of s_i and t_j for $1 \le i \le n$ and $1 \le j \le m$ by (x_i^s, y_i^s, l_i^s) and (x_j^t, y_j^t, l_j^t) respectively. The candidate positions of the TSVs are determined by the following theorem.

THEOREM 1. Given $X = \{x_1^s, \dots, x_n^s, x_1^t, \dots, x_m^t\}$ and $Y = \{y_1^s, \dots, y_n^s, y_1^t, \dots, y_m^t\}$, the optimal TSV positions can be chosen from the set $\{(x, y) : x \in X, y \in Y\}$, i.e., the Hanan grid [10] formed by union of masters and slaves.

PROOF. For simplicity, we only consider the case when $B = 1$. The proof can be extended to the case of larger B value. By the additivity of rectilinear distance for horizontal and vertical directions, we can determine the x and y coordinates of TSVs separately. Without loss of generality, we only consider x coordinates of TSV locations. Suppose $x_1^v, x_2^v, \dots, x_{L-1}^v$ is an optimal assignment of TSVs where x_k^v denotes the x coordinate of a (set of) TSV(s) connecting adjacent layers M_k and M_{k+1}, for $1 \le k \le L - 1$. Let k_0 be the smallest layer index such that $x_{k_0}^v \notin X$. We choose $x^1 = \max\{x : x < x_{k_0}^v \text{ and } x \in X\}$ and $x^2 = \min\{x : x > x_{k_0}^v \text{ and } x \in X\}$. Using the property of linear length metric, we can show that there exists $x' \in \{x^1, x^2, x_{k_0+1}^v\}$ so that moving $x_{k_0}^v$ to x' will not increase the objective value. If $x' = x^1$ or x^2, we achieve an optimal solution where the positions of the first k_0 TSVs are chosen from the Hanan grid instead of $k_0 - 1$. Otherwise if $x' = x_{k_0+1}^v$, we move $x_{k_0}^v$ to x' and repeat the previous process by moving $x_{k_0}^v$ and $x_{k_0+1}^v$ together. By induction, we have that there exists an optimal solution in which all TSVs are located on the Hanan grid. ∎

As noted in [9], restriction to Hanan grids may be suboptimal for minimizing the maximum delay between source-terminal pairs. We assume that timing constraints can still be met by adding buffers in buses. In the following two

subsections, we present an efficient dynamic programming algorithm to determine the optimal TSV location assignment when $B = 1$ and a local search heuristic to determine the approximately optimal TSV location assignment when $B > 1$.

3.1 One TSV between Adjacent Layers

In the case where $B = 1$, we can determine the x and y coordinates of TSVs separately by coordinate-wise additivity of rectilinear distance. Here, we give the algorithm for finding x coordinates of TSVs. We again represent an assignment of TSV locations by $x_1^v, x_2^v, \dots, x_{L-1}^v$ where x_k^v denotes the x coordinate of the single TSV between M_k and M_{k+1}. Now our objective can be expressed as

$$\min_{x_1^v, \dots, x_{L-1}^v} \sum_{i=1}^{n} \sum_{j=1}^{m} c_{i,j} d(s_i, t_j)$$

$$= \min_{x_1^v, \dots, x_{L-1}^v} (\sum_{k=1}^{L-2} w_k^v |x_{k+1}^v - x_k^v|$$

$$+ \sum_{i=1}^{n} w_i^s |x_i^s - x_{l_i^s-1}^v| + \sum_{i=1}^{n} w_{n+i}^s |x_i^s - x_{l_i^s}^v|$$

$$+ \sum_{j=1}^{m} w_j^t |x_j^t - x_{l_j^t-1}^v| + \sum_{j=1}^{m} w_{m+j}^t |x_j^t - x_{l_j^t}^v|$$

$$+ \sum_{i=1}^{n} \sum_{j=1}^{m} \chi(l_i^s = l_j^t) c_{i,j} |x_i^s - x_j^t|), \qquad (1)$$

where $d(s_i, t_j)$ is the length of a shortest path from s_i to t_j; w_i^s, w_j^t and w_k^v indicate constant weights corresponding to each term; and χ is a 0-1 indicator function. We explain the meaning of each term of (1) as follows.

- $|x_{k+1}^v - x_k^v|$: The segment between two TSVs connecting M_{k+1} with M_{k+2} and M_k with M_{k+1}, respectively.

- $|x_i^s - x_{l_i^s-1}^v|$: The segment between s_i and the TSV connecting the layer of s_i with the layer above.

- $|x_i^s - x_{l_i^s}^v|$: The segment between s_i and the TSV connecting the layer of s_i with the layer below.

- $|x_j^t - x_{l_j^t-1}^v|$: The segment between t_j and the TSV connecting the layer of t_j with the layer above.

- $|x_j^t - x_{l_j^t}^v|$: The segment between t_j and the TSV connecting the layer of t_j with the layer below.

- $\chi(l_i^s = l_j^t)|x_i^s - x_j^t|$: The segment between s_i and t_j if they are on the same layer.

If the shortest path from s_i to t_j passes through a segment listed above, it will contribute $c_{i,j}$ to the corresponding weight constant. Figure 4 gives an example where the shortest path between s_3 and t_2 passes through four segments whose corresponding weight constants are labeled below. Therefore, the shortest path will contribute $c_{3,2}$ to w_3^s, w_2^v, w_1^v and w_{m+2}^t, respectively.

Since the last term of (1) is constant, we remove it from our objective for simplicity. Let $\vec{x} = (x_1^s, \dots, x_n^s, x_1^t, \dots, x_m^t)$ and let $\vec{x}(r)$ be the rth coordinate of $\vec{x}$ for $1 \le r \le n+m$. Let $S(k)$ and $T(k)$ respectively denote the sets of masters and slaves in layer M_k, for $1 \le k \le L$. To derive our dynamic programming algorithm, we use a function OPT to indicate partial solutions, defined as

$$
\begin{aligned}
OPT(k_c, r_c) &= \min_{x_1^v, \ldots, x_{k_c}^v = \vec{x}(r_c)} \left(\sum_{k=1}^{k_c-1} w_k^v |x_{k+1}^v - x_k^v| \right. \\
&+ \sum_{k=1}^{k_c} \sum_{i \in S(k)} (w_i^s |x_i^s - x_{k-1}^v| + w_{n+i}^s |x_i^s - x_k^v|) \\
&+ \left. \sum_{k=1}^{k_c} \sum_{j \in T(k)} (w_j^t |x_j^t - x_{k-1}^v| + w_{m+j}^t |x_j^t - x_k^v|) \right), \\
&\quad 1 \le k_c \le L-1; 1 \le r_c \le n+m. \quad (2)
\end{aligned}
$$

The initial values of function OPT can be evaluated as

$$
\begin{aligned}
OPT(1, r_c) &= \sum_{i \in S(1)} w_{n+i}^s |x_i^s - \vec{x}(r_c)| \\
&+ \sum_{j \in T(1)} w_{m+j}^t |x_j^t - \vec{x}(r_c)|, 1 \le r_c \le n+m. (3)
\end{aligned}
$$

The value of $OPT(k_c, r_c)$ indicates the minimal total length of weighted shortest paths between masters and slaves in layers $\{M_1, \ldots, M_{k_c}\}$ plus weighted shortest paths from masters/slaves in $\{M_1, \ldots, M_{k_c}\}$ to the set of TSVs with position $\vec{x}(r_c)$ between M_{k_c} and M_{k_c+1}. The key step for computing OPT is expressed by the following recursive formula:

$$
\begin{aligned}
OPT(k_c, r_c) &= \min_{1 \le r \le n+m} (OPT(k_c-1, r) + w_{k_c-1}^v |\vec{x}(r_c) - \vec{x}(r)| \\
&+ \sum_{i \in S(k_c)} (w_i^s |x_i^s - \vec{x}(r)| + w_{n+i}^s |x_i^s - \vec{x}(r)|) \\
&+ \sum_{j \in T(k_c)} (w_j^t |x_j^t - \vec{x}(r)| + w_{m+j}^t |x_j^t - \vec{x}(r)|)), \\
&\quad 2 \le k_c \le L-1; 1 \le r_c \le n+m. \quad (4)
\end{aligned}
$$

Intuitively, since all communications from masters/slaves in $\{M_1, \ldots, M_{k_c-1}\}$ to masters/slaves in M_{k_c} and the TSV between M_{k_c} and M_{k_c+1} must go through the TSV between M_{k_c-1} and M_{k_c}, we only need to enumerate all possible positions $\vec{x}(r)$ between M_{k_c-1} and M_{k_c} and choose the best one to achieve $OPT(k_c, r_c)$. Finally, our objective (1) will be computed as

$$
\begin{aligned}
\min_{1 \le r \le n+m} \quad &(OPT(L-1, r) + \sum_{i \in S(L)} w_i^s |x_i^s - \vec{x}(r)| \\
&+ \sum_{j \in T(L)} w_j^t |x_j^t - \vec{x}(r)|). \quad (5)
\end{aligned}
$$

THEOREM 2. The time complexity of the dynamic programming algorithm for determination of TSV locations is $O((n+m)^2 L)$.

PROOF. As in recursive formula (4), each $OPT(k_c, r_c)$ can be evaluated in time $O(n+m)$ where the summation term can be gradually updated in constant time as r is increasing. Since there are $O((n+m)L)$ number of OPT values, we can obtain all of them in time $O((n+m)^2 L)$. Computing the final solution by (5) takes additional $O(n+m)$ time. Therefore, the total running time of our algorithm is $O((n+m)^2 L) + O(n+m) = O((n+m)^2 L)$. ∎

3.2 Multiple TSVs between Adjacent Layers

When multiple TSVs between adjacent layers are allowed (i.e., $B > 1$), we cannot determine TSV locations for x and y coordinates separately. Therefore, we represent the TSVs' locations by $\{(x_{kb}^v, y_{kb}^v) : 1 \le k \le L-1, 1 \le b \le B\}$ where (x_{kb}^v, y_{kb}^v) denotes the location of the bth TSV between layers M_k and M_{k+1}. We describe a local search heuristic to

determine locations of TSVs in Algorithm 1. It first finds a best assignment of TSVs on a coarse grid Z by exhaustive search and then updates it locally to find better solutions until there is no improvement of the objective function. If Z is a candidate set of positions for TSVs, each element V in the set $Z^{(L-1)B}$ corresponds to an assignment of TSVs on Z. We use $V(g)$ to denote the gth element of V by taking V as a sequence of elements in Z with length $(L-1)B$. If V is the assignment of TSVs, the function $TotShortestPaths(V)$ returns the total length of shortest paths weighted by communication frequencies between each pair of masters and slaves, .

Algorithm 1: Finding multiple TSVs by local search

Input: $n, m, L, B, \{(x_i^s, y_i^s, l_i^s) : 1 \le i \le n\}, \{(x_j^t, y_j^t, l_j^t) : 1 \le j \le m\}$

Output: TSV positions (x_{kb}^v, y_{kb}^v) where $1 \le k \le L-1, 1 \le b \le B$

1   prt = a small integer (e.g., 5);
2   $d_x = W/prt$;
3   $d_y = H/prt$;
4   $Z = \{0, d_x, 2d_x, \ldots, W\} \times \{0, d_y, 2d_y, \ldots, H\}$;
5   V_0 = the best assignment $V \in Z^{(L-1)B}$ such that $TotShortestPaths(V)$ is smallest;
6   $X = \{x_i^s : 1 \le i \le n\} \cup \{x_j^t : 1 \le j \le m\}$;
7   $Y = \{y_i^s : 1 \le i \le n\} \cup \{y_j^t : 1 \le j \le m\}$;
8   **while** *true* **do**
9    $update\_flag = false$;
10   **for** $1 \le g \le (L-1)B$ **do**
11   V_1 = the best assignment V with $V(g') = V_0(g')$ for $g' \ne g$ and $V(g) \in X \times Y$ such that $TotShortestPaths(V)$ is smallest;
12   **if** $V_1 \ne V_0$ **then**
13      $update\_flag = true$;
14   **end**
15     $V_0 = V_1$;
16   **end**
17   **if** $update\_flag = false$ **then**
18   break;
19   **end**
20 **end**
21 **return** V_0;

We describe the details of Algorithm 1 as follows:

- Lines 1-5: We partition the rectangle $W \times H$ into $prt \times prt$ parts with the size of each part to be $d_x \times d_y$. Let Z be the set of lattice points in this partition. We first assume TSVs can only take positions from Z and obtain the best assignment V_0 by exhaustive search in Line 5. V_0 will be updated to a better solution afterward in Lines 8-20.

- Lines 6-7: We define the Hanan grid formed by positions of masters and slaves as $X \times Y$.

- Lines 8-20: In each iteration of the while loop, if there exists a better solution V_1 than V_0, we update V_0 to be V_1 and continue the loop. Otherwise, we stop and output V_0 as the final solution.

- Lines 10-16: For each TSV g, let V_1 be the best assignment by moving TSV g in V_0 to a new position as in Line 11. If V_1 is better than V_0, we update V_0 in Lines 12-15.

Based on our experimental results, the local search algorithm achieves nearly optimal solution by comparing with the lower bound obtained from the case where arbitrary TSVs are allowed between adjacent layers (i.e., $B = \infty$).

Algorithm 2: Greedy RSSG heuristic

Input: A set of n masters $s_1, \ldots, s_n$ and m slaves $t_1, \ldots, t_m$ in rectilinear plane.

Output: A subgraph $G(V, E)$ of Hanan grid H_g formed by masters and slaves so that G contains a rectilinear shortest path for each master-slave pair.

1 $V = \{s_1, \ldots, s_n, t_1, \ldots, t_m\}$;
2 $E = \emptyset$;
3 $P = $ set of vertices in Hanan grid H_g;
4 $D_p = \{1, 2, \ldots, n\}$ for $p \in \{t_1, \ldots, t_m\}$;
5 $D_p = \emptyset$ for $p \in P$ and $p \notin \{t_1, \ldots, t_m\}$;
6 **while** $\exists p, q \in P$ *s.t.* $D_p \cap D_q \neq \emptyset$ **do**
7    $BestBenefit = 0$;
8   **for** $(p, q \in P$ *s.t.* $D_p \cap D_q \neq \emptyset)$ and $d \in \{0, 1\}$ **do**
9      $Benefit = \sum_{i \in D_p \cap D_q} GetBenefit(s_i, p, q, d)$;
10   **if** $Benefit > BestBenefit$ **then**
11      $BestBenefit = Benefit$;
12      $p^* = p, \ q^* = q, \ d^* = d$;
13   **end**
14   **end**
15   **for** $i \in D_{p^*} \cap D_{q^*}$ **do**
16      $v_i = GetSteinerPoint(s_i, p^*, q^*, d^*)$;
17      $V = V \cup \{v_i\}$;
18      $D_{p^*} = D_{p^*} \backslash \{i\}, \ D_{q^*} = D_{q^*} \backslash \{i\}$;
19      $D_{v_i} = D_{v_i} \cup \{i\}$;
20   **end**
21    $r = GetMergePoint(p^*, q^*, d^*)$;
22    $E = E \cup ShortestPath(p^*, r)$;
23    $E = E \cup ShortestPath(q^*, r)$;
24 **end**
25 **for** $p \in P$ *s.t.* $D_p \neq \emptyset$ **do**
26   **for** $i \in D_p$ **do**
27      $E = E \cup ShortestPath(s_i, p)$;
28   **end**
29 **end**
30 $RemoveRedundantEdges(G)$;
31 **return** $G(V, E)$;

---

4. PROBLEM TWO: APPROXIMATION ALGORITHMS FOR GENERATING RECTILINEAR A SHORTEST-PATH STEINER GRAPH

In this section, we build a network to connect masters, slaves and TSVs in each device layer. Since signals are delivered on one path with distributed multiplexers/demultiplexers, we formulate our problem as the RSSG synthesis first considered in [16]. We assume that locations of a set of masters $s_1, \ldots, s_n$ and slaves $t_1, \ldots, t_m$ in a fixed layer are given. Notice that TSVs connected with this layer are considered as both master and slave. A solution of RSSG is a rectilinear routing containing all master-to-slave rectilinear shortest paths, with total wire length as small as possible. RSSG is a generalization of the minimum rectilinear Steiner ar-

borescence (RSA) problem [8][13] and a relaxation of the Minimum Manhattan network (MMN) problem [3], both of which have been proven to be NP-Complete in [14] and [7] respectively. The RSA problem corresponds to the case where there is only one master s. We will introduce two approximation algorithms for RSSG in the following subsections. One is a greedy heuristic which extends the insight of a 2-approximation algorithm given in [13] for solving the RSA problem. It requires only $O(nm)$ extra memory beyond inputs and easily handles large-scale RSSG instances. Our algorithm uses linear programming (LP) relaxation and rounding to solve small cases with higher accuracy. In practice, we observe that the solution of LP relaxation and rounding is within a factor 1.0005 of optimal. However, the LP formulation of RSSG has $O(nm(n + m)^2)$ number of variables and constraints which makes it suitable only for solving instances with < 1000 master-slave pairs on a computer with 4GB RAM.

4.1 Greedy Heuristic

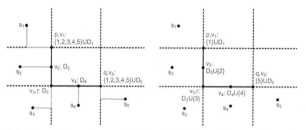

(a) Illustration for evaluating $GetBenefit(s_i, p, q, 0)$.
(b) Updated demand set for (a).

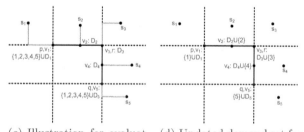

(c) Illustration for evaluating $GetBenefit(s_i, p, q, 1)$.
(d) Updated demand set for (c).

Figure 5: Contributions of two possible cases for merging p and q.

The greedy heuristic starts with m slaves as m subtrees and iteratively merges a pair of subtree roots p^* and q^* with merging point r is farthest from masters. We can think of this heuristic as obtaining the largest possible benefit, i.e., the sum of Manhattan distances from masters to p^* and q^*, in each iteration. Algorithm 2 provides the details of the greedy heuristic, with explanations as follows.

- Lines 1-3: The subgraph G and the set of vertices P of Hanan grid H_g are initialized.

- Lines 4-5: For each slave $p \in \{t_1, \ldots, t_m\}$, we define a demand set D_p initialized to be $\{1, \ldots, n\}$ which means $s_i - p$ rectilinear shortest path needs to be constructed in graph G for every $i \in D_p$. We also set the demand set to be empty for other non-slave points in P.

- Lines 6-14: In Line 6, if there exist two slaves or Steiner points p and q such that their demand sets have non-empty intersection, we will choose one such pair, i.e., p^* and q^*, with largest benefit for merging them using direction d^* between Lines 7 and 13. The function $GetBenefit(s_i, p, q, d)$ returns the benefit we can obtain in terms of s_i by merging p and q using direction d. Without loss of generality, we assume that p is above and to the left of q as illustrated in Figure 5. Two possible directions of merging p and q are shown in Figures 5(a) and 5(c). The benefit obtained by $GetBenefit(s_i, p, q, d)$ is the length of the blue line connecting s_i with a slave or Steiner point v_i. For other positions of s_i other than those shown in Figure 5, the function $GetBenefit(s_i, p, q, d)$ will return zero. Notice that after p and q are merged, the rectilinear shortest paths from s_i to p and from s_i to q can share the path from s_i to v_i; this is the benefit we achieve compared with connecting s_i with p and q separately.

- Lines 15-20: We update the demand sets after merging p^* and q^* using direction d^*. For each master s_i such that i belongs to the intersection of D_{p^*} and D_{q^*}, the slave or Steiner point v_i is found by the function $GetSteinerPoint$ and we henceforth only need to connect s_i with v_i by a rectilinear shortest path, instead of connecting with p^* and q^*. The corresponding update of demand sets is described in Lines 18-19 and illustrated in Figures 5(b) and 5(d).

- Lines 21-23: We get the merging point r and update the graph G according to the merging operation. The function $ShortestPath$ gives a rectilinear shortest path between two points using smallest extra edges not in G. It is implemented by a simple dynamic programming algorithm.

- Lines 25-29: When there are no intersecting demand sets, we fulfill the remaining connection requirements.

- Lines 30: We delete all redundant edges in G, i.e., as long as removing them does not change the length of any $s - t$ shortest path.

4.2 LP Relaxation and Rounding

In this section, we first give an integer linear programming (ILP) formulation of the RSSG problem and relax it into an instance of general linear programming (LP). Then, we devise a rounding technique to obtain an approximation solution of RSSG from the optimal solution of the LP relaxation. By comparing with the objective value of the LP relaxation, our results show that this approach achieves nearly optimal solutions.

The ILP formulation for the RSSG problem is given in (6) and the details are described as follows. Suppose there are Q pairs of masters and slaves where $Q = n \times m$. We use (s^l, t^l) to denote the lth master-slave pair. To transform the RSSG problem into an ILP, we construct a directed graph N_l as in Figure 6 for (s^l, t^l) on the Hanan grid formed by masters and slaves. Without loss of generality, assume s^l is to the left of t^l. Figures 6(a) and 6(b) show the directed graph N_l for the cases where s^l is below and above t^l, respectively. We notice that an $s^l - t^l$ rectilinear shortest path exists

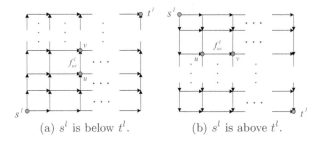

(a) s^l is below t^l.   (b) s^l is above t^l.

Figure 6: Directed network N_l on the Hanan grid.

if and only if an integer $s^l - t^l$ flow with value one exists in N_l. Let E_l be the set of arcs of N_l and let f_{uv}^l be a variable denoting the flow from u to v where $(u, v) \in E_l$. We use the first three constraints in (6) to guarantee the existence of a valid flow for each (s^l, t^l) pair. Now let E_H be the set of undirected edges in the Hanan grid and let the binary variable x_{uv} denote whether the edge $(u, v) \in E_H$ is selected in the RSSG solution. If a rectilinear shortest path from s^l to t^l includes the edge (u, v) for any $1 \leq l \leq Q$, then $x_{uv} = 1$. We use the fourth constraint in (6) to denote such constraints. Finally, if d_{uv} denotes the length of edge (u, v), our objective is to minimize the total wire length which can be described by the objective function of (6). Since there are $O((n + m)^2)$ edges in the Hanan grid and $O(nm)$ master-slave pairs, it is easy to see that the ILP formulation contains $O(nm(n + m)^2)$ variables and $O(nm(n + m)^2)$ constraints.

$$
\begin{aligned}
\min \quad & \sum_{(u,v) \in E_H} d_{uv} x_{uv} \\
s.t. \quad & \sum_{(u,v) \in E_l} f_{uv}^l - \sum_{(v,u) \in E_l} f_{vu}^l \geq 0, \\
& 1 \leq l \leq Q \text{ and } v \neq s^l, t^l; \\
& \sum_{(s^l,u) \in E_l} f_{s^l u}^l \geq 1, \quad 1 \leq l \leq Q; \\
& \sum_{(u,t^l) \in E_l} f_{ut^l}^l \geq 1, \quad 1 \leq l \leq Q; \\
& x_{uv} - f_{uv}^l \geq 0, \\
& (u,v) \in E_H \text{ and } l \in \{l_0 : (u,v) \in E_{l_0}\}; \\
& f_{uv}^l \in \{0,1\}, \quad 1 \leq l \leq Q \text{ and } (u,v) \in E_l; \\
& x_{uv} \in \{0,1\}, \quad (u,v) \in E_H. \qquad (6)
\end{aligned}
$$

Since solving ILP (6) is time-consuming and even intractable when n and m are large, we propose a LP relaxation and rounding technique described in Algorithm 3. We begin by relaxing the binary constraints on variables f_{uv}^l and x_{uv} so that they can take any nonnegative real values. We thus obtain an LP instance that can be solved efficiently. The variables x_{uv} in the optimal solution of the LP relaxation are in the range $[0, 1]$ based on other constraints, but may be fractional, which does not yield a feasible RSSG solution. To construct an RSSG from the LP solution, we adopt usual approach of viewing each x_{uv} as the probability of selecting edge (u, v). Hence, we start with a graph G containing all edges in the Hanan grid and try to delete edges as long as the remaining graph still contains a rectilinear shortest path for each master-slave pair. The order of deletion depends on the optimal x_{uv} values. An edge with smaller x_{uv} will be deleted from graph G earlier if possible. Finally, the remaining graph G will be a feasible RSSG solution. The optimal objective function value of the LP relaxation lower-bounds the optimal objective function value of the ILP (6), and hence optimal RSSG solution. We can evaluate the quality of the rounding solution by comparing it with the lower bound; experimental results below show that the largest ratio between them in practice is 1.0005.

Algorithm 3: LP relaxation and rounding for RSSG
____

Input: A set of n masters $s_1, \ldots, s_n$ and m slaves
$\quad$ $t_1, \ldots, t_m$ in the rectilinear plane.
Output: A subgraph $G(V, E)$ of the Hanan grid H_g
$\quad$ with edge set E_H, such that G contains a
$\quad$ rectilinear shortest path for each master-slave
$\quad$ pair.
1 $V = \{s_1, \ldots, s_n, t_1, \ldots, t_m\}$;
2 $E = E_H$;
3 $n_e = |E_H|$;
4 Solve LP relaxation of (6) to obtain x_{uv} for every
$\quad$ $(u, v) \in E_H$;
5 $(e_1, e_2, \ldots, e_{n_e})$ = edges in E_H sorted from smallest to
$\quad$ largest according to the value of x_{uv};
6 **for** $e = e_1, e_2, \ldots, e_{n_e}$ **do**
7 $\quad$ **if** $E - e$ *contains a rectilinear shortest path for each*
$\quad\quad$ *master-slave pair* **then**
8 $\quad\quad$ $E = E - e$;
9 $\quad$ **end**
10 **end**
11 **return** $G(V, E)$;

---

5. EXPERIMENTAL RESULTS

5.1 TSV Location

We first show the TSV location results with different communication frequencies and different upper bounds B on the number of available TSVs between adjacent layers. Let L be the number of device layers. We randomly distribute N masters and slaves in each layer. Figure 7 shows two optimal assignments of TSVs with different communication frequencies and the same distribution of masters and slaves. TSV locations are plotted by red lines. Blue dots and green dots denote masters and slaves respectively. For clarity, we simplify each layer as a straight line. Here we assume $B = 1$ and the dynamic programming approach is adopted to obtain optimal solutions. We notice that in Figure 7(b), TSV locations connecting the first three layers are to the left of those in Figure 7(a). At the same time, masters in bottom two layers are farther away from slaves in the top layer in Figure 7(b). These observations reflect the fact that master-slave pairs in the first two layers communicate more frequently.

Table 1 compares the power consumption results for various bounds B. We assume that communication frequencies for all master-slave pairs are the same in this experiment. The power consumption is defined as the total length of shortest paths normalized by the number of master-slave pairs. Notice that when $B = \infty$, every master-slave pair can be reached by a path with shortest rectilinear distance which gives us a lower bound on power consumption. In each cell of

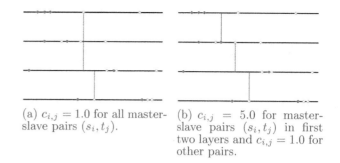

(a) $c_{i,j} = 1.0$ for all master-slave pairs (s_i, t_j).

(b) $c_{i,j} = 5.0$ for master-slave pairs (s_i, t_j) in first two layers and $c_{i,j} = 1.0$ for other pairs.

Figure 7: Single TSV assignments for $(L, N) = (4, 5)$.

the table, the first value represents the power consumption and the second value is the increase of (as a percentage) of power consumption relative to the case $B = \infty$ in the same row. We can see that the lower bound can be almost achieved by only using three TSVs between adjacent layers. Figure 8 gives an example of the TSV assignment results along with their power consumption for bound $B = 1, 2, 3$ on the same setup of masters and slaves.

Table 1: Power consumption with varying B.

(L, N)	$B = \infty$	$B = 1$	$B = 2$	$B = 3$
(3,10)	346.36	456.46 (31.79%)	375.93 (8.54%)	360.86 (4.19%)
(3,20)	292.39	384.34 (31.45%)	336.13 (14.96%)	305.50 (4.49%)
(3,50)	337.27	453.09 (34.34%)	393.07 (16.55%)	355.22 (5.32%)
(4,20)	343.93	476.15 (38.45%)	413.28 (20.16%)	367.62 (6.89%)
(5,20)	346.68	492.00 (41.92%)	431.04 (24.33%)	367.39 (5.97%)
(5,50)	323.76	452.50 (39.77%)	401.23 (23.93%)	346.31 (6.96%)

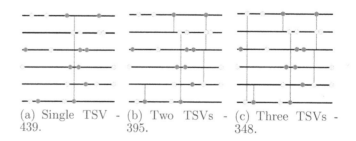

(a) Single TSV - 439.

(b) Two TSVs - 395.

(c) Three TSVs - 348.

Figure 8: Wire length with varying TSV budgets.

5.2 RSSG Construction

Table 2 gives the total wire length improvement of our algorithms when compared with the RSSG result in [16]. The column "LP(obj)" denotes the optimal objective value of the LP relaxation in (6). The column "LP(Round)" denotes the total wire length of the LP relaxation and rounding result. Notice that in most cases, the rounding result and the optimal LP objective are the same, which means that we already obtain the optimal RSSG (since LP relaxation gives a

Table 2: Total wire length results.

(n, m)	RSSG	Greedy	LP(Obj)	LP(Round)	Ratio
(3, 16)	5388	5068	4882	4882	9.39%
(5, 15)	7024	6586	6342	6342	9.71%
(12, 6)	6698	6306	5915	5915	11.69%
(6, 12)	7127	6160	5575	5575	21.78%
(12, 12)	11559	10385	10319	10319	10.73%
(20, 20)	16236	15921	13847	13847	14.71%
(30, 30)	302619	287042	251841	251968	16.78%

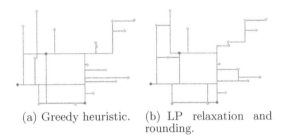

(a) Greedy heuristic.   (b) LP relaxation and rounding.

Figure 9: Comparison of RSSG results.

lower bound of smallest total wire length). In practice during our experiments, the largest gap between "LP(round)" and "LP(obj)", i.e., LP(round)/LP(obj), is 1.0005 which indicates that our rounding method can achieve nearly optimal solutions. The column "Ratio" gives the improvement of our RSSG generated by LP relaxation and rounding relative to RSSG generated by the algorithm in [16]. We can see that our algorithm can reduce the total wire length by up to 21.78%. Figure 5.2 plots examples of our RSSG solutions generated by the greedy heuristic and by LP relaxation and rounding. There are three masters and sixteen slaves, which are represented by blue dots and green dots, respectively. The LP relaxation and rounding method for this example achieves an optimal RSSG with total wire length of 5683, 5.38% better than the greedy solution whose total wire length is 6006. Table 3 shows the time and memory complexity of our LP relaxation and rounding algorithm. We use the Gurobi Optimizer [1] running on a machine with Intel Core i3 2.4GHz CPU and 4GB memory to solve LP relaxation of (6). Based on the memory limit, we can solve RSSG instances with up to 1000 master-slave pairs by LP relaxation and rounding.

Table 3: Scaling of LP relaxation and rounding.

(n, m)	# Variables	# Constraints	Time	Memory
(3, 16)	4326	6209	<1s	<50MB
(5, 15)	8019	11458	<1s	<50MB
(20, 20)	155396	239037	3m43s	234MB
(30, 30)	778358	1175974	4h22m	1.2GB

6. CONCLUSION

In this paper, we construct a framework and algorithms for synthesis of gated buses in 3D ICs. The power consumption of on-chip communication is considered to be the first objective and we achieve it by optimizing TSV locations between device layers. In each device layer, we build a rectilinear shortest-path Steiner graph to connect masters and slaves whose objective is to minimize the total wire length.

Experimental results show that our dynamic programming (resp. local search) algorithm efficiently determines the optimal (resp. nearly optimal) TSV locations, and that the approximation algorithms for generating RSSG reduce the total wire length by up to 22% compared with [16].

7. ACKNOWLEDGMENTS

The authors would like to acknowledge the support of NSF CCF-1017864 and CCF-1116667, the DARPA/MARCO Gigascale Systems Research Center, as well as SRC and DOE support. C.K. Cheng would like to acknowledge the support of National Science Council of Taiwan Grant No.: NSC 100-2811-E-002-034. We thank Prof. Ion Mandoiu of the University of Connecticut for pointers to the MMN problem and related references.

8. REFERENCES

[1] Gurobi optimizer 4.0. http://www.gurobi.com/.
[2] K. Bazargan, R. Kastner and M. Sarrafzadeh. 3-D floorplanning: simulated annealing and greedy placement methods for reconfigurable computing systems. *IEEE Int. Workshop on Rapid System Prototyping*, pp. 38–43, 2002.
[3] M. Benkert, A. Wolff, F. Widmann and T. Shirabe. The minimum Manhattan network problem: approximations and exact solutions. *Computational Geometry: Theory and Applications*, 35(3), pp. 188–208, 2006.
[4] B. Black, D. W. Nelson, C. Webb and N. Samra. 3D processing technology and its impact on IA32 microprocessors. *Int. Conference on Computer Design*, pp. 316–318, 2004.
[5] J. Y. Chen, W. B. Jone, J. S. Wang, H. I. Lu and T. Chen. Segmented bus design for low power system. *IEEE Trans. on VLSI Systems*, 7(1), pp. 25–29, 1999.
[6] L. Cheng, L. Deng and M. Wong. Floorplanning for 3-D VLSI design. *Asia and South Pacific Design Automation Conference*, pp. 405–411, 2005.
[7] F. Y. Chin, Z. Guo and H. Sun. Minimum Manhattan network is NP-complete. *Proc. of the 25th Annual Symposium on Computational Geometry*, pp. 393–402, 2009.
[8] J. Cong, A. B. Kahng and K. S. Leung. Efficient algorithms for the minimum shortest path Steiner arborescence problem with applications to VLSI physical design. *IEEE Trans. on Computer-Aided Design*, 17(1), pp. 24–39, 1998.
[9] K. D. Boese, A. B. Kahng, B. A. McCoy and G. Robins. Near-optimal critical sink routing tree constructions *IEEE Trans. on Computer-Aided Design of Integrated Circuits and Systems*, 14(12), pp. 1417–1436, 1995.
[10] M. Hanan. On Steiner's problem with rectilinear distance. *SIAM Journal on Applied Mathematics*, 14, pp. 255–265, 1966.
[11] C.-T. Hsieh and M. Pedram. Architectural power optimization by bus splitting. *Design, Automation and Test in Europe*, pp. 612–616, 2000.
[12] M. Pathak, Y. Lee, T. Moon and S. Lim. Through-silicon-via management during 3D physical design: when to add and how many? *Int. Conference on Computer-Aided Design*, 2010.
[13] S. K. Rao, P. Sadayappan, F. K. Hwang and P. W. Shor. The rectilinear Steiner arborescence problem. *Algorithmica*, 7, pp. 277–288, 1992.
[14] W. Shi and C. Su. The rectilinear Steiner arborescence problem is NP-complete. *Proc. of the 11th Annual ACM-SIAM Symposium on Discrete Algorithms*, pp. 780–787, 2000.
[15] V. V. Vazirani. *Approximation algorithms*. Springer, 2010.
[16] R. Wang, N. C. Chou, B. Salefski and C. K. Cheng. Low power gated bus synthesis using shortest-path Steiner graph for system-on-chip communications. *Design Automation Conference*, pp. 166–171, 2009.

TSV-Constrained Micro-Channel Infrastructure Design for Cooling Stacked 3D-ICs*

Bing Shi and Ankur Srivastava
University of Maryland, College Park, MD, USA
{bingshi, ankurs}@umd.edu

ABSTRACT

Micro-channel based liquid cooling has significant capability of removing high density heat in 3D-ICs. The conventional micro-channel structures investigated for cooling 3D-ICs use straight channels. However, the presence of TSVs which form obstacles to the micro-channels prevents distribution of straight micro-channels. In this paper, we investigate the methodology of designing TSV-constrained micro-channel infrastructure. Specifically, we decide the locations and geometry of micro-channels with bended structure so that the cooling effectiveness is maximized. Our micro-channel structure could achieve up to 87% pumping power savings compared with the structure using straight micro-channels.

Categories and Subject Descriptors

B.7.2 [**Integrated Circuits**]: Design Aids

General Terms

Design, Algorithm

Keywords

3D-IC, micro-channel, liquid cooling, power, TSV

1. INTRODUCTION AND MOTIVATION

The three-dimensional integrated circuit (3D-IC) consists of two or more layers of active electronic components which are stacked vertically. Despite its significant performance improvement over 2D circuits such as fast on-chip communications, 3D-IC also exhibits thermal issues due to its high power density caused by the stacked architecture.

While the conventional air cooling might be not enough for stacked 3D-ICs, the micro-channel based liquid cooling provides a better option to address this problem. In the 3D-IC, active (silicon) layers consist of functional units such as cores and memories which dissipate power and are stacked vertically. Micro-channel heat sinks are embedded below each silicon layer and the coolant fluid is pumped through the micro-channels, and takes away the heat generated in the silicon layer [4]. Micro-channels have significant capability of cooling high heat density (as much as $700W/cm^2$ [13]) and therefore are very appropriate for cooling 3D-ICs.

Many works have investigated the thermal modeling of 3D-ICs with micro-channels heat sinks [4][12]. Some other

*This work is partly supported by NSF grants CCF 0937865 and CCF 0917057

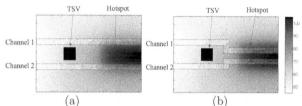

Figure 1: Example of silicon layer thermal profile with TSV and (a) straight, (b) bended micro-channels

works try to find the best dimensional parameters such as channel width and height so as to improve the overall cooling effectiveness of the micro-channel system [5][13].

The conventional micro-channel structures investigated for cooling 3D-ICs use straight channels that spread on the whole chip or in areas that demand high cooling capacity. If the spatial distribution of micro-channels is unconstrained then such an approach results in the best cooling efficiency with the minimum cooling energy (power dissipated to pump the fluid). However 3D-ICs impose significant constraints on how and where the micro-channels could be located due to the presence of TSVs, which allow different layers to communicate. A 3D-IC usually contains thousands of TSVs which are incorporated with clustered or distributed topologies [7]. These TSVs form obstacles to the micro-channels since the channels cannot be placed at the locations of TSVs. Therefore the presence of TSVs prevents distribution of straight micro-channels. This results in the following problems.

1. As illustrated in figure 1(a), micro-channels would fail to reach thermally critical areas thereby resulting in thermal violations and hotspots.

2. To fix the thermal hotspots in areas where microchannels cannot reach, we need to increase the fluid flow rate resulting in a significant increase in cooling energy.

To address this problem, we investigate micro-channel with bends as illustrated in figure 1(b). With bended structure, the micro-channels can reach those TSV-blocked hotspot regions which straight micro-channels cannot reach. This results in better coverage of hotspots and therefore better cooling efficiency and reduced cooling energy. While micro-channels with bends (or serpentine organization of micro-channels) have been investigated in the past [9], our work is the first one to investigate this structure from the context of 3D-ICs and more specifically address the constraint imposed by TSVs towards spreading of straight micro-channels. In this paper, we investigate the methodology of designing TSV-constrained micro-channel infrastructure. Specifically, we decide the locations and geometry of micro-channels with bended structure so that its cooling effectiveness is maximized. Our micro-channel structure could achieve up to 87% pumping power savings compared with the micro-channel structure using straight channels.

The organization of this paper is as follows. In section 2, we introduce the thermal and power model of 3D-IC with micro-channel cooling. We investigate the TSV-constrained micro-channel infrastructure design methodology in section 3. The experimental result is given in section 4.

2. THERMAL AND POWER MODEL FOR 3D-IC WITH MICRO-CHANNELS

2.1 Thermal modeling

The thermal behavior of a 3D-IC with micro-channels could be modeled as an RC network as illustrated in [12]. In the RC network, the resistance corresponds to thermal conduction and capacitance corresponds to heat capacity. The power profile represents current sources in this RC network. In several cases, we are mostly interested in the steady state thermal behavior of the 3D-IC, hence, enabling us to capture the thermal behavior as a pure resistive network [4]. In this case, for a given 3D-IC power profile, the thermal profile could be estimated by solving a system of linear equations of the form $GT = Q$ where G is the thermal conductivity matrix and Q is the power profile. The thermal conductivity matrix G depends on many factors including the material properties, location of channels and TSVs, fluid flow rate etc. Interested readers are referred to [4][11][13] for details.

2.2 Micro-channel power consumption

The power used by micro-channels for performing chip cooling comes from the work done by the fluid pump to push the coolant fluid into micro-channels. The pumping power Q_{pump} is decided by the pressure drop and volumetric flow rate of micro-channels: $Q_{pump} = \sum_{n=1}^{N} f_n \Delta P_n$, where N is the total number of channels, ΔP_n and f_n are the pressure drop and fluid flow rate of the n-th micro-channel. In this paper, we use single-phase laminar liquid flow as the working fluid. Pressure drop and fluid flow rate are interdependent and also related to other micro-channel parameters such as length and width. The pressure drop in a **straight** micro-channel is decided by:

$$\Delta P = \frac{2\gamma\mu L v}{D_h^2} \tag{1}$$

Here L is the length of micro-channel, D_h is hydraulic diameter, v is fluid velocity, μ is fluid viscosity and γ is determined by the micro-channel dimension (given in [5]).

Usually fluid pumps are designed to work such that all the micro-channels experience the same pressure drop ΔP. For a given ΔP that the pump delivers across all the channels, fluid velocity v could be estimated by equation 1. The fluid flow rate $f = v w_a w_b$ could be estimated as well (w_a, w_b are micro-channel width and height). Also, flow rate could be controlled by changing the pressure drop. **Higher pressure drop results in higher flow rate and better cooling.**

2.3 Modeling Micro-channels with bends

Consider the channel structure shown in figure 2. The existence of a bend causes a change in the flow properties which impact the cooling effectiveness and pressure drop. An otherwise fully developed laminar flow in the straight part of the channel, when comes across a 90° bend becomes turbulent/developing around the corner and settles down after traveling some distance downstream into laminar fully developed again (see figure 2). So a channel with bends has three distinct regions, 1) fully developed laminar flow region 2) the bend corner 3) the developing/turbulent region after the bend [3][9]. The length of flow developing region is [10]:

$$L_d = (0.06 + 0.07\beta - 0.04\beta^2) Re D_h \tag{2}$$

where $\beta = w_b/w_a$ is the channel aspect ratio. Re is the Raynolds number defined by $Re = \rho v D_h/\mu$, where ρ, μ and v are the fluid density, viscosity and velocity.

The rectangular bend impacts the pressure drop. Due to the presence of bends, the pressure drop in the channel is greater than an equivalent straight channel with exactly the same dimensions. The total pressure drop in a channel with bends is the sum of the pressure drop in the three regions described above (which finally depend on how many bends the channel has). Assuming L is the total channel length, and m is the bend count. Hence $m \cdot L_d$ is the total length that has developing/turbulent flow and $m \cdot w_a$ is the total length attributed to corners (see figure 2). Hence the effective channel length attributed to fully developed laminar

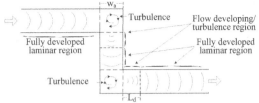

Figure 2: Micro-channel with bends

flow is $L - m \cdot L_d - m \cdot w_a$. The pressure drop in the channel is the sum of the pressure drop in each of these regions.

Pressure drop in fully developed laminar region: The total pressure drop in fully developed laminar region is [5]:

$$\Delta P_f = \frac{2\gamma\mu(L - m \cdot L_d - m \cdot w_a)v}{D_h^2} = \frac{2\gamma\mu L_f v}{D_h^2} \tag{3}$$

Here $L_f = L - m \cdot L_d - m \cdot w_a$ is the total length of the fully developed laminar region which is explained above, the other parameters are the same as in equation 1.

Pressure drop in flow developing region: The pressure drop in each flow developing region is: $\Delta p_d = \frac{2\mu v}{D_h^2} \int_0^{L_d} \gamma(z) dz$ [6]. Here $\gamma(z)$ is given by $\gamma(z) = 3.44\sqrt{(ReD_h)/z}$, where z is the distance from the entrance of developing region in the flow direction. Assuming there are a total of m corners in a given micro-channel, so there are m developing regions with the same length L_d in this channel. By putting the expression of $\gamma(z)$ and L_d into the equation of Δp_d and solving the integration, we can get the total pressure drop of the developing region in this micro-channel:

$$\Delta P_d = m\Delta p_d = mK_d\rho v^2 \tag{4}$$

where $K_d = 13.76(0.06 + 0.07\beta - 0.04\beta^2)^{\frac{1}{2}}$ is a constant associated with the aspect ratio β. Please refer to [3][6] for details.

Pressure drop in corner region: The total pressure drop at all the 90° bends in a micro-channel is decided by:

$$\Delta P_{90°} = m\Delta p_{90°} = m\frac{\rho}{2}K_{90}v^2 \tag{5}$$

where m is the number of corners in the channel, $\Delta p_{90°}$ is the pressure drop at each bend corner and K_{90} is the pressure loss coefficient for 90° bend whose value can be found in [3].

Total pumping power: The total pressure drop in a micro-channel with bends is the sum of the pressure drop in the three regions discussed above:

$$\Delta P = \Delta P_d + \Delta P_f + \Delta P_{90°} = \frac{2\gamma\mu L_f}{D_h^2}v + m(K_d + \frac{K_{90}}{2})\rho v^2 \tag{6}$$

From equations 6, the total pressure drop of a micro-channel is a quadratic function of the fluid velocity v. For a given pressure difference applied on a micro-channel, we can calculate the associated fluid velocity by solving equation 6. With the fluid velocity, we can then estimate the fluid flow rate f, and thus estimate the thermal resistance and pumping power for this channel. Hence the pumping power as well as cooling effectiveness of micro-channels with bends is a function of 1) number of bends, 2) location of channels 3) pressure drop across the channel.

Comparing equations 1 and 6, due to the presence of bends, if the same pressure drop is applied on a straight and a bended micro-channel of the same length, the bended channel will have lower fluid velocity, which leads to a lower cooling capability. Therefore, to provide sufficient cooling, we will need to increase the overall pressure drop that the pump delivers, which results in increase of pumping power. But bends allow for better coverage in the presence of TSVs.

3. MICRO-CHANNEL INFRASTRUCTURE DESIGN: ALGORITHMS

Designing 3D-IC micro-channel infrastructure is a very complex problem. For example there are exponentially many ways to incorporate micro-channels with bends whose impact on the silicon temperature requires us to solve complex system of thermal equations. The specific problem formulation is as follows.

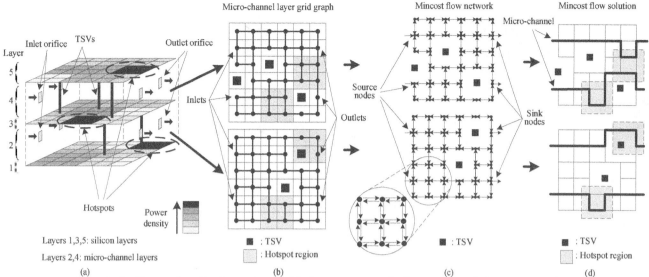

Figure 3: Example of micro-channel infrastructure design using minimum cost flow

$$ min \quad Q_{pump}(g_{i,j}^l, \Delta P) $$

$$ s.t. \sum_{\forall j \in I(i)} g_{i,j}^l = 1, \forall \text{grid } i \in \{CI, CO\}, \forall \text{channel layer } l $$

$$ \sum_{\forall j \in I(i)} g_{i,j}^l = k \in \{0,2\}, \forall \text{grid } i \notin \{CI, CO, TSV\}, \forall \text{channel layer } l $$

$$ g_{i,j}^l = 0, \text{if grid } i \text{ or } j \in \{TSV\}, \forall \text{channel layer } l $$

$$ T_i^l(g_{i,j}^l, \Delta P) \le T_{max}, \forall \text{grid } i, \forall \text{channel layer } l $$

$$ g_{i,j}^l \in \{0,1\}, \forall \text{grids } i, j, \forall \text{channel layer } l $$

$$ g_{i,j}^l = g_{j,i}^l, \forall \text{grids } i, j, \forall \text{channel layer } l $$

$$ (7) $$

Figure 3 represents the problem formulation graphically. Given a set of stacked silicon layers, some of the intermediate layers between silicon layers would have micro-channels (as shown in figure 3(a), two intermediate layers comprise of micro-channels). The locations of input and output orifices for the micro-channels are assumed known. We would like to find micro-channel routes from one side to the other such that the routes do not intersect, avoid TSVs and provide sufficient cooling at minimum pumping energy.

We impose a graph on each micro-channel layer as indicated in figure 3(b). In the graph, each grid is represented by a node, and the edges define the immediate neighbors of a node. The micro-channel routing would be performed on this graph. If there is a TSV located on a grid, then its corresponding neighborhood edges are removed since micro-channels cannot be routed through TSVs. Let $g_{i,j}^l = 1$ represents the fact that there is a channel connecting grids i and j in the l-th micro-channel layer of the 3D-IC (so i and j must be neighboring nodes in the grid graph and $g_{i,j}^l = g_{j,i}^l$). Neither i nor j should have a TSV (because TSVs will not allow channels to go through them). In the first constraint, $\{CI, CO\}$ represents the set of input and output orifice nodes, $I(i)$ represents the set of i's neighboring nodes. So the first constraint imposes that the input and output orifice nodes must have a neighboring grid they are connected to so that their incoming/outgoing fluid can be pushed into/out-of the micro-channel layer. The next constraint imposes that either a channel goes through a grid (and therefore $\sum_{\forall j \in I(i)} g_{i,j}^l = 2$) or it does not (and therefore $\sum_{\forall j \in I(i)} g_{i,j}^l = 0$). In the third constraint, $\{TSV\}$ represents the set of grids containing TSVs, so micro-channels cannot be routed through these nodes. The following constraint imposes that the temperature is within acceptable limits and the objective tries to minimize the pumping power.

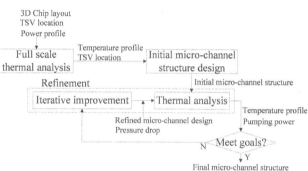

Figure 4: Micro-channel infrastructure design flow

3.1 Overall micro-channel design flow

This is a very complex problem since 1) the variables need to be discrete, and 2) the thermal and pumping power models are highly nonlinear. In this paper we investigate such a methodology as illustrated in figure 4. Our methodology follows a sequence of logical steps. First the severity of the thermal problem and the need for having micro-channels is evaluated by performing a full scale thermal analysis. Based on the severity of the thermal problem (location, intensity of hotspots) an initial micro-channel design is developed. This design is further improved for reducing the cooling power footprint and improving the thermal effectiveness using iterative methods. Now we go into the details of these individual steps.

3.2 Mincost flow based micro-channel design

The full scale 3D thermal analysis would identify locations of hotspots in different layers which cannot be removed by conventional package/air cooling based approaches. These are the areas which require sufficient proximity to the micro-channels. Since solving the formulation in equation 7 is intractable, we use simple models to come up with a *sufficiently good* initial micro-channel infrastructure which is iteratively improved subsequently. In order to develop this initial solution we use the minimum cost flow formulation.

Initialization of the minimum cost flow network: Consider the 3D-IC and the corresponding grid graph of each micro-channel layer as illustrated in figure 3(a)(b). For each micro-channel layer, we instantiate a minimum cost flow problem as follows (see figure 3(c) for illustration). The nodes corresponding to the input/output orifices for the given micro-channel layer are assigned a supply/demand of one flow unit. All nodes in the grid graph have a capacity one. The edges have unlimited capacity and are bi-directional (can take fluid flow in either direction). As indicated earlier the edges between two neighboring nodes exist only if neither of the nodes has a TSV. This enforces the routing constraint imposed by TSVs. Figure 3(c) indicates the flow network for the two micro-channel layers.

Each node has a cost whose assignment would be discussed subsequently. We would like to send flow from inlet nodes to outlet nodes such that the capacity constraints are not violated and the cost is minimum. Assigning the node capacity to be 1 would ensure that all the flow from inlet to outlet follows simple paths (non-intersecting and non-cyclic). A minimum cost flow formulation with a well defined node capacity could be solved using very similar methods as a formulation with edge capacity alone [8]. It is noteworthy that because there is an edge between each pair of neighboring nodes, the flow path could take several bends if necessary.

Cost assignment: The cost assignment should be such that the minimum cost flow formulation develops an initial infrastructure that distributes the micro-channels with higher density in areas that demand more cooling. The chip scale thermal analysis would identify locations of grids in the silicon layers that are in dire need of cooling (see figure 3(a)). A silicon layer would be cooled by the micro-channels both above and below (unless the silicon layer is at the very top or very bottom of the stack). For example, the middle silicon layer in figure 3(a) could be cooled by two micro-channel layers unlike the top and bottom silicon layers.

As illustrated in figure 3(b), each micro-channel layer is represented as a grid graph. The amount of cooling required at a certain node in this graph is a function of how hot the top and bottom grids in the silicon layers are. It also depends on how we chose to distribute the cooling demand at a certain location in the silicon layer between the micro-channel layers just above and just below. Let us suppose a certain location in the silicon layer has temperature $T \geq T_{max}$ and requires cooling (estimated by full scale thermal analysis). Let uT (with $0 \leq u \leq 1$) represent the fraction of this cooling demand assigned to the micro-channel grid right above and $(1 - u)T$ represent the cooling demand from the micro-channel grid just below. If u is set too low then most of the cooling will be done by the channel layer below and vice versa for large u. Let u_i^l be the heat load partitioning factor of grid i in silicon layer l, it is assigned as follows.

Case 1: If l is the topmost (bottommost) layer, then $u_i^l = 0(u_i^l = 1)$ so that all the cooling demand goes to the micro-channel layer right below (above) l, which is layer $l-1$ $(l+1)$.

Case 2: If l is neither top nor bottom layer, $0 \leq u_i^l \leq 1$, implying that the heat generated in grid i of silicon layer l needs to be distributed in the two micro-channels layers right above and below. If the channel layers above and below (layers $l + 1$ and $l - 1$) have the same number of TSVs then $u_i^l = 1/2$, else it is scaled linearly such that more cooling demand is assigned to the micro-channel layer with lesser TSVs.

Given the partitioning factor u_i^l, the cost is assigned as follows. (See figure 5 for an illustration.) Let $cost(i,l)$ denote the cost for node i in micro-channel layer l (hence layers $l - 1$ and $l + 1$ correspond to silicon layers just below and above the micro-channel layer l), three cases are considered depending on whether there is hotspot below and above this node in the silicon layers $l - 1$ and $l + 1$.

Case 1: Hotspots on both sides. When the grid i in both silicon layers $l - 1$ and $l + 1$ are in hotspot regions ($T_i^{l-1} > T_{max}$ and $T_i^{l+1} > T_{max}$), the micro-channel should provide cooling to both sides (above and below), so the cost is:

$$cost(i,l) = -[(1 - u_i^{l+1})T_i^{l+1} + u_i^{l-1}T_i^{l-1}] \quad (8)$$

Here the first component inside the square bracket indicates the cooling demand from the silicon grid above and the second component corresponds to the cooling demand from the silicon grid just below. Higher demand leads to lower cost since we would like micro-channels to pass through high cooling demand regions. See figure 5 for an illustration.

Case 2: Hotspot in one side. When the silicon grid i on

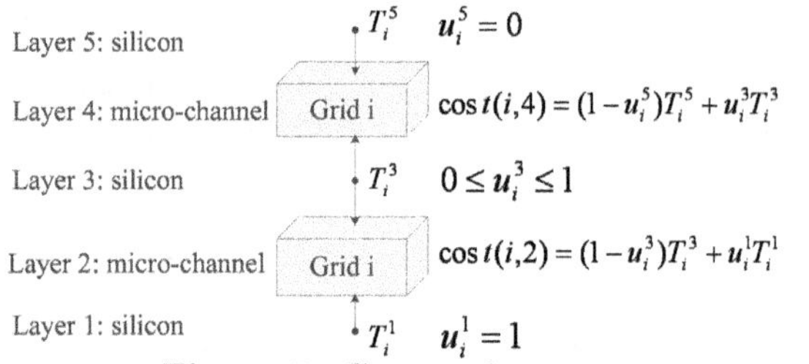

Figure 5: Cost assignment

only one side ($l - 1$ or $l + 1$) is in hotspot region (but not both), the cost is assigned as

$$cost(i,l) = \begin{cases} -(1 - u_i^{l+1})T_i^{l+1}, & \text{if } T_i^{l+1} \geq T_{max} \\ -u_i^{l-1}T_i^{l-1}, & \text{if } T_i^{l-1} \geq T_{max} \end{cases} \quad (9)$$

Case 3: No hotspot in either side. When there is no hotspot in either side, then the node cost is assigned to a small positive value $cost(i,l) = \epsilon > 0$.

The minimum cost flow formulation would therefore route flows such that maximum number of high cooling demand grids are touched by the channels. The non-hotspot regions are assigned a small positive cost. This would enable the minimum cost flow formulation to avoid areas that do not demand high cooling.

3.3 Micro-channel refinement

The primary objective of the minimum cost flow formulation is to come up with an initial micro-channel design that carries cooling in sufficient proximity of hot areas. This is not enough to guarantee effective cooling. For example, some channels have several bends and/or may be routed over disproportionately large number of hotspots. Both of these situations cause a degradation in the overall cooling quality. In this section we present approaches for iteratively refining the design for improved cooling effectiveness. The micro-channel infrastructure refinement process works as illustrated in figure 4.

3.3.1 Temperature and Pumping Power Analysis

The impact of micro-channels on the 3D-IC thermal profile is a function of how the micro-channels are routed and also how much fluid flow they carry. The initial design generated using minimum cost flow technique does not prescribe the pressure drop and the fluid flow rate that the channels need to work at. Hence given the micro-channel design, we then need to estimate the *smallest* pressure drop that the pump needs to work at such that thermal constraints are satisfied. Given the micro-channel design, the smallest pressure drop value results in the smallest pumping energy. As indicated earlier, we assume that all channels are subjected to the same pressure drop by the pump, hence the minimum pressure drop can be determined by linearly increasing ΔP and calculating the thermal profile for each value until the thermal constraints are met. For a given pressure drop across the pump and a given micro-channel design, equation 6 could be used to determine the velocity (fluid flow rate) in each channel. Note that because each channel has different number of bends and total length, the flow rate would be different too. Based on the flow rate information which is computed for a given pressure drop, the associated thermal conductance matrix G could be computed. This information could be used to estimate the thermal profile of the 3D-IC for a given pressure drop. After finding the minimum required ΔP, we could calculate the required pumping power. This technique is highlighted in Algorithm 1.

3.3.2 Iterative micro-channel optimization

The objective of minimum cost flow formulation did not capture cooling energy and/or number of bends in the channels. Figure 6 illustrates typical situations that can occur. In figure 6, the two micro-channels have significantly different cooling demands (figure 6(a)) and number of bends (figure 6(b)). Such imbalance (in cooling demand and bend

Algorithm 1 Finding the minimum required pumping power

1. $\Delta P = \Delta P_{min}$, and repeat steps 2-6:
2.   Calculate the fluid velocity using equations 6;
3.   Calculate thermal conductance matrix G;
4.   Estimate temperature profile;
5.   If thermal violation occurs, $\Delta P = \Delta P + \delta P$;
6.   Else break;
7. Calculate pumping power.

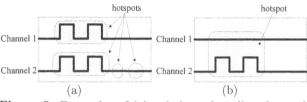

Figure 6: Examples of (a) unbalanced cooling demand, (b) different number of bends

count) leads to increase in the required pressure drop and thereby increasing the pumping energy. The basic idea is that all the channels should have similar levels of heat load, length and number of bends. Hence if a channel has too many bends or goes through many hotspots while others are shorter, then other channels could be made longer thereby more uniformly distributing the heat load and also reducing the number of bends in the most *critical* micro-channel.

Based on these considerations, we try to refine the initial design by 1) balancing the heat loads among micro-channels and 2) reducing unnecessary bends.

Micro-channel heat load balancing:

Starting from the initial design we identify the micro-channels which have disproportionately high heat removal load and spread their heat load into neighboring channels.

Algorithm 2 highlights the iterative pairwise micro-channel cooling load balance process. In the first iteration of pairwise micro-channel cooling workload balance, we start from the channel with the highest cooling workload. Here the cooling workload is measured by the total heat absorbed by the micro-channel, which could be calculated using $q = (T_{out} - T_{in})/R_{io}$. Here T_{in} is the fluid supply temperature at micro-channel inlet, and T_{out} is the fluid temperature at micro-channel outlet, R_{io} is the total thermal resistance between the fluid inlet and outlet of that specific channel. Given the pressure drop, power profile of the 3D-IC and the location and dimensions of the micro-channels, these parameters could be easily calculated (see discussion in section 2 and reference [11]). Assuming i is the channel with the highest cooling workload, we then pick one of i's neighbors (either left or right) with lower cooling workload, say channel k, and balance the workload between channels i and k.

To balance the workload of channels i and k, we firstly partition the hotspot regions covered by channels i and k. Basically, we would like the resultant two parts have similar total amount of heat load (cooling demand). As indicated earlier, the cost of a node i at the l-th micro-channel layer (defined in section 3.2) signifies the degree of cooling desired there. Therefore the total cooling needed in the region covered by channels i and k is simply the sum total of the cost in all the associated grids. We would like each channel to be assigned half of this total cooling load in that region. Hence we partition this region into two subregions with the same total cooling load (that is, same total grid cost). To find the exact route of the micro-channels we can remove the edges connecting the two subregions and solve the minimum cost flow formulation once again. This would ensure that channels i and k do not encroach on each others regions.

The min-cost flow gives a refined micro-channel design. We then redo the temperature analysis and find the required pumping power for the new design using algorithm 1.

In the next iteration of optimization, we find the currently highest workload micro-channel in the new design and balance workload for this channel using the new graph updated in the previous iteration. We repeat this process iteratively

Algorithm 2 Pairwise micro-channel cooling load balance

Repeat:
1. Pick the micro-channel with highest cooling load i;
2. Pick a micro-channel k from i's neighbor with smaller cooling load, that is, $k = argmin_{k \in \{i-1, i+1\}}(load(k))$;
3. Equally divide the hotspot region covered by channels i and k, and assign one of the region to channel i, the other to channel k;
4. Remove some edges on the boundary between these two regions from the grid graph;
5. Resolve the minimum cost flow based on new graph;
6. Temperature analysis and calculating minimum required pumping power using algorithm 1;
7. If no further pumping power saving could be achieved, stop.

until no further pumping power saving could be achieved.

Bend Elimination

As shown in section 2.3, the corners/bends in the micro-channel will introduce considerable pressure drop, which increases the pumping power. Bends in micro-channels allow us to reach areas which cannot be directly connected by straight channels due to the presence of TSV obstacles. But unnecessary bends which have been incorporated due to the heuristic nature of our algorithm provide little benefit while impacting the cooling quality. As a final refinement step, we identify all unnecessary bends and replace them with equivalent straight channels or patterns with lesser corners. Note that removing corners in the hotspot region might lead to reduction in the micro-channel cooling performance since it reduces the level of coverage. Hence we only remove those corners in the non-hotspot regions which can easily be identified by the thermal analysis.

4. EXPERIMENTAL RESULTS

We test our method on a two-tier stacked 3D-IC, and each tier contains a micro-channel layer. For the sake of experiment, each tier contains a four-core CPU. We assume a typical floor plan for each tier. To obtain the power data for each core, we simulated a high performance out-of-order processor with SPEC 2000 CPU benchmarks [2] (using Wattch [1]). For each benchmark, we simulated a representative 250M instructions and sampled the chip power dissipation values using uniform time intervals. We simulated 20 such benchmarks and the resultant power data gave us the power profile for each core of the CPU on each tier. To generate the power profile of a four-core CPU, we randomly choose 4 of these profiles and arrange them according to the typical floor plan, so each of the resultant power profile represents the power profile on one tier. In the experiment on this two-tier 3D chip, we choose 2 of these power profiles and each of them represents the power profile on one tier. That is, the benchmark we use in this experiment is a combination of two of these power profiles, each power profile corresponds to one tier. The area of each chip stack is $1.2 \times 1.2cm^2$, and the grid size is $200 \times 200\mu m^2$ (so 60×60 grids in each layer). The channel dimensions are $w_a = 100\mu m$, $w_b = 400\mu m$. The maximum temperature constraint T_{max} is 85°C. The maximum available pressure drop is $500kPa$.

4.1 Comparison with straight channels

We evaluate our method by comparing our micro-channel design with the micro-channel structure with straight channels. In our design, we use 20 micro-channels. For each benchmark, we tested on different number of TSVs and these TSVs are randomly distributed across each layer of the chip. We find the minimum pressure drop required to cool the 3D chip below thermal constraint for our design and then calculate the associated pumping power Q_{pump}. While in the straight channel design we place as many channels as possible to maximize its cooling efficiency. Note that in the micro-channel system with only straight channels, the number and location of channels are constrained by the TSVs. We also find the minimum required pressure drop and associated pumping power for the straight channel design. The comparison is shown in table 1. Note that for the cases when the cooling demand is so large that even using the

Table 1: Comparison of our design with straight channel structure (temperature: °C, power: W)

Percentage of grids containing TSV	Bench mark	P_{chip}	Straight channel		Our design						Q_{pump} saving
					Initial design		With pairwise balance		With bend elimination		
			T_{peak}	Q_{pump}	T_{peak}	Q_{pump}	T_{peak}	Q_{pump}	T_{peak}	Q_{pump}	
0.3%	1	149.86	84.86	6.02	84.75	5.25	84.78	4.83	84.62	4.52	25%
	2	222.76	84.82	15.16	84.69	13.65	84.66	13.42	84.70	12.65	17%
	3	298.88	84.92	24.16	84.77	23.85	84.75	22.28	84.79	21.08	13%
0.7%	1	149.86	84.62	10.40	84.72	5.26	84.89	4.92	84.62	4.65	55%
	2	222.76	84.44	66.76	84.90	14.08	84.85	13.69	84.96	12.92	81%
	3	298.88	violation	NA	84.97	24.11	84.95	23.68	84.91	22.59	NA
1.4%	1	149.86	84.81	43.96	84.74	6.55	84.98	6.23	84.98	5.94	87%
	2	222.76	violation	NA	84.85	14.28	84.92	13.89	84.96	13.40	NA
	3	298.88	violation	NA	84.92	39.23	84.94	38.78	84.96	38.11	NA
2.1%	1	149.86	violation	NA	84.96	7.12	84.94	6.98	84.93	6.65	NA
	2	222.76	violation	NA	84.95	18.89	84.98	18.52	85.00	18.10	NA
	3	298.88	violation	NA	84.88	46.64	84.86	46.44	84.89	46.03	NA

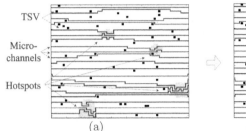 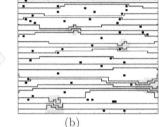 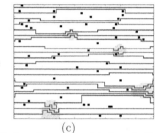

(a)   (b)   (c)

Figure 7: Resulting micro-channel structure, (a) initial design, (b) with pairwise balancing, (c) with bend elimination

maximum pressure drop will lead to temperature violation, we label "violation" in these cases.

In the table, Q_{chip} is the total power dissipation for each benchmark, and Q_{pump} is the pumping power, T_{peak} is the maximum temperature achieved under the given pumping power. In the results of our design, we show the results of both the initial design (after solving minimum cost flow), and also the design after pairwise balancing and bend elimination refinement.

As we can see from the table, our algorithm could achieve higher cooling effectiveness compared with the straight channel (with pumping power savings from 13% to 87%). This is because, the presence of TSV constrained the count and locations of straight channels. Especially when the number of TSVs increases, the available locations for straight micro-channels reduce dramatically. Therefore, to provide sufficient cooling to the hotspots blocked by TSVs, the flow rate should increase significantly. Note that, even though the percentage of grids containing TSVs are small (no more than around 2.1%), its impact on straight micro-channel design can be significant since the whole row will become unavailable even when there is only one TSV in this row. However, the presence of TSVs will have much less impact on our design. So we could use a smaller pressure drop (flow rates) to provide sufficient cooling. Moreover, as the number of TSVs increases, the improvement of our design over straight channel becomes more significant.

Figure 7 shows the micro-channel infrastructures for benchmark 1 generated by our algorithm. TVSs, which are represented as black squares in the figure, are placed in 1.4% of the grids. The gray area are hotspot regions. Figure 7(a) shows the initial design generated by minimum cost flow, and figures 7(b) and 7(c) are the refined design after pairwise balancing and bend elimination.

5. CONCLUSION

In this paper, we investigated the methodology of designing TSV-constrained micro-channel infrastructure. We decide the locations and geometry of micro-channels with bends so that sufficient cooling could be provided using minimum pumping power. Our design could achieve up to 87% pumping power saving compared with the micro-channel structure using straight channels.

6. REFERENCES

[1] D. Brooks, V. Tiwari, and M. Martonosi. Wattch: A framework for architectural-level power analysis and optimizations. In *27th International Symposium on Computer Architecture*, 2000.

[2] G. Hamerly, E. Perelman, J. Lau, and B. Calder. Simpoint 3.0: Faster and more flexible program analysis. In *Journal of Instruction Level Parallelism*, 2005.

[3] S. Kandlikar, S. Garimella, and et al. Heat transfer and fluid flow in minichannels and microchannels. *Elsevier*, 2005.

[4] Y. J. Kim, Y. K. Joshi, and et al. Thermal characterization of interlayer microfluidic cooling of three dimensional integrated circuits with nonuniform heat flux. *ASME Trans. Journel of Heat Transfer*, 2010.

[5] R. W. Knight, D. J. Hall, and et al. Heat sink optimization with application to microchannels. *IEEE Trans. on Components, Hybrids, and Manufacturing Technology*, 1992.

[6] Y. S. Muzychka and M. M. Yovanovich. Modelling friction factors in non-circular ducts for developing laminar flow. In *2nd AIAA Theoretical Fluid Mechanics Meeting*, 1998.

[7] M. Pathak, Y.-J. Lee, T. Moon, and S. K. Lim. Through-silicon-via management during 3d physical design: When to add and how many? In *IEEE/ACM International Conference on Computer-Aided Design*, 2010.

[8] K. A. Ravindra, L. M. Thomas, and J. B. Orlin. Network flows: Theory, algorithms and applications. *Prentice Hall*, 1993.

[9] S. Senn and D. Poulikakos. Laminar mixing, heat transfer and pressure drop in tree-like microchannel nets and their application for thermal management in polymer electrolyte fuel cells. *Journal of Power Sources, Vol. 130*, 2004.

[10] R. K. Shah and A. L. London. Laminar flow forced convection in ducts: A source book for compact heat exchanger analytical data. *Academic*, 1978.

[11] B. Shi, A. Srivastava, and P. Wang. Non-uniform micro-channel design for stacked 3d-ics. In *Design Automation Conference (DAC'11)*, 2011.

[12] A. Sridhar, A. Vincenzi, M. Ruggiero, T. Brunschwiler, and D. Atienza. 3D-ICE: Fast compact transient thermal modeling for 3D ICs with inter-tier liquid cooling. In *IEEE/ACM Intl. Conf. on Computer Aided Design*, 2010.

[13] D. B. Tuckerman and R. F. W. Pease. High-performance heat sinking for VLSI. *IEEE Electron Device Letters*, 1981.

Construction of Minimal Functional Skew Clock Trees

Venky Ramachandran
Mentor Graphics
Fremont, CA, USA
Venky_Ramachandran@mentor.com

Abstract

Power is the number one implementation challenge for many consumer SoCs, especially in the telecommunications and mobile space. A large portion of that, 30% or more, can be directly attributed to the clock tree. 10-20% additional power can additionally be attributed to increased logic area on account of skew in the clock tree. Increasingly, we find designers trying to reduce power both by manual specifications of multiple voltage and power domains, block and core level clock gating as well as more automated techniques like leaf-level combinational and sequential clock gating. However, this tends to increase the problem complexity for the clock tree synthesis tool. Furthermore, these numbers are only likely to get worse with larger designs and more complex clocking scenarios. Already it is not uncommon to see over 200 clocks defined for a modern SoC; several of them asynchronous to each other in the timing sense; but which nevertheless overlap physically. This creates a nightmare problem instance for traditional clock tree synthesis tools attempting to balance path delays to all sinks. Further, due to DFT and other design considerations, the clock tree network is increasingly resembling more a regular logic network of gates rather than a simple buffer tree with the occasional clock gates.

 The premise of this paper, backed by experimental data on real designs, is that the clock tree synthesis problem for complex SoCs needs to be done in a true timing driven fashion where the delay requirements are automatically derived from the timing constraints themselves. One key observation is that global skew constraints is not only unnecessary, but could indeed be harmful to the goals of building reduced power but timing friendly clock trees. Power reduction is achieved through lower buffer counts brought upon by the relaxed balancing requirements. By considering only the true timing paths in the design, the real impact of functional skew in the clock tree can be more appropriately determined and considered during the tree building process itself. As such the CTS solution could borrow from some of the timing driven techniques of the physical synthesis world that have been developed for reducing total negative slack. Moreover, this needs to happen in a tight incremental loop, where the timing impact of refining or rebuilding some portion of the clock tree can be immediately analyzed and acted upon. Care should be exercised in this process not to over-buffer to make some marginal timing improvement. For instance some initial clustering can often be done to group related registers, and the clock to all such registers can be shifted together.

Categories & Subject Descriptors: B.7.2 Placement and routing

General Terms: Clock tree synthesis, Algorithms, Design.

Bio

The speaker is a Place and Route Architect at Mentor Graphics and Technical Leader of Olympus-SoC advanced clock tree synthesis team. He has over 15 years experience in EDA industry working on problems in physical synthesis and clock tree optimization. He has coauthored novel algorithms for automatic generation of clock gating functions including those involving sequential reasoning. More recently, his focus is on leveraging the timing optimization capabilities of physical synthesis for solving complex clock tree synthesis problems. He received a Ph.D in Computer Science and Electrical Engineering from the University of Michigan.

Novel Pulsed-Latch Replacement
Based on Time Borrowing and Spiral Clustering

Chih-Long Chang
Dept. of Electronics Eng. &
Inst. Of Electronics
National Chiao Tung University

paralost.ee96@nctu.edu.tw

Iris Hui-Ru Jiang
Dept. of Electronics Eng. &
Inst. Of Electronics
National Chiao Tung University

huiru.jiang@gmail.com

Yu-Ming Yang
Dept. of Electronics Eng. &
Inst. Of Electronics
National Chiao Tung University

yuming.yyang@gmail.com

Evan Yu-Wen Tsai
Faraday Technology Corp.
Hsinchu, Taiwan

ywtsay@faraday-tech.com

Aki Sheng-Hua Chen
Faraday Technology Corp.
Hsinchu, Taiwan

akiyama@faraday-tech.com

ABSTRACT

Flip-flops are the most common form of sequencing elements; however, they have a significantly higher sequencing overhead than latches in terms of delay, power, and area. Hence, pulsed-latches are promising to reduce power for high performance circuits. In this paper, we propose a novel pulsed-latch replacement approach to save power and satisfy timing constraints. We fully utilize the intrinsic time borrowing property of pulsed-latches and develop a spiral clustering method with clock gating consideration. In addition, spiral clustering works well for both rectangular and rectilinear shaped layouts; the latter are popular in modern IC design. Experimental results show that our approach can generate very power efficient results.

Categories and Subject Descriptors

B.7.2 [**Integrated Circuits**]: Design Aids – *Placement and Routing*

General Terms

Algorithms, Performance, Design, Theory.

Keywords

Pulsed-latch, pulsed-register, time borrowing, clock power.

1. INTRODUCTION

Clock power has become the major contributor of total chip power consumption, and a large portion of it is consumed by sequencing elements [1][2]. Due to deep pipeline and high clock frequency, how to minimize the sequencing overhead is a crucial issue for high performance circuits.

Flip-flops are the most common form of sequencing elements because of its simple timing model. Since a conventional flip-flop contains two cascaded latches (master and slave) triggered by a

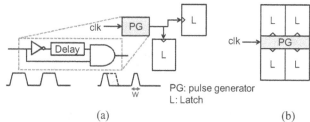

Figure 1. (a) Generic pulsed-latch: a pulse generator (PG) drives a group of latches (L). (b) Multi-bit pulsed-latch.

clock signal, it has a significantly higher sequencing overhead than a latch in terms of delay, power, and area. A *pulsed-latch* is a latch synchronized by a pulse clock as shown in Figure 1(a) [3].[1] Since the pulse is short, a pulsed-latch can be approximated as a fast, low-power, and small flip-flop taking advantages of both flip-flops and latches. Therefore, migration from a flip-flop-based design to a pulsed-latch-based counterpart can effectively reduce the sequencing overhead.

Recent research endeavors have been devoted to EDA solutions for pulsed-latch-based circuits [4–13]. Most of these works adopt the generic pulsed-latch structure illustrated in Figure 1(a) and flip-flop-like timing analysis. To minimize clock period, Lee *et al.* in [5][6] and Paik *et al.* in [7] apply aggressive time borrowing techniques, e.g., clock skew scheduling, pulse width allocation, and retiming. These skews, retimed values and widths are derived based on a logic synthesized netlist instead of placement. Based on predefined pulse widths, Chuang *et al.* propose a pulsed-latch-aware analytical placer in [8]. They control pulse distortion by limiting the number of latches and total wirelength of the pulse clock net driven by each pulse generator, but they do not consider timing slack constraints. On the other hand, to reduce power consumption, Shibatani and Li demonstrate a post-placement methodology for pulsed-latch replacement and pulse generator insertion in [9]. Kim *et al.* discuss how to generate clock gating functions of pulse generators in [10]. Without clock gating consideration, Lin *et al.* try to minimize pulse generator usage during clock tree synthesis in [11]. By integrating [8] and [11], Chuang *et al.* perform placement and clock-network co-synthesis

---

[1] In the generic form of pulsed-latches, the pulse generator generates the pulse clock by ANDing the clock signal with a delayed inverted clock. The delay along the inverted clock determines the pulse width.

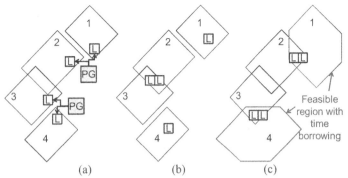

Figure 2. Pulsed-latch replacement. (a) Generic pulsed latches without time borrowing may incur pulse distortion. (b) MBPL without time borrowing. (c) MBPL with time borrowing.

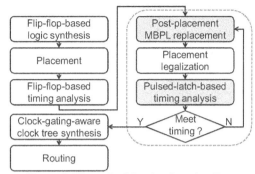

Figure 3. The pulsed-latch migration flow.

in [12]. Paik *et al.* minimize both the number of pulse generators and the wirelength of the pulse clock net in [13].

Since the pulse generator and latches are placed apart in the generic form, pulses can easily be distorted, and thus the load capacitance of each pulse generator should be delicately constrained. In addition, clock gating can be integrated into the pulse generator, e.g., [2][13][14], but this integration almost doubles the hardware cost of the pulse generator and forces clock gating to be applied around the clock sinks. Hence, we adopt *multi-bit pulsed-latch* (MBPL) design as the primitive cell of sequencing element in this paper [15][16]. (see Figure 1(b)) In a MBPL cell, the pulse generator and latches are placed and hard-wired together in a compact and symmetric form. The pulse distortion and clock skew can be well controlled in this structure. Moreover, clock gating is not included into the cell to keep the cell small and to preserve the flexibility to clock-gating-aware clock network synthesis.

Under flip-flop-like timing analysis, prior works [5–7] use aggressive time borrowing techniques; however, various pulse widths, clock skew scheduling, and retiming may induce some difficulties on timing closure and functional verification. In fact, we can utilize only the intrinsic time borrowing[2] of latches to provide flexibility to relocate pulsed-latches as well as tolerate timing violations [3]. In addition, the commercial timing analysis tools are mature and can handle time borrowing very well.

Hence, in this paper, based on the multi-bit pulsed-latch structure and time borrowing offered by the pulse width, we apply post-placement pulsed-latch replacement to minimize power consumption subject to timing constraints. According to the timing slacks obtained from a placed flip-flop-based design, we extract feasible moving regions (timing safe regions) with time borrowing consideration to replace flip-flops with pulsed-latches. Interestingly, the feasible regions with time borrowing consideration are very irregular. Directly manipulating these irregular shapes is complicated, and we show that they can be efficiently handled by four interval graphs. We propose spiral clustering with clock gating consideration to form MBPLs. For high performance circuits, the feasible regions without time borrowing could be very small. As shown in Figure 2, with time borrowing, flip-flops can be replaced by multi-bit pulsed-latches

to reduce power and satisfy timing requirements. We summarize our contributions as follows.

- We reveal irregular feasible regions: We derive timing analysis formulae with time borrowing consideration and reveal that the feasible regions can be very irregular. We adopt an efficient representation—four interval graphs—to manipulate the irregular feasible regions.

- We propose spiral clustering: Considering the distribution and proximity of flip-flops, we propose spiral clustering. Our clustering method is suitable for not only rectangular but also rectilinear shaped layouts; the latter are popular in modern IC design due to macros.

- We consider clock gating patterns: Since clock gating is widely used for clock power reduction, we incorporate clock gating consideration into pulsed-latch replacement to gain double benefits from clock gating and pulsed-latch.

- Experimental results show that with time borrowing, spiral clustering, and clock gating consideration, our approach can effectively reduce clock power consumed by sequencing elements.

The remainder of this paper is organized as follows. Section 2 introduces the pulsed-latch migration flow and the timing model and gives problem formulation; Section 3 derives the feasible regions based on time borrowing; Section 4 describes our algorithm; Section 5 shows experimental results; finally, Section 6 concludes this paper.

2. PRELIMINARIES AND PROBLEM FORMULATION

In this section, we introduce the pulsed-latch migration flow, describe the timing model, and give problem formulation.

2.1 Pulsed-Latch Migration Flow

To gain the power benefits of pulsed latches, we adopt a simple yet effective pulsed-latch migration flow shown in Figure 3 to convert a flip-flop-based circuit into a pulsed-latch-based one. Given a placed flip-flop-based circuit, we replace flip-flops by multi-bit pulsed-latches based on their timing slacks and the available amount of time borrowing. We fix pulsed-latches and their fanin/fanout gates and do legalization to remove placement overlap if needed. Then, we apply clock-gating-aware clock tree synthesis (same as flip-flop-based circuits) [17]. It can be seen that our pulsed-latch replacement can easily be integrated to the conventional flip-flop-based design flow.

---

[2] A robust pulse width is process dependent but frequency independent. Typically, a safe value is around the delay of one or more inverters. The amount of time borrowing offered by the pulse width is significant for high performance circuits.

2.2 Timing Model

In this subsection, we briefly describe the timing models of flip-flop-based and pulsed-latch-based circuits used in this paper.

Figure 4 illustrates flip-flop-based and pulsed-latch-based timing analysis. In the sequel, let t_{su} (t'_{su}) denote the setup time, t_{hd} (t'_{hd}) denote the hold time, t_{cq} (t'_{cq}) denote clock-to-Q delay for flip-flops (pulsed-latches). T denotes cycle period, and w denotes pulse width. $(t_{fo}(i) + D_{ij} + t_{fi}(j))$ is the maximum delay of combinational circuit between sequencing elements i and j, and $(t_{fo}(i) + d_{ij} + t_{fi}(j))$ is the minimum. In addition, based on Synopsys' Liberty library, $t_{fo}(i)$ (respectively $t_{fi}(j)$) means the wire delay contributed by the wirelength between sequencing element i (respectively j) and its fanout (respectively fanin).

As shown in Figure 4(a), for flip-flops, the setup time constraint between i and j is:

$$t_{cq} + t_{fo}(i) + D_{ij} + t_{fi}(j) + t_{su} \le T, \qquad (1)$$

while the hold time constraint is:

$$t_{cq} + t_{fo}(i) + d_{ij} + t_{fi}(j) - t_{hd} \ge M, \qquad (2)$$

where M is referred to as the hold time margin.

As shown in Figure 4(b), when we replace flip-flops with pulsed-latches, the data can depart the launching latch on the rising edge of the clock, but does not have to set up until the falling edge of the clock on the receiving latch. If the total delay from i to j is larger than a cycle period, it can borrow time from the delay from j to k. Time borrowing can accumulate across multiple timing windows. However, in designs with feedback, the timing windows with longer delays must be balanced by those with shorter delays so that the overall loop completes in the time available [3]. To guarantee successful time borrowing, in this paper, time borrowing is allowed between two adjacent timing windows. The setup time constraint is:

$$t'_{cq} + t_{fo}(i) + D_{ij} + t_{fi}(j) + t'_{su} \le T + w, \qquad (3)$$

$$t'_{cq} + t_{fo}(i) + D_{ij} + t_{fi}(j) + \\ t_{fo}(j) + D_{jk} + t_{fi}(k) + t'_{su} \le 2T, \qquad (4)$$

while the hold time constraint is:

$$t'_{cq} + t_{fo}(i) + d_{ij} + t_{fi}(j) - t'_{hd} \ge M_0 + w. \qquad (5)$$

To facilitate pulsed-latch replacement, the hold time margin[3] M (see Equation (2)) used at flip-flop-based logic synthesis, placement, and flip-flop-based timing analysis is set to $M = M_0 + w$, where M_0 is the original hold time margin, and w is the pulse width defined by the MBPL library.

We can convert the timing slacks for $t_{fo}(i)$ and $t_{fi}(j)$ obtained by flip-flop-based timing analysis into pulsed-latch-based slacks without time borrowing as follows.

$$S_{fo}(i) = S_{fi}(j) = \frac{T - (D_{ij} + t_{cq} + t_{su}) + \epsilon_1}{2}, \qquad (6)$$

$$H_{fo}(i) = H_{fi}(j) = \frac{d_{ij} + t_{cq} - t_{hd} - \epsilon_2 - M_0 - w}{2}, \qquad (7)$$

where $S_{fo}(i), S_{fi}(j), H_{fo}(i), H_{fi}(j)$ are setup and hold slacks for $t_{fo}(i)$ and $t_{fi}(j)$, and ϵ_1 and ϵ_2 are the differences on

---

[3] Paik and Shin in [4] reveal that the area overhead to pad short paths to achieve this extra hold time margin is very small.

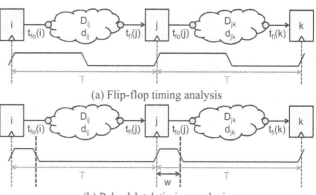

(a) Flip-flop timing analysis

(b) Pulsed-latch timing analysis

Figure 4. Timing analysis.

$(t_{cq} + t_{su})$ and $(t_{cq} - t_{hd})$ between flip-flops and pulsed-latches, $\epsilon_1 = (t_{cq} + t_{su}) - (t'_{cq} + t'_{su})$, $\epsilon_2 = (t_{cq} - t_{hd}) - (t'_{cq} - t'_{hd})$. We define D'_{ij} and d'_{ij} as follows.

$$D'_{ij} = D_{ij} + t_{cq} + t_{su} - \epsilon_1,$$
$$d'_{ij} = d_{ij} + t_{cq} - t_{hd} - \epsilon_2 - M_0.$$

Then, the slacks given in Equations (6) and (7) can be simplified.

$$S_{fo}(i) = S_{fi}(j) = \frac{T - D'_{ij}}{2}, \qquad (8)$$

$$H_{fo}(i) = H_{fi}(j) = \frac{d'_{ij} - w}{2}. \qquad (9)$$

In Section 3, we will combine these slack values with time borrowing consideration to derive feasible regions.

2.3 Problem Formulation

The multi-bit pulsed-latch replacement problem is formulated as follows.

The Multi-Bit Pulsed-Latch Replacement (mPLR) problem: Given a multi-bit pulsed-latch library, the timing slacks and clock gating patterns of flip-flops in a placed design, replace flip-flops by multi-bit pulsed-latches with time borrowing to minimize power on pulsed-latches subject to timing slack and placement density constraints.

The timing slacks are extracted based on the timing model described in Section 2.2, while the clock gating patterns can be obtained by simulations [18].

3. OUR FEASIBLE REGION ANAYLSIS

In this section, we derive the feasible region with time borrowing consideration for pulsed-latches. It can be seen that the feasible regions can be very irregular.

3.1 Basics

Based on Synopsys' Liberty library, wire delays $t_{fo}(i)$ and $t_{fi}(j)$ can be approximated by piece-wise linear functions with the Manhattan distances $L_{fo}(i)$ and $L_{fi}(j)$ as follows.

$$t_{fo}(i) = \tau L_{fo}(i),$$
$$t_{fi}(j) = \tau L_{fi}(j),$$

where τ is calibrated by the delay table of the pulsed-latch library. If i has multiple fanouts, then each fanout corresponds to one constraint on $L_{fo}(i)$.

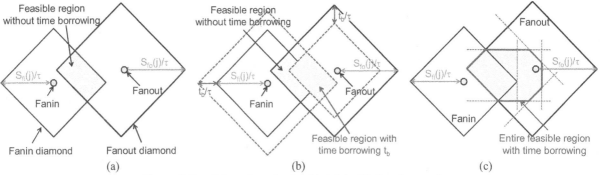

Figure 5. Feasible region of pulsed-latch j with time borrowing.

3.2 Feasible Regions with Time Borrowing

Considering the setup and hold time constraints with time borrowing described in Section 2.2, we derive the feasible region to place pulsed-latch j as shown in Figure 5. Time borrowing makes pulsed-latch replacement flexible.

As mentioned in Section 2.2, flip-flop-based synthesis and placement have considered the extra hold time margin w. Therefore, we consider the input design satisfies setup and hold time constraints in this paper, $S_{fi}(j), S_{fo}(j) \geq 0$; $H_{fi}(j), H_{fo}(j) \geq 0$. The hold time are still safe after including $t_{fo}(i)$ and $t_{fi}(j)$. Based on Inequalities (3), (4), (8), and (9), we have

$$t_{fi}(j) = \tau L_{fi}(j) \leq S_{fi}(j) + t_b, \qquad (10)$$

$$t_{fo}(j) = \tau L_{fo}(j) \leq S_{fo}(j) - t_b, \qquad (11)$$

$$t_b \leq w, \qquad (12)$$

where t_b is the amount of time borrowed from the timing window j-k to window i-j. Initially, the fanin and fanout setup time slacks define two diamonds centered at the fanin and fanout gates of pulsed-latch j. The overlap area is the initial feasible region without time borrowing as shown in Figure 5(a). When we borrow some time t_b, then the fanin diamond is expanded by t_b/τ, while

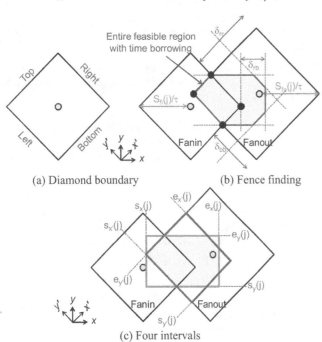

(a) Diamond boundary   (b) Fence finding

(c) Four intervals

Figure 6. Feasible region extraction. Any feasible region can be represented by four intervals.

the fanout diamond is shrunk by t_b/τ. The overlap area slides horizontally or vertically as shown in Figure 5(b). When we keep borrowing, the fanin or fanout diamond would reach the middle lines of the boundaries of fanin/fanout diamonds, and the overlap area are truncated. The entire feasible region is irregular, as illustrated in Figure 5(c). In the worst case, the feasible region could be an octagon.

4. OUR APPROACH

In this section, we present our post-placement pulsed-latch replacement framework. First of all, we extract feasible regions and represent them by four interval graphs. Second, we use spiral clustering to form multi-bit pulsed-latches; meanwhile, we consider clock gating during MBPL extraction. Finally, we relocate the newly formed multi-bit pulsed-latches.

4.1 Feasible Region Extraction

Based on Section 3.2, the feasible regions are determined by the pulse width and the differences between boundaries of fanin/fanout diamonds.

The fanin diamond expands, while the fanout diamond shrinks with time borrowing; if some fanout boundary is *outer* of the corresponding fanin one, we can find a fence to constrain the feasible region sliding. Given the initial feasible region, the entire feasible region with time borrowing can be extracted by finding eight fences. Using these eight fences, we can handle all irregular feasible regions.

To facilitate our feasible region extraction, we adopt a simple and fast coordinate transformation proposed in [19].[4] The fanin/fanout diamonds in Cartesian coordinate system C become squares in C', obtained by rotating by 45-degree. As shown in Figure 6(a), we define the four boundaries of a fanin/fanout diamond as right, bottom, left, and top boundaries.

Let δ_{rr}, δ_{bb}, δ_{ll}, and δ_{tt} denote the differences between fanin and fanout boundaries in top, right, bottom, left directions.

$$\delta_{rr}(j) = \max\left(0, x'_{r,fo}(j) - x'_{r,fi}(j)\right),$$

$$\delta_{bb}(j) = \max\left(0, y'_{b,fi}(j) - y'_{b,fo}(j)\right),$$

$$\delta_{ll}(j) = \max\left(0, x'_{l,fi}(j) - x'_{l,fo}(j)\right),$$

$$\delta_{tt}(j) = \max\left(0, y'_{t,fo}(j) - y'_{t,fi}(j)\right). \qquad (14)$$

---

[4] Consider a new coordinate system C' with origin at $(0, 0)$ in Cartesian coordinate system C. Considering a scaling factor as $\sqrt{2}$, the coordinate transformation between point (x, y) in C and its counterpart (x', y') in C' is defined as follows. $x' = y + x$, $y' = y - x$; $x = \frac{x'-y'}{2}$, $y = \frac{x'+y'}{2}$.

$x'_{r,fo}(j)$ means the x' coordinate of the right fanout boundary of j. We define the rest of coordinates similarly. For example, in Figure 6(b), fanout's right (bottom) boundary is outer of fanin's right (bottom) boundary, $\delta_{tt} = \delta_{ll} = 0, \delta_{rr}, \delta_{bb} > 0$.

First of all, we have four fences in C' located at:

$$x' = x'_{r,fi}(j) + \min(w, \delta_{rr}(j)/2) = e_{x'}(j),$$
$$y' = y'_{b,fi}(j) - \min(w, \delta_{bb}(j)/2) = s_{y'}(j),$$
$$x' = x'_{l,fi}(j) - \min(w, \delta_{ll}(j)/2) = s_{x'}(j),$$
$$y' = y'_{t,fi}(j) + \min(w, \delta_{tt}(j)/2) = e_{y'}(j). \quad (15)$$

We also have four fences in C.

$$x = x_{rb,fr}(j) + \min(w, \delta_{rr}(j)/2, \delta_{bb}(j)/2) = e_x(j),$$
$$y = y_{lb,fr}(j) - \min(w, \delta_{ll}(j)/2, \delta_{bb}(j)/2) = s_y(j),$$
$$x = x_{lt,fr}(j) - \min(w, \delta_{ll}(j)/2, \delta_{tt}(j)/2) = s_x(j),$$
$$y = y_{rt,fr}(j) + \min(w, \delta_{rr}(j)/2, \delta_{tt}(j)/2) = e_y(j). \quad (16)$$

$x_{rb,fr}(j)$ means the x coordinate of the right-bottom corner of the initial feasible region of j. We define the rest of coordinates similarly.

As shown in Figure 6(c), the four fences in C' defined by (15) form a rectangle, while the four fences in C defined by (16) form another. The entire feasible region is the overlap area of these two rectangles. Projecting these two rectangles to x'-, y'-, x-, y-axes, we obtain four intervals.

Based on the setting for successful time borrowing described in Section 2.2, Equations (14), (15) and (16) are applicable to loops. Moreover, if j has multiple fanouts, Equation (14) can be generalized to compare the boundaries of all fanouts together, and eight fences can then be obtained.

4.2 Spiral Clustering

After extracting all feasible regions, we encode them into four interval graphs (i.e., record by four sequences). As shown in Figure 6(c), projecting the two rectangles formed by eight fences to x'-, y'-, x-, y-axes, we obtain four intervals. Hence, we use four sequences X', Y', X, Y to record the starting and ending coordinates of x', y', x, and y intervals in ascending order.

Several pulsed-latches can form a multi-bit pulsed-latch if their feasible regions overlap, i.e., forming a clique in x'-, y'-, x-, y-intervals at the same time. We propose spiral clustering to form multi-bit pulsed-latches in this paper.

Figure 7 compares one-way clustering with spiral clustering, where the shaded area are occupied by macros. One-way clustering finds cliques along the diagonal of the chip [19]. Since pulsed-latches around corners have fewer choices to be merged, one-way clustering may leave many pulsed-latches alone (orphans) around the corners. Hence, spiral clustering is proposed. We find cliques from four corners towards the center. Moreover, spiral clustering is suitable for rectangular and rectilinear shaped layouts; the latter are popular in modern IC design.

Prior work [19] proposes a fast way to find maximal cliques in *one* interval graph by the following definition and theorem.

Definition 1 [19]: If there exist two consecutive points x'_k and x'_{k+1} in X', where $x'_k = s_{x'}(i)$, $x'_{k+1} = e_{x'}(j)$, $1 \leq i, j \leq n$, a decision point is the coordinate of x'_{k+1}, i.e., $e_{x'}(j)$.

Theorem 1 [19]: Consider sequence X' (i.e., sorted x' intervals). A maximal clique can be retrieved at some decision point.

First of all, we extend the above theorem as follows.

Theorem 2: Consider sequence X' (i.e., sorted x' intervals). Maximal cliques retrieved by scanning X' from left to right are the same as those retrieved by scanning X' from right to left.

Theorem 2 allows us to scan X' from two endpoints toward the middle. In addition, to find a maximal clique in four interval graphs, we may scan X' to find a maximal clique candidate and verify its ingredients in the other three interval graphs. Similarly, we can scan Y' and verify in the other three interval graphs, too. Based on Theorem 2, we develop spiral clustering.

In spiral clustering, we simultaneously maintain X' and Y'. We iteratively scan X' from left, scan Y' from bottom, scan X' from right and finally scan Y' from top. At each iteration, we handle one decision point to find a maximal clique in each direction. Multi-bit pulsed latches are extracted from the found maximal cliques (see Section 4.3). When a multi-bit pulsed latch is formed, its members are removed from X' and Y'. The process is repeated until all latches are processed. Figure 8 shows an example, where each block indicates the feasible region of some pulsed-latch. It can be seen that spiral clustering with time borrowing delivers the best result.

4.3 MBPL Extraction Considering Clock Gating

Since a maximal clique found during spiral clustering may not fit a MBPL cell, we need to extract a subset from this clique.

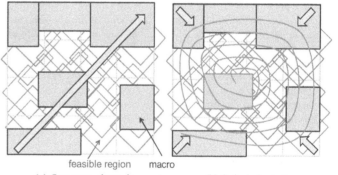

(a) One-way clustering   (b) Spiral clustering
Figure 7. Spiral clustering.

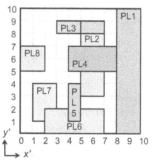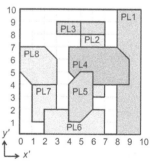

(a) Without time borrowing   (b) With time borrowing
Figure 8. One-way clustering vs. spiral clustering with and without time borrowing. (a) Without time borrowing: One-way: {8}, {6, 7}, {2, 3}, {5}, {1, 4}; spiral: {8}, {6, 7}, {1, 4}, {2, 3}, {5}. (b) With time borrowing: One-way: {7, 8}, {2, 3}, {4, 5}, {6}, {1}; spiral: {7, 8}, {5, 6}, {1, 4}, {2, 3}.

Since the latches inside one MBPL cell share the pulse clock, their clock gating functions are integrated by logic OR operations. If we merge pulsed-latches with very different clock gating patterns, we may not reduce power consumption. For example, assume that one 2-bit pulsed-latch consumes only 74% power compared with two single-bit latches. If the feasible regions of two latches overlap, but the active rate of their integrated clock gating function increases 50%, then the power consumption increases 11%, 0.74*1.5=1.11. Hence, we should consider clock gating during MBPL extraction. The effective power ratio of a multi-bit pulsed latch is

$$P_r = \alpha\beta, \qquad (17)$$

where α is the power ratio of the MBPL cell w.r.t. separate single-bit pulsed-latches, while β is the activity ratio of the shared clock gating pattern w.r.t. the average of original patterns.

To reduce power, our strategy is to extract a subset of feasible bit number and with minimum effective power ratio $P_r \leq 1$ from a found maximal clique. We try to keep a balance between the number of clock sinks and the pulsed-latch power reduction.

4.4 Pulsed-Latch Relocation

It can be shown that given a MBPL, if it is located within the bounding box formed by the median x- and y-coordinates of its fanin/fanout gates, the total wirelength on $L_{fo}(i)$ and $L_{fi}(j)$ is minimum.

Since the feasible region is irregular, it is non-trivial to find the point inside the feasible region which is nearest to the bounding box. We determine the point by two steps: First of all, we consider the rectangle constructed by x' and y' intervals. We find some point on the rectangle which is nearest to four corners or the center of the bounding box. If this point is also inside the rectangle constructed by x and y intervals, we place the MBPL at this point. Otherwise, we compute the distance between each corner of the feasible region to the bounding box and place the MBPL at the nearest one. If this point has been occupied or conflicts with the placement density constraint, we check the vacancy around this point; if we still cannot successfully relocate this MBPL, we will generate a smaller MBPL from its originating maximal clique.

5. EXPERIMENTAL RESULTS

We implemented our algorithm in the C programming language and executed the program on a platform with an Intel Xeon 3.8 GHz CPU and with 16 GB memory under Ubuntu 10.04 OS. Experiments are conducted on five industrial cases.

Table 1 lists the normalized power and area ratio of the used MBPL library, while Table 2 lists benchmark statistics. We employ a MBPL library containing 1-/2-/4-/8-bit MBPL cells based on 55-nm technology; the maximum allowable time borrowing from the pulse width is 100ps. All sequencing elements in input designs are single-bit flip-flops. First of all, we replace them by single-bit pulsed-latches. Then, we apply our pulsed-latch replacement approach. The active rates of clock gating patterns range from 0.13 to 0.69; the setup slacks for these circuits range from 10ps to 200ps.

Table 3 compares one-way clustering with spiral clustering, Table 4 displays the impact of pulse widths, and Table 5 shows the impact of clock gating strategy. 'Power Ratio' means the total power contributed from MBPL library cell (considering only α in

TABLE 1. MBPL LIBRARY: POWER VS. AREA

Bit Number	Normalized power	Normalized area
1	1.00	1.00
2	1.48	1.92
4	2.45	3.85
8	4.60	7.58

TABLE 2. STATISTICS OF THE BENCHMARK

Circuit	#FFs	#Bins	#Grids	Avg. activity
Industry1	120	6×6	600×600	0.25
Industry2	120	6×6	600×600	0.13
Industry3	60,000	100×300	2,000×3,000	0.69
Industry4	5,524	100×200	2,000×2,000	0.44
Industry5	953	30×160	600×1,600	0.25

Remark: avg. activity is the average active rate of clock gating functions.

Equation (17)) over the total power from single-bit pulsed-latch cells. 'Pattern-Aware Power Ratio' means the total *effective* power of MBPLs (considering both α and β in Equation (17)) over the total *effective* power of single-bit pulsed-latches. '#Sinks' means the number of clock sinks. In Table 3, we focus on power reduction contributed from the MBPL library during spiral clustering. Compared with one-way clustering without time borrowing [19], it can be seen that spiral clustering with time borrowing can effectively reduce the power consumption (see 'Power Ratio'). In Industry3, the initial feasible regions strongly overlap with each other; hence, we do not gain many benefits from time borrowing. Table 4 compares spiral clustering with time borrowing 150ps and with time borrowing 200ps. In general, if the pulse width increases, the power saving can be further improved.

From Table 3, it can be seen that even if power reduction from MBPL library is significant, the effective power saving may be not as good as expected. In Table 5, we further consider clock gating during spiral clustering. Without clock gating consideration, spiral clustering generates very few clock sinks but consumes more power, especially under very random clock gating patterns, e.g., Industry1 and Industry5. Our clock gating strategy generally delivers low power consumed by pulsed-latches but generates more clock sinks. Even so, we still can effectively reduce the number of sinks by 44.23%. We can keep the balance between the effective power reduction on pulsed-latches and the number of sinks. It can be seen that using time borrowing, spiral clustering, and clock gating consideration, we can achieve very power efficient results.

6. CONCLUSION

Pulsed-latches are promising to reduce power for high performance circuits. In this paper, we propose a novel pulsed-latch replacement approach to save power and satisfy timing constraints. We derive the irregular feasible regions under time borrowing offered by the pulse width and develop a spiral clustering method with clock gating consideration. Experimental results show that our approach can generate very power efficient results. Future work includes the generalization of our method to compensate setup and/or hold violations.

7. ACKNOWLEDGMENTS

The authors greatly appreciate constructive comments and suggestions from anonymous reviewers. This work was partially supported by Faraday Technology Corporation and NSC of Taiwan under Grant No. NSC 100-2220-E-009-047.

TABLE 3. COMPARISON ON POWER, THE NUMBER OF CLOCK SINKS, AND RUNTIME: ONE-WAY VS. SPIRAL CLUSTERING

Circuit	One-Way Clustering [19]				Spiral Clustering with Time Borrowing w=100ps w/o Clock Gating			
	Power Ratio	Pattern-Aware Power Ratio	#Sinks (1/2/4/8-bit PLs)	Runtime (s)	Power Ratio	Pattern-Aware Power Ratio	#Sinks (1/2/4/8-bit PLs)	Runtime (s)
Industry1	74.93%	130.67%	62(18/37/7/0)	< 0.01	69.34%	140.38%	49 (4/32/13/0)	< 0.01
Industry2	75.78%	101.22%	64 (20/38/6/0)	< 0.01	72.36%	104.30%	56 (14/31/11/0)	< 0.01
Industry3	57.54%	79.53%	7,558 (10/35/46/7,467)	3.36	57.50%	79.49%	7,500 (0/0/0/7,500)	3.07
Industry4	62.98%	96.61%	1,520 (52/432/920/116)	0.41	60.84%	99.33%	1,233 (16/182/784/251)	0.39
Industry5	65.36%	113.79%	311 (27/123/152/9)	0.04	62.33%	121.02%	246 (9/62/145/30)	0.05
Avg.	67.32%	104.36%	35.55%	-	64.47%	108.90%	29.63%	-

TABLE 4. COMPARISON ON POWER, THE NUMBER OF CLOCK SINKS, AND RUNTIME: W=150PS VS. W=200PS

Circuit	Spiral Clustering with Time Borrowing w=150ps w/o Clock Gating				Spiral Clustering with Time Borrowing w=200ps w/o Clock Gating			
	Power Ratio	Pattern-Aware Power Ratio	#Sinks (1/2/4/8-bit PLs)	Runtime (s)	Power Ratio	Pattern-Aware Power Ratio	#Sinks (1/2/4/8-bit PLs)	Runtime (s)
Industry1	68.07%	142.54%	46 (4/26/16/0)	< 0.01	67.64%	144.35%	45 (4/24/17/0)	< 0.01
Industry2	70.22%	101.35%	51 (10/27/14/0)	< 0.01	69.79%	103.56%	50 (10/25/15/0)	< 0.01
Industry3	57.50%	79.53%	7,500 (0/0/0/7,500)	3.20	57.50%	79.47%	7,500 (0/0/0/7,500)	3.23
Industry4	60.52%	99.68%	1,184 (14/157/727/286)	0.41	60.46%	99.95%	1,170 (14/163/690/303)	0.40
Industry5	62.00%	121.95%	239 (7/55/145/32)	0.05	62.12%	122.86%	240 (7/63/135/35)	0.04
Avg.	63.66%	109.01%	27.97%	-	63.50%	110.04%	27.61%	-

TABLE 5. COMPARISON ON POWER, THE NUMBER OF CLOCK SINKS, AND RUNTIME: WITHOUT VS. WITH CLOCK GATING (W=100PS)

Circuit	Spiral Clustering with Time Borrowing w=100ps w/o Clock Gating				Spiral Clustering with Time Borrowing w=100ps w/ Clock Gating			
	Power Ratio	Pattern-Aware Power Ratio	#Sinks (1/2/4/8-bit PLs)	Runtime (s)	Power Ratio	Pattern-Aware Power Ratio	#Sinks (1/2/4/8-bit PLs)	Runtime (s)
Industry1	69.34%	140.38%	49 (4/32/13/0)	< 0.01	95.68%	95.68%	110 (104/4/2/0)	< 0.01
Industry2	72.36%	104.30%	56 (14/31/11/0)	< 0.01	78.38%	78.38%	70 (32/32/6/0)	< 0.01
Industry3	57.50%	79.49%	7,500 (0/0/0/7,500)	3.07	63.59%	68.78%	15,033 (8,578/25/17/6,413)	5.20
Industry4	60.84%	99.33%	1,233 (16/182/784/251)	0.39	73.33%	73.99%	2,633 (1,584/328/621/100)	0.45
Industry5	62.33%	121.02%	246 (9/62/145/30)	0.05	77.46%	77.59%	535 (337/102/89/7)	0.05
Avg.	64.47%	108.90%	29.63%	-	77.69%	78.88%	55.77%	-

8. REFERENCES

[1] R. S. Shelar, "An efficient clustering algorithm for low power clock tree synthesis," in *Proc. Int. Symp. on Physical Design (ISPD)*, Mar. 2007, pp. 181–188.

[2] S.D. Naffziger, G. Colon-Bonet, T. Fischer, R. Riedlinger, T.J. Sullivan, T.S. Grutkowski, "The implementation of the Itanium 2 microprocessor," *IEEE Journal of Solid-State Circuits (JSSC)*, vol. 37, no. 11, Nov. 2002, pp. 1448–1460.

[3] N. H. E. Weste and D. M. Harris, *Integrated Circuit Design*, 4th ed., Addison Wesley, 2011.

[4] S. Paik and Y. Shin, "Pulsed-latch circuits to push the envelope of ASIC design," in *Proc. Int'l SoC Design Conf. (ISOCC)*, pp. 150–153, Nov. 2010.

[5] H. Lee, S. Paik, and Y. Shin, "Pulse width allocation with clock skew scheduling for optimizing pulsed-latch-based sequential circuits," in *Proc. Int'l Conf. on Computer-Aided Design (ICCAD)*, pp. 224–229, Nov. 2008.

[6] S. Lee, S. Paik, and Y. Shin, "Retiming and time borrowing: optimizing high-performance pulsed-latch-based circuits," in *Proc. Int'l Conf. on Computer-Aided Design (ICCAD)*, pp. 375–380, Nov. 2009.

[7] S. Paik, L.-E. Yu, and Y. Shin, "Statistical time borrowing for pulsed-latch circuit designs," in *Proc. Asia South Pacific Design Automation Conf. (ASPDAC)*, pp. 675–680, Jan. 2010.

[8] Y.-L. Chuang, S. Kim, Y. Shin, and Y.-W. Chang, "Pulsed-latch-aware placement for timing-integrity optimization," in *Proc. Design Automation Conf. (DAC)*, pp. 280–285, June 2010.

[9] S. Shibatani and A. H.C. Li, "Pulse-latch approach reduces dynamic power," *EE Times*, July 2006.

[10] S. Kim, I. Han, S. Paik, and Y. Shin, "Pulser gating: a clock gating of pulsed-latch circuits," in *Proc. Asia South Pacific Design Automation Conf. (ASPDAC)*, pp. 190–195, Jan. 2011.

[11] H.-T. Lin, Y.-L. Chuang, and T.-Y. Ho, "Pulsed-latch-based clock tree migration for dynamic power reduction," in *Proc. Int'l Symp. on Low Power Electronics and Design (ISLPED)* Aug. 2011, pp. 39–44.

[12] Y.-L. Chuang, H.-T. Lin, T.-Y. Ho, Y.-W. Chang, and D. Marculescu, "PRICE: Power reduction by placement and clock-network co-synthesis for pulsed-latch designs, " in *Proc. Int'l Conf. on Computer-Aided Design (ICCAD)*, pp. 85–90, Nov. 2011.

[13] S. Paik, G.-J. Nam, and Y. Shin, "Implementation of pulsed latch and pulsed register circuits to minimize clocking power, " in *Proc. Int'l Conf. on Computer-Aided Design (ICCAD)*, pp. 640–646, Nov. 2011.

[14] S. Kozu, M. Daito, Y. Sugiyama, H. Suzuki, H. Morita, M. Nomura, K. Nadehara, S. Ishibuchi, M. Tokuda, Y. Inoue, T. Nakayama, H. Harigai, and Y. Yano, "A 100 MHz 0.4W RISC processor with 200 MHz multiply-adder, using pulse-register technique," in *Proc. IEEE Int. Solid-State Circuits Conf. (ISSCC)*, pp. 140–141, Feb. 1996.

[15] H.R. Farmer, D.E. Lackey, and S.F. Oakland, "Pipeline array," US patent 6856270 B1, Feb. 2005.

[16] A. Venkatraman, R. Garg, and S. P. Khatri, "A robust, fast pulsed flip-flop design," in *Proc. Great Lakes Symp. VLSI (GLSVLSI)*, pp. 119–122, May 2008.

[17] W. Shen, Y. Cai, X. Hong, and J. Hu, "An effective gated clock tree design based on activity and register aware placement," *IEEE Trans. Very Large Scale Integration Systems (TVLSI)*, vol. 18, no. 12, Dec. 2010, pp. 1639–1648.

[18] E. Arbel, C. Eisner, and O. Rokhlenko, "Resurrecting infeasible clock-gating functions, " in *Proc. Design Automation Conf. (DAC)*, pp. 160–165, July 2009.

[19] C.-L. Chang, I. H.-R. Jiang, Y.-M. Yang, E. Y.-W. Tsai and L. S.-F. Chen, "INTEGRA: Fast multi-bit flip-flop clustering for clock power saving based on interval graphs," in *Proc. Int'l Symp. on Physical Design (ISPD)*, pp. 115–121, Mar. 2011.

On Constructing Low Power and Robust Clock Tree via Slew Budgeting

Yeh-Chi Chang
Institute of Electronics
National Chiao Tung University
Hsinchu, Taiwan
yehchi0604@gmail.com

Chun-Kai Wang
Institute of Electronics
National Chiao Tung University
Hsinchu, Taiwan
oldkai.ee90@gmail.com

Hung-Ming Chen
Institute of Electronics
National Chiao Tung University
Hsinchu, Taiwan
hmchen@mail.nctu.edu.tw

ABSTRACT

Clock skew resulted by process variation becomes more and more serious as technology shrinks. In 2010, ISPD held a high performance clock network synthesis contest; it considered supply-voltage variation and wire manufacturing variation. Previous works show that the main issue of variation induced skew is on supply-voltage variation. To trade off power and supply-voltage variation induced skew more effectively, we adapt a tree topology which use a timing model independent symmetrical tree at top level to drive the bottom level non-symmetry trees. Our method gives top tree more power budget to reduce supply-voltage variation induced skew and greedily saves power consuming in bottom level. Experimental results are evaluated from the benchmarks of ISPD contest 2010. Compared with state-of-the-art cross link work, the proposed technique reduces 10% of power consumption on average and also improves the run time.

Categories and Subject Descriptors

B.7.2 [**Design Aids**]: Placement and routing

General Terms

Algorithms, Design

Keywords

clock synthesis, slew

1. INTRODUCTION

In very large scale integration (VLSI) systems, low power designs attracted great attention. Sequential circuit accounts for much power consumption owing to clock network. Furthermore, a well-designed clock network must minimize the skew because the circuit speed is limited by it. As the process technology scales below 65nm, variation in the manufacture, temperature or power-supply noise induce noticeable skew. The issue to cope with unpredictable variations becomes more and more important. ISPD held a high performance clock network synthesis contest considering power-supply and wire variation problem. The objective is to minimize power consumption under skew and slew constraint. And if there is a blockage, the inserted buffers can not overlap with it.

1.1 Previous Works

To synthesize a low power clock network, the commonly used structure is a tree due to its common path from root to sinks. On the other hand, non-tree structures like mesh or spine providing more robust timing yield than tree structure while consume much more power in connections. In this paper, we focus on the low-power tree design. We review three important previous works and then conclude on our strategy.

Later Fine-tuning with Two-stage Synthesis: This type of clock synthesis first generates the topology of the tree then performs buffer insertion and wire sizing. Since the power consumption of a clock network comes from wire and inserted buffers, and more wire using needs more buffers to drive it; this two-stage method tends to generate a simple topology in the first step. However, for a fixed topology, inserting a new buffer or resizing a wire changes latency of corresponding sinks. Contango [1], the first place of the ISPD 2010 contest solves this problem by a later stage fine-tuning which includes wire snaking and delay buffer insertion.

Interleaving Topology Generation and Buffer Insertion: The other strategy, as proposed in [2], did not separate the topology generation and buffer insertion. During the bottom up pahse of Deferred-Merging Embedding (DME) [3, 4, 5] buffers are inserted simultaneously. And an early stage skew estimation is used to ensure valid skew. The estimation scheme decides a suitable inserted buffer size. However, in [2], the latency variation model is oversimplified so that the inserted buffer cannot reduce supply-voltage variation induced skew very effectively. We will discuss it later in this paper.

Timing Model Independent Tree: Two aforementioned works both need a lot of run time for SPICE simulations because there is no efficient yet accurate timing model for the strict skew constraint of GHz high performance design. In [6], it proposed a timing model independent method. The skew-free problem is based on the symmetry structure, where all the paths from the clock source to sinks are similar. In other words, there are the same wire-length, the same number of branched and buffers, and identical type of buffer in all paths from root to all sinks in the clock tree. The disadvantage of the symmetry structure

Table 1: Comparison of four clock tree works. Early skew estimation means the skew estimation is before buffer insertion. Run time and power are in rank. (TG/BI: topology generation and buffer insertion)

Work	[1]	[2]	[7]	Ours
Skew Estimation	later	early	none	early
Symmetry	-	-	+ (bottom mesh)	top: + bottom: -
TG/BI separation	+	-	+	+
run time (SPICE)	4 +	3 +	1 -	2 +
power	3	2	4	1

is its higher power consumption, since a symmetry tree always has longer total wire length than an asymmetry tree. Furthermore, in case of non-uniform sink distribution, dense in some region and empty in others, the symmetrical topology results in explosive wire length in leaf levels, and the too long path results valid skew cannot be conformed. The author improved this deficiency in a modified version [7], which used a symmetry tree to drive a bottom level mesh. But there is still a power performance gap compared with non-symmetry tree works. On power minimization, there is a potential of symmetry structure has not been shown. Its nature of separation between topology generation and buffer insertion wire sizing would make one plans buffer insertion more efficiently.

1.2 Our Contributions

In this work, we adopt a hybrid structure that a symmetrical top tree drives bottom level asymmetry trees. The bottom level asymmetry trees prevents the explosive wire length. We greedily save power in bottom level and give top level tree more power budget to insert buffers that helps reduce supply-voltage variation induced skew. A early skew estimation concept was used on buffer insertion in top level tree, it ensures timing yield and makes buffer insertion more power efficient.

Table 1 shows a comparison between three previous works and our work. Without skew estimation, [7] needs a large margin to prevent skew fail so that a bottom level mesh is used but cost most power among works. Early skew estimation is preferred to has better power performance. Symmetry structure has obvious advantage on run time. Our bottom level asymmetry trees still need some SPICE simulation to ensure small nominal skew. Although a symmetry tree always has longer total wire length than a asymmetry tree, a better buffer plan can save more power. For asymmetry tree, the separate topology generation and buffer insertion wire sizing needs a lot of run time on later stage fine-tuning.

Even compared with [8] which is a cross link work and had the best power performance on ISPD 2010 benchmark, we have 10% power reduction and less run time.

The contributions are summarized as follows:

- We check the latency variation model in [2]. The over-simplification in this model misleads to minimize the number of buffer levels. It may result in bad slew and worse supply-voltage variation induced skew.

- We adopt a hybrid structure to construct clock tree,

this structure makes power optimization easier than the structures other works propose.

The remainder of this paper is organized as follows: Section 2 illustrates the problem description. Section 3 describes the early skew estimation and power optimization for buffer insertion. Section 4 introduces the sub-tree generation and our flow. Experimental results are demonstrated in Section 5. Finally, Section 6 concludes this paper.

2. PROBLEM DESCRIPTION

The problem is provided by the ISPD 2010 High Performance Clock Network Synthesis Contest [9]. The problem description is given as follows: the inputs are

1. A set of clock pins of Flip-Flops $\{s_1, s_2, ..., s_n\}$.

2. A set of blockages $\{b_1, b_2, ..., b_m\}$.

3. A $W \times H$ layout region.

4. A library of buffer and wire.

The objective is minimizing total capacitance of wire and buffer under the constraints described as follows. The constraints are:

1. No overlap between buffers and blockages.

2. No slew-rate violation: input slew of buffer/sink must less than 100ps.

3. No local clock skew violation which is defined as the 95% skew under 500 times SPICE simulation should not exceed a given skew threshold, and only the sinks within a given local distance need be considered. We want to construct a clock tree satisfying all constraints. Moreover, our experimental results are verified by Monte Carlo simulations under wire and power-supply variations.

3. EARLY SKEW ESTIMATION FOR POWER OPTIMIZATION

In this section, we review the previous skew estimation method used in [2] and introduce the power optimization.

3.1 Early Skew Estimation in CTS

[10] proposed a skew upper bound model. It can be utilized when we know the number of sinks and the value of delay standard deviation from clock root to a sink. [2] utilized this method to estimate the 95% skew as shown in (1) to (4).

$$95\% R_I \approx NCS + \alpha \cdot \left[E(R_I) + 2\sqrt{Var(R_I)} \right] \quad (1)$$

$$E(R_I) = \sigma \left[\frac{4\ln N - \ln\ln N - \ln 4\pi + 2C}{(2\ln N)^{1/2}} + O\left(\frac{1}{\log N}\right) \right] \quad (2)$$

$$Var(R) = \frac{\sigma^2}{\ln N} \frac{\pi^2}{6} + O\left(\frac{1}{\log^2 N}\right) \quad (3)$$

$$\sigma^2 = {\sigma_0}^2 B \quad (4)$$

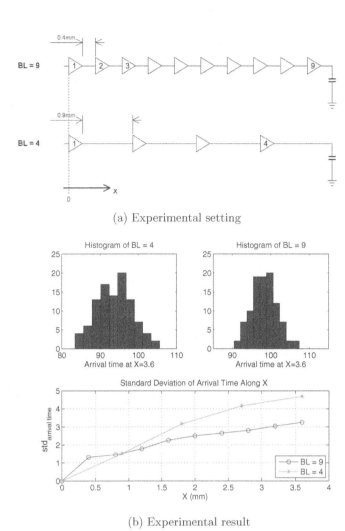

(a) Experimental setting

(b) Experimental result

Figure 1: Buffer level experiment. We set two kinds of buffer levels to drive the same loading. The total power usage for both is set the same. (a) One is 9 buffer levels with 12x inv-1, the other is 4 buffer levels with 27x inv-1. The wires are all in type of wire-0. (b) The 9 buffer levels design has smaller arrival time variance. Along the path, at 1.8, 2.7 and 3.6mm, the 9 buffer levels design has better standard deviation than 4 buffer levels design.

where the random variable R_I refers to the supply-voltage variation resulted skew, NCS is the nominal skew, α is a correction coefficient for metric-free tree model, N is the number of sinks, σ_0 is a constant value represents the standard deviation of a buffer stage delay, B is the number of buffer levels, and $C(= 0.5772)$ is the Euler's constant. [2] estimates 95% bound of a random variable by 2 times standard deviation in (1) and substitutes the variance of total clock latency by the multiplication of σ_0^2 and B in (4). In (2) and (3), both expected value of variation induced skew and variance of variation induced skew have positive relations with the number of buffer levels. It seems less buffer levels would reduce supply-voltage variation induced skew. But in our experience, more buffer levels may result in better skew.

We experiment on the effect of different number of buffer

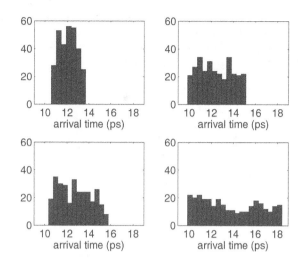

Figure 2: Histogram of buffer stage delay under different input signal. Slower input slew amplifies the supply-voltage variation induced clock skew. In each case, we run 300 SPICE simulations under ISPD10 supply-voltage variation. The circuit is two 12x inv-1 in parallel with 0.4mm wire-0 connection.

levels. Figure 1(a) shows the experimental setting. We use two kinds of buffer levels to drive a 100pF capacitance connected with 3.6mm wire-0(a wire type in ISPD 2010 benchmarks). The first one has 9 buffer levels, and each buffer level is a 12x inv-1(12 inverters of type1 in parallel, the type is also from ISPD 2010 benchmarks). The second has 4 buffer levels, and each of them is a 27x inv-1. The numbers of total buffer usage in these two cases are the same. Result was shown in Figure 1(b). Under ISPD supply-voltage variation setting and 100 times SPICE simulation, we probe the arrival time on nodes of each buffer input and capacitance load. In Figure 1(b), the histograms shows that arrival time of 9 buffer levels design has narrower distribution than 4 buffer levels design.

Actually, the standard deviation of a buffer stage delay should be considered more carefully. It is not a constant value. The input slew affects the supply-voltage variation induced skew tremendously. In our experiment, no matter how fast the input slew is, after a 27x inv-1 with 0.9mm wire-0, the slew value is about 50 to 60ps. But in 9 buffer levels design, since the wire is shorter, the slew is about 20 to 30ps. We examine how different input slew effects latency variation by a single stage in 9 buffer levels design that is two 12x inv-1 in series with 0.4mm wire connection. If the input signal is rising with slew equal to 30ps(left-top of Figure 2), the standard deviation of a buffer stage delay is 0.7899. But if the input slew is 50ps(left-bottom), the standard deviation would raise to 1.4588. In falling edge(right), the gap still exists.

3.2 Power Optimizaion

[2] neglects the slew factor that misleads a conclusion: less buffer levels help to reduce supply-voltage variation induced skew. To reduce supply-voltage variation induced skew, satisfying slew constraint only is not sufficient. In our clock tree

Table 2: Power Optimization from skew estimation. For a target σ derived by the skew estimation, we move power usage from leaf to top, like case1 to case2. It would effectively reduce power.

	Case1		Case2	
	σ	Power	σ	Power
Lv1	1	1×2^1	0.89	2×2^1
Lv2	1	1×2^2	1	1×2^2
Lv3	1	1×2^3	1	1×2^3
Lv4	1	1×2^4	1.1	0.8×2^4
Total	2	31	2	26.8

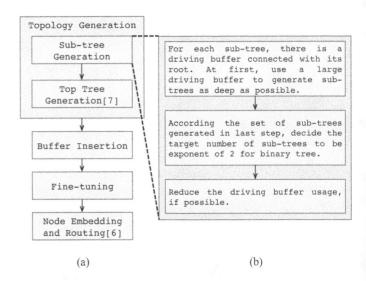

(a)   (b)

Figure 3: The flow of methodology. (a)Overall flow. (b)Sub-tree generation strategy

synthesis, we try to keep the input slew of buffers much better than slew constraint. In this work, a buffer distance was defined, that is the distance between two buffers in series. This distance is shorter than other works to ensure better slew. Although, it causes more buffer levels, the better slew lets us use less buffers in parallel. That is why we have lower power consumption.

To estimate supply-voltage variation induced skew, we adopt the symmetry tree structure and build a look-up-table(LUT) for standard deviation of a buffer stage delay. The symmetry structure reduces complexity of buffer insertion. We only need to focus buffer insertion on a single path. On this single path, we look up the standard deviation of each buffer level by its input slew and its RC network topology. Since every standard deviation value in table is extracted from time-consuming Monte Carlo simulation under ISPD 2010 supply-voltage variation, we do not want the table to be large.

We firstly insert buffers on every branch node. In this way a buffer stage will be either two buffers in series or a buffer driving two buffers of next level and the branch node is just at the output pin of driving buffer. The confined RC network topology would reduce table size but lose some solution space at the same time. But this smaller solution space would not harm too much. The reason is that, when a branch node is far behind an output pin of driving buffer, the RC network topology would be more like a large resistance in front of a capacitance. The large resistance comes from the wire between buffer output and branch node. Compared with the branch node on the output pin of a driving buffer, in front of branch there is only output resistance of driving buffer. The RC network topology with large capacitance in front leads bad slew soon. vie local viewpoint, it is not a good solution.

And a preprocess for metric-free tree [10] correction is adopted. We use a binary tree to simulate variation induced skew. The binary tree is symmetrical, and among a level, all nodes are i.i.d random variables refers to stage delay variation. Variation induced skew can be easily calculated by gathering latency of leaves. A correction function for metric-free tree model can be built by regressing skew and standard deviation of random variables on the path.

Another problem is to distribute the total standard deviation of clock latency to buffer levels. [2, 7] both insert buffers when the slew rate is about to violate. That means they equally distribute the standard deviation of clock latency to all buffer levels. It is not a power efficient strategy. [1] at first inserts buffers to minimize the clock latency and

then insert delay buffer to pass a LUT skew estimation. It does not directly tackle the standard deviation distribution. Table 2 is a simple example to distribute standard deviation more efficient and result in power-saving. Since a parent has two children in a H-tree, it makes a level has double power dissipation compared with its parent level. Therefore, for a target standard deviation of clock latency, reducing delay variation in top level is more efficient than reducing it in bottom level.

4. BALANCED SUB-TREE GENERATION FOR LOW POWER AND ROBUST CLOCK TREE

The overall clock tree synthesis flow is illustrated in Figure 3(a). The two-phase method utilizing the notion of DME algorithm is presented. First, all sinks are clustered into several sub-trees driven by identical buffers. Moreover, the size of driving buffer and the number of sub-trees are determined for greedy power minimization. Then these buffers are regarded as leaves, and we construct a symmetrical obstacle avoiding tree topology at top-level. After topology generation, buffers will be inserted considering supply-voltage variation. As discussed in Section 3, we ensure that the input slew of inserted buffers has better slew rate than only satisfying slew constraint. We embed the SPICE simulation in the last step to ensure that the nominal skew is smaller than 1 ps. Finally, in the top-down phase node embedding will be performed.

The symmetrical top tree implementation and final node embedding is from [7, 6]. We depict our sub-tree generation, buffer insertion and fine-tuning in this section.

4.1 Sub-Tree Generation

During sub-tree generation, $subtreeGen(b, nS)$ is a procedure to decide the number of sub-trees and the size of driving buffer. It is a DME-based function and has two

Algorithm 1 $subtreeGen(b, nS)$

Input: driving buffer size b, and
Output: a set of subtrees V
1: $V \leftarrow \phi$
2: $U \leftarrow allsinks$
3: $N \leftarrow \|U\|$
4: **while** $N > nS$ and $\|U\| \neq 0$ **do**
5:      $n_1 \leftarrow$ the subtree with min. delay in U
6:      $n_2 \leftarrow$ nearest neighbor of n_1 in U
7:      $n_{new} \leftarrow merge(n_1, n_2)$
8:      $U \leftarrow U \setminus \{n_1, n_2\}$
9:      **if** $slewcheck(b, n_{new})$ fail **then**
10:        $V \leftarrow V \cup \{n1, n2\}$
11:      **else**
12:        $U \leftarrow U \cup \{n_{new}\}$
13:        $N \leftarrow N - 1$
14:      **end if**
15: **end while**
16: $V \leftarrow V \cup U$
17: **return** V

Algorithm 2 $subtreeGen\_all(allsinks, B)$

Input: $allsinks$ and buffer library B
Output: drving buffer size $b_{subtree}$, and
1: sort buffer in B in increase order of size
2: **for** $i \leftarrow 1$ to $\|B\|$ **do**
3:      **if** $b_i \in B$ drives the max loading sink without slew violation **then**
4:        $idx \leftarrow i + \delta$
5:        break;
6:      **end if**
7: **end for**
8: $V = subtreeGen(b_{idx}, 0)$
9: $nS_{target} \leftarrow 2^{\lceil \log \|V\| \rceil}$
10: **for** $j \leftarrow i$ to idx **do**
11:      $V = subtreeGen(b_j, nS_{target})$
12:      **if** $\|V\| = ns_{target}$ **then**
13:        $b_{subtree} \leftarrow b_j$
14:        break;
15:      **end if**
16: **end for**
17: **return** $b_{subtree}$ and nS_{target}

parameters: b is the buffer size to drive a sub-tree, nS is a target number of sub-trees in each run. We show the detail of $subtreeGen(b, nS)$ in Algorithm1. Every time we invoke $subtreeGen(b, nS)$, it generates a set of sub-trees. In $subtreeGen(b, nS)$, we firstly initialize two containers U and V, where U contains sub-trees those are waiting to be merged, and V is the container which collects sub-trees we do not want to merge them any more, because those sub-trees are about to violate slew constraint. We pick up a merging pair as suggested by [11] - merging two nodes with similar delay will reduce excessive wire-snacking. Therefore, the first node n_1 we pick is the one with smallest delay, and n_2 is the nearest neighbor of n_1. After merging, n_1 and n_2 are removed from U, and there is a slew check for the new sub-tree. This slew check is referenced from bottom up merge in [2]: if a sink of new sub-tree has slew violation, we give up this merge and push n_1 and n_2 into V. If slew check is passed, we push the new sub-tree into U. Once the target number of sub-trees is achieved or there is no more sub-tree to be merged, $subtreeGen(b, nS)$ terminates. Since sinks have zero delay at first, and we pick up the smallest delay node in each iteration, orphan sinks will not result large number of sub-trees in this method, only if a orphan sink has too large capacitance loading to merge with the other sink without slew violation.

Figure 3(b) illustrates our strategy of sub-tree generation. The method to decide our final number of sub-trees and the size of driving buffer is shown in Algorithm2. In line 1 to 7, we select a buffer size, b_{idx}, which mush be large enough to drive maximum sink loading without slew violation. In line 4, if δ is large, it would give more chance to synthesis less number of sub-trees. Since the major clock power dissipation is in bottom level, less sub-trees usually has better power performance. In our experiment, δ is set to 10. Then we run a $subtreeGen(b_{idx}, nS = 0)$ to generate a set of sub-trees. The zero nS asks $subtreeGen$ to merge as deep as possible. Our target number of sub-trees should be the exponent of 2 for top symmetry tree, so we set the least exponent of 2 larger than $\|V\|$ as nS_{target}. This is our final target number of sub-trees. Finally, we minimize the buffer usage by sweeping smaller buffer size for $subtreeGen(b_j, nS_{target})$. If

a smaller buffer size matches the target number of sub-trees, nS_{target}, we set it as final sub-tree driving buffer, $b_{subtree}$.

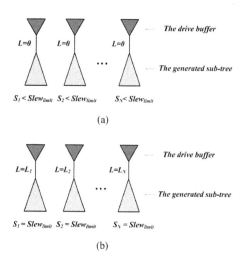

Figure 4: When sub-trees are just generated, slew rate of sinks has no violation but are in different value, and the driving buffer are connected to its sub-tree with zero wire length, like (a) shows. Then in (b), the wire length is adjusted to make the max value slew among leaves in sub-tree equals slew limit.

After the sub-trees are generated, the buffer is inserted with connected wire between this buffer and the root of sub-tree, as shown in Figure 4(a). Then we adjust the length of wire so that max value slew among leaves in sub-tree equals a given slew limit in Figure 4(b). We use the binary search to get the wire-length with the particular slew limit. In our experiments, the slew limit is set to 75ps(The slew constraint is 100ps in benchmark). Since delay has a strong correlation with slew in dc-connected RC tree, the sub-trees have similar latency from buffer input to sinks when they have similar slew. Moreover, the skew within each sub-tree is close to zero. Therefore, we can easily construct delay-balance cluster by slew evaluation instead of getting accurate latency by SPICE simulation.

4.2 Buffer Insertion

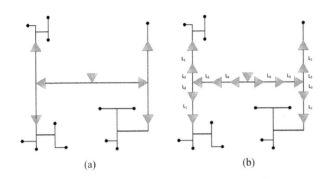

Figure 5: Buffer Insertion. (a)Symetrical buffers are inserted on all branches.(b)Symetrical buffers are inserted into the path.

Based on the symmetrical structure, identical buffers are inserted at the same level and the same distance away from the tree root. As discussed in Section 3.2, buffers are first inserted on all branches. The procedure of buffer insertion is illustrated in Figure 5. In Figure 5(a), buffers are inserted on all branches. Then we insert buffers with a fixed distance along path as Figure 5(b) shows. The decision criterion of buffer sizing is based on the skew estimation described in Section 3.1. We manually try different sizes of buffer and according the skew estimation result to decide it. Besides, to get a power efficient distribution of clock latency standard deviation, we use the narrower wire type in bottom level (sub-trees) and wider wire type in top level.

4.3 Fine-Tuning

In previous steps, we match sink input slew to a slew limit. When sinks have similar capacitance loading, the similar slew goes with similar latency and results small skew. Nevertheless, skew would increases when sinks are with wide range of capacitance. To ensure that the nominal skew is close to zero, we embed the SPICE simulation in the last step.

In our clock topology, nominal skew can be easily improved by adjusting length of the wire previously mentioned in Figure 4. Unlike to match sink slew in previous sub-tree generation step, what to match here is latency. In this step, SPICE simulation first extracts the latency information, and then we set a target latency and calculate the adjusting wire length by Elmore delay. The SPICE simulation procedure repeats until the nominal skew is smaller than 1 ps.

5. EXPERIMENTAL RESULTS

The proposed approach is implemented in the C++ programming language, compiled with gcc v4.1.2 and run on a 2.4GHz Intel workstation. We obtain our result using

Table 3: Comparison of our results with [7, 1, 2, 8] on ISPD 2010 Benchmark

ISPD 2010 Benchmark	Method	95%LCS (ps)	Cap (fF)	Runtime (s)
cns01	[7]	7.16	445331	0.4
	Contango 2.0 [1]	7.01	198300	12015
	[2]	5.79	177460	2790
	[8]	7.32	142644	1092
	Our work	6.48	137971	472
cns02	[7]	7.33	933574	2.42
	Contango 2.0 [1]	7.34	375900	25006
	[2]	6.69	329920	7787
	[8]	7.42	265207	4314
	Our work	7.38	268289	1450
cns03	[7]	4.88	183702	1.57
	Contango 2.0 [1]	4.18	55860	3840
	[2]	3.46	50810	2094
	[8]	4.49	36609	383
	Our work	4.76	34167	79
cns04	[7]	4.01	196337	0.27
	Contango 2.0 [1]	4.46	71840	6075
	[2]	3.79	57440	2763
	[8]	6.70	51070	934
	Our work	7.14	42773	110
cns05	[7]	3.81	89094	0.10
	Contango 2.0 [1]	4.41	37690	2406
	[2]	3.68	28930	1110
	[8]	4.78	25129	278
	Our work	5.88	22133	40
cns06	[7]	7.40	160447	0.28
	Contango 2.0 [1]	6.05	47810	2660
	[2]	4.01	36120	1142
	[8]	6.41	32680	285
	Our work	5.61	28548	61
cns07	[7]	6.24	228243	0.30
	Contango 2.0 [1]	4.58	72660	2351
	[2]	5.65	57930	2968
	[8]	5.86	48316	818
	Our work	6.62	43910	133
cns08	[7]	7.64	228243	0.28
	Contango 2.0 [1]	5.15	52490	1987
	[2]	4.24	40430	1497
	[8]	5.07	32699	327
	Our work	6.50	28413	54
Average comparison	[7]		**4.94**	
	Contango 2.0 [1]		**1.63**	
	[2]		**1.33**	
	[8]		**1.10**	
	Our work		**1.00**	

the benchmarks from ISPD contest [9]. The benchmark includes information of circuit, inverter library, wire library and process variations. Wire variation has ±5% and power-supply variation has ±7.5%. The 95% local clock skew(LCS) obtained through 500 times of NGSPICE based on Monte Carlo simulations.

Table 3 compares our solutions with [7, 1, 2, 8][1]. The results show that our approach achieves averagely 394%, 63%, 33% and 10% better capacitance compared with [7], Contango 2.0, [2] and [8] respectively. Moreover, the average runtime is faster than Contango 2.0, [2] [8] but slower than [7]. Though it is not fair to compare the run-time on the different workstation, we believe that our approach can

---

[1] [7] is performed on a 2.6GHz AMD-64; [1] is on 2.5GHZ Intel Quad Core; [2] is on 2GHz Intel with 3GB memory; [8] is on 2.2GHz Intel with 2.9GB memory

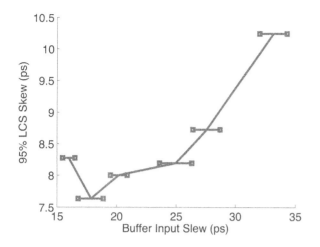

Figure 6: Slew vs. skew. When slew is less than 20ps, improving slew rate may not be effective strategy to reduce skew. All designs in this experiment are set to have equal power, the deviation is less than 0.6% .

reduce the number of times of SPICE simulation compared with others.

Figure 6 shows an experiment on benchmark ispdcns08. In this experiment, we sweep buffer input slew and observe its resulted skew. To get different slew value, we do not insert buffer on branches in this experiment. That is because, in some levels, wire lengths (between two connecting branches) are much shorter than other levels. Inserting buffer on branches results divergent input slew among levels. Instead, we insert buffers by a fixed distance. Although the branches makes the buffers have different output RC network, but input slew varies little. We manually adjust the fixed distance and buffer size to make all clock designs in this experiment have equal level power and different input slew. The power deviation is less than 1%, and the total capacitance is about 28.5pF. When slew is larger than 25ps, it is clear that the larger slew results in larger supply-voltage variation induced skew. When slew is less than 20ps, continuing to improve slew may not be a good strategy, since slew value is hard to improve in this range. And the effect of buffer level gradually dominates the skew. According these equal power designs, it is clear that to synthesize a low power clock tree, selecting a reasonable input slew value for inserted buffers in different levels is influential. Our methodology effectively selects the slew rate of inserted buffers for clock tree synthesis and results in lower power than other works.

6. CONCLUSIONS

In this paper, we point out that for power optimization of clock network under supply-voltage variation, searching solutions along with slew constraint boundary will not get the best solution. Although the solutions on the boundary of slew constraint have less number of variation source, its slow slew amplifies the variation more. We adopt a hybrid structure, a symmetry top tree driving bottom level non-symmetry tree, which keeps wire-saving in bottom level and

reduce the complexity to estimate supply-voltage induced skew. By keeping the slew rate better than constraint, the modification on equal distribution of clock latency variation among buffer levels, and greedily saving power in bottom level; we reduce the 10% power compared with state-of-the-art clock network synthesis on ISPD 2010 benchmarks.

In addition, the buffer size decision still needs some manual work in this paper, only one buffer size is used in each design, and spatial correlation of supply-voltage variation is not addressed; we will pursue these topics in our future work.

7. REFERENCES

[1] D.J. Lee, M.C. Kim, and I.L. Markov. "Low-Power Clock Trees for CPUs". In *International Conference on Computer-Aided Design*, pages 444–451, 2010.

[2] S. Bujimalla and C.-K. Koh. "Synthesis of Low Power Clock Trees for Handling Power-Supply Variations". In *International Symposium on Physical Design*, pages 37—44, 2011.

[3] Y.-C. Hsu T.-H. Chao and J.-M. Ho. "Zero Skew Clock Net Routing". In *Design Automation Conference*, pages 518—523, 1992.

[4] K.D. Boese and A.B. Kahng. "Zero-Skew Clock Routing Trees with Minimum Wirelength". In *ASIC Conference and Exhibit*, pages 17—21, 1992.

[5] R.S. Tsay. "Exact Zero Skew". In *International Conference on Computer-Aided Design*, pages 336—339, 1991.

[6] X.W. Shih and Y.W. Chang. "Fast Timing-Model Independent Buffered Clock-Tree Synthesis". In *Design Automation Conference*, pages 80–85, 2010.

[7] X.W. Shih, H.C. Lee, K.H. Ho, and Y.W. Chang. "High Variation-Tolerant Obstacle-Avoiding Clock Mesh Synthesis with Symmetrical Driving Trees". In *International Conference on Computer-Aided Design*, pages 452–457, 2010.

[8] T. Mittal and C.-K. Koh. "Cross Link Insertion for Improving Tolerance to Variations in Clock Network Synthesis". In *International Symposium on Physical Design*, pages 29—36, 2011.

[9] C.N. Sze. "ISPD 2010 High Performance Clock Network Synthesis Contest: Benchmark Suite and Results". In *International Symposium on Physical Design*, pages 143–143, 2010.

[10] S.D. Kugelmass and K. Steighlitz. "An Upper Bound on Expected Clock Skew in Synchronous Systems". *IEEE Transactions on Computers*, 39(12):1475–1477, 1990.

[11] A. Rajaram and D.Z. Pan. "Variation Tolerant Buffered Clock Network Synthesis with Cross Links". In *International Symposium on Physical Design*, pages 157—164, 2006.

Optimizing the Antenna Area and Separators in Layer Assignment of Multi-Layer Global Routing

Wen-Hao Liu, Yih-Lang Li

Department of Computer Science, National Chiao-Tung University, Hsin-Chu,Taiwan
dnoldnol@gmail.com, ylli@cs.nctu.edu.tw

ABSTRACT - Traditional solutions to antenna effect, such as jumper insertion and diode insertion peformed at post-route stage may produce extra vias and degrade circuit performance. The work in [1] suggests combining layer assignment, jumper insertion and diode insertion together to achieve a better design quality with less additional cost. Based on our observations on *global* and *local* antenna violations, this work proposes a dynamic-programming based single-net layer assignment called NALAR, which first enumerates all antenna-violation-safe layer assignment solutions of a net, and then extracts the minimum-cost one for the net. NALAR can minimize via count and separators as well. In addition, an antenna avoidance layer assignment algorithm (ANLA) adopting NALAR as its kernel not only avoids global antenna violations, but also eliminates local antenna violations. Experimental results reveal that, in 11 benchmarks, ANLA can yield 5 violation-free assignments while the algorithms of other works yield no violation-free assignment. As for the total number of antenna violations in all benchmarks, this work and the works in [2], [3] and [4] yield 21, 43506, 41261 and 29671 antenna violations, respectively. However, ANLA performs about 7 times slower than other antenna-aware layer assignment [4].

Categories and Subject Descriptors

B.7.2 [**Integrated Circuits**]: Design Aids - Placement and Routing.

General Terms: Algorithms, Design.

Keywords

Layer assignment, Global Routing, Antenna Effect, Separator, Via

1. INTRODUCTION

Owing to the time consuming nature of 3D routing, the conventional approach condenses the 3D grid graph into a 2D grid graph. 2D routing results are then obtained using 2D global routing. Finally, layer assignment assigns each net edge to the corresponding metal layer in order to obtain the final 3D routing results. Figure 1 displays the common design flow adopted by most state-of-the-art global routers [5-13].

The layer assignment in 3D global routing minimizes the via count without altering the routing topology or increasing the wire overflow, which is called constrained via count minimization problem and has been proven to be NP-complete [14]. This problem has been extensively studied [2][3][12][15]. The antenna effect is another relevant topic that must be considered in routing to improve circuit reliability. Some investigations [1][4][16] have studied the antenna effect during layer assignment stage. The antenna effect is charge collection by plasma during lithography, possibly resulting in gate oxide damages and ultimately the

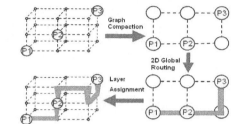

Fig. 1. modern 3D global routing flow.

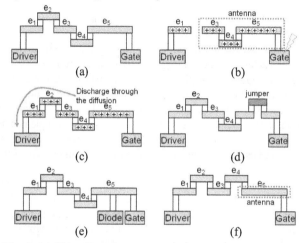

Fig. 2. (a) Side view of a routing result connecting a driver to a gate; (b) the circumstance in which metal 1 and metal 2 are etched; (c) the circumstance after manufacturing metal 3; (d) solving antenna violation by jumper insertion; (e) solving antenna violation by diode insertion; (f) a routing result as layer assignment obeys the antenna rule.

circuit reliability problem. When exposed to plasma, a wire segment functioning as an antenna may gather charging current. If the wire segment connects only to the gate oxides (but not diffusions) and collects significantly more charges than a threshold value does, Fowler-Nordheim tunneling current discharges through the thin oxide to the substrate, incurring the gate oxide damage [17][18]. For instance, Fig. 2(a) shows the side view of a routing result connecting a driver (diffusion) to a gate, wire segment e_4 located on metal 1, e_1, e_3 and e_5 located on metal 2, and e_2 located on metal 3. Figure 2(b) shows the circumstance, while metals 1 and 2 are etched; since metal 3 has not yet been built, the collected charges of e_3, e_4 and e_5 discharge through the thin gate oxides. In Fig. 2(b), wire segments e_3, e_4 and e_5 function as an antenna. Also, the gate oxide damage may occur if the collected charges of the antenna exceed a threshold, which is referred to as **antenna violation**. Figure 2(c) shows the circumstance after manufacturing metal 3, in which the collected charges can be released through the diffusion.

Jumper insertion and diode insertion are generally adopted to solve the antenna violation problem. Figure 2(d) illustrates an example of solving the antenna violation problem by using jumper insertion. The wire segments with antenna violations are split and then routed to the top-metal layer; the wire on the top-metal layer is called a jumper. The jumper cuts a long antenna into a shorter one, yet consumes additional vias. On the other hand, by placing diodes near the gates with antenna violations (Fig. 2(e)), the diode insertion approach can protect the gates from current discharge through the thin oxide to the substrate by restraining the charge voltage level. Also, search-and-repair is performed for antenna violation during detailed routing or post optimization stages by using jumper insertion and diode insertion. However, the additional vias and inserted diodes consume additional routing resources, and degrade the circuit performance and manufacturing yield.

In contrast to the above mentioned search-and-repair approaches during the post-route stage, considering the antenna effect during early stages can prevent antenna violations at a lower cost. Previous works [1][4][16] have developed layer assignment algorithms while considering the **antenna rule** to avoid antenna violation. Figure 2(f) shows a routing result as layer assignment complies with the antenna rule. Although Figs. 2(a) and 2(f) have the same via count, the antenna area in Fig. 2(f) is shorter than that in Fig. 2(a). In [1], layer assignment considering antenna rule is regarded as a complementary approach of jumper insertion and diode insertion. These three approaches can be integrated to achieve an improved design solution.

Recent works [1][4] have applied a tree-partitioning algorithm to facilitate antenna avoidance layer assignment. Those algorithms treat each net as a tree, and break each one into sub-trees by the tree-partitioning algorithm. The wire segments of each sub-tree are then assigned to the corresponding layer under the antenna rule. Although efficient, these tree-partitioning-based algorithms may fall into the local optimal. This work presents a novel antenna avoidance layer assignment algorithm called ANLA, capable of avoiding antenna violations and minimizing the via count for an entire net. The proposed algorithm provides a more global view and achieve a higher quality than previous tree-partitioning-based algorithms [1][4].

The rest of paper is organized as follows: Section 2 describes the problem formulation. Section 3 then introduces previous works. Section 4 presents the proposed antenna avoidance layer assignment algorithm. Additionally, Section 5 summarizes the experimental results. Conclusions are finally drawn in Section 6.

2. PROBLEM FORMULATION
2.1. Preliminaries

The grid graph model is generally applied in the global routing and layer assignment problems. According to Fig. 3(a), a k-layer routing region can be partitioned into an array of tiles and modeled by a k-layer grid graph $G^k(V^k, E^k)$ such as that shown in Fig. 3(b), where V^k denotes the set of 3D grid cells; each grid cell represents a tile; and E^k refers to the set of 3D grid edges, in which each grid edge is termed by the adjacency of the related tile of its two end nodes. The capacity of a grid edge e ($cap(e)$) indicates the number of

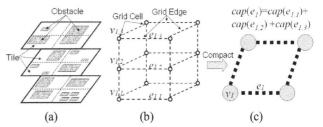

Fig. 3. Mapping a 3D routing region to the grid graph model. (a) Three-layer routing region; (b) 3D grid graph; (c) compacted 2D grid graph from (b);

wire segments that can pass through e. The overflow of e is defined as the amount of demand in excess of capacity.

The layer assignment problem is formulated as follows. Given a 3-tuple(G^k, G, S), $G^k(V^k, E^k)$ denotes a k-layer 3D grid graph; $G(V, E)$ represents a 2D grid graph that is compressed from G^k; and S refers to a 2D global routing result on G. The 3D grid edge $e_{i,z}$, $1 \leq z \leq k$, is called the *corresponding edge* of 2D grid edge e_i, and the 3D grid node $v_{i,z}$, $1 \leq z \leq k$, is called the *corresponding node* of 2D grid node v_i. Layer assignment assigns the wire segments of S to the corresponding edges in order to obtain a 3D global routing result S^k. For instance, in Figs. 3(b) and 3(c), $e_{1,1}$, $e_{1,2}$ and $e_{1,3}$ are the corresponding edges of e_1; by assuming that a 2D global router identifies a wire segment passing through e_1 in Fig. 3(c), layer assignment assigns the wire segment to one of $e_{1,1}$, $e_{1,2}$ and $e_{1,3}$ in Fig. 3(b). With the imposed wire congestion constraints and antenna rule, layer assignment can ensure the feasibility of assignment results.

2.2. Wire Congestion Constraints

Given a 2D global routing result S, layer assignment identifies the k-layer assignment result S^k. To ensure its legality and routability, S^k must satisfy the following wire congestion constraints:

$$TWO(S^k) \leq TWO(S) \quad (1)$$

$$MWO(S^k) \leq \lceil MWO(S) * (2/k) \rceil \quad (2)$$

, where TWO and MWO denote the total wire overflow and maximum wire overflow, respectively. The first constraint ensures that the wire overflow in 3D graph does not exceed that in 2D graph, while the second constraint ensures that the congestion is averagely distributed to each layer for overflowed regions.

2.3. Antenna Rule

Given a 3D global routing result of a net with a driver and a set of output pins (gates). For each output pin p, let $L_{top}(p)$ denote the top layer containing at least one wire segment in the routing path between the driver and p. The antenna of p is defined as the maximum sub-tree that contains p and consists of a set of wire segments on the metal layers lower than $L_{top}(p)$. The wire segments surrounding antennas are called *separators* [1]. For instance, Fig. 4(a) shows a 3D routing result of a net, the notation next to each wire segment represents the 3D grid edge that the wire segment is assigned. The layers of $L_{top}(p_1)$, $L_{top}(p_2)$, $L_{top}(p_3)$ and $L_{top}(p_4)$ are 3, 3, 4 and 6, respectively. The associated antennas of p_1, p_2, p_3 and p_4 include the sets of wire segments $\{e_{2,2}, e_{3,1}, e_{5,2}\}$, $\{e_{2,2}, e_{3,1}, e_{5,2}\}$, $\{e_{6,3}\}$ and $\{e_{8,5}, e_{9,4}\}$, respectively. Notably, p_1 and p_2 share the same antenna. The blue wire segments in Fig. 4(a) are separators. A separator is formally defined as follows:

Definition 1. *separator*: a wire segment $e_{a,al}$ is regarded as a separator if $e_{a,al}$ is the nearest wire segment to pin p in a path from p to the driver such that the layer of $e_{a,al}$ is $L_{top}(p)$.

To avoid gate oxide damage, [4] attempts to make the antenna ratio of each antenna less than a given threshold A_{max}. This limitation is the antenna rule imposed on the layer assignment problem in [4] which adopted the cumulative antenna model to define the antenna ratio as follows:

$$\text{antenna ratio} = \frac{\text{total exposed antenna area}}{\text{total gate oxide area}} \quad (3)$$

To simplify the antenna ratio calculation, [4] assumed a uniform wire width and a uniform gate oxide area. Thus, the antenna ratio calculation can be simplified as L/n, where L denotes the antenna length and n represents the number of output pins connected by this antenna. In Fig. 4(a), the antenna ratio of the associated antennas of p_1, p_2, p_3 and p_4 are 1.5, 1.5, 1 and 2, respectively.

Complying with the antenna rule of [4] can reduce the antenna violations in subsequence stages. However, in some circumstances, layer assignment results that obey the abovementioned antenna rule still damage gate oxides. For instance, $e_{2,3}$ is the separator of p_1 and p_2 (Fig. 4(b)) and assume that A_{max} is 2. The p_1 and p_2 share an antenna $\{e_{3,1}, e_{4,1}, e_{5,1}, e_{6,2}\}$. The antenna ratio of $\{e_{3,1}, e_{4,1}, e_{5,1}, e_{6,2}\}$ is 2, which does not exceed A_{max}, explaining why Fig. 4(b) conforms to the above antenna rule. However, during manufacturing, while metal 1 is etched and metal 2 and metal 3 have not yet been built, the collected charges of e_3, e_4 and e_5 may discharge through p_1 and then damage the gate oxide of p_1. This circumstance is referred to herein as ***local-antenna-violation.*** Conversely, the circumstance in which the antenna ratio of an entire sub-tree exceeds A_{max} is defined to ***global-antenna-violation.*** The work [4] addressed the feasibility of eliminating global-antenna-violation, yet was unaware of local-antenna-violation. Thus the experiment in section 5 reveals that the layer assignment results of [4] contain a significant amount of antenna-violations. In this work, local-antenna-violation is avoided using a strict antenna rule; the strict antenna rule limits a situation in which the antenna length rather than the antenna ratio is less than A_{max}. If an assignment result of a net conforms to this strict antenna rule, the assignment result is regarded as ***antenna-violation-safe.*** This finding implies that the net never incurs global-antenna-violation and local-antenna-violation.

2.4. Objectives

In the antenna avoidance layer assignment problem, minimizing the number of nets with the antenna violations is of priority concern, while minimizing the via count is the secondary objective. Moreover, the fact that the separators must be fixed at a specified layer explains why too many separators may degrade the flexibility of the subsequent detailed routing. Thus, minimizing separators is the third objective. The final assignment result must conform to wire congestion constraints. In this work, antenna effect is regarded as an objective rather than a constraint,

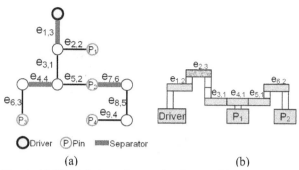

O Driver (P) Pin ▰ Separator

(a)   (b)

Fig. 4. (a) 3D global routing result of a net; (b) example with local-antenna-violation.

since antenna violations in the final assignment result can be resolved by jumper or diode insertion at detailed routing stage.

3. PREVIOUS WORKS

The following sub-section briefly describes a recent layer assignment work [3] that focuses on minimizing the via count, yet fails to consider the antenna effect. Sub-section 3.2 then introduces current antenna avoidance layer assignment works [1][4] and discusses the potential limitations of [1][4].

3.1. Negotiation-Based Via Count Minimization

[3] developed a negotiation-based single-net layer assignment algorithm called *NANA*, which can always identify the minimum-cost assignment result for a net $N(V_N, E_N)$. NANA is a 2-phase dynamic-programming algorithm. During the first phase, NANA enumerates all possible 3D trees starting at leaf nodes in a bottom-up manner until reaching the root. During the second phase, the minimum-cost assignment solution for the entire tree is extracted from a set of possible 3D trees. Each net edge is then assigned to the corresponding layer based on the minimum-cost assignment solution in a top-down manner. The following equation defines the cost of a 3D tree t:

$$\text{cost}(t) = viaCost \times \text{numVia}(t) + \sum_{e \in t} \text{congCost}(e) \quad (4)$$

, where the cost of a 3D tree equals the sum of the congestion cost and the via cost; *viaCost* refers to a user defined constant for the cost of a single via; numVia(t) represents the via number of t and congCost(e) denotes the congestion cost of the grid edge e.

The framework of [3] consists of three stages: initial layer assignment, wire overflow reduction and post optimization. Each stage adopts NANA with different congestion cost formulae to assign nets for different objectives. The initial layer assignment stage first identifies a minimal via count solution without considering wire congestion constraints for each net by fixing the congestion cost at zero. During the wire overflow reduction stage, if a grid edge e overflows, the nets passing through e are ripped up and re-assigned; the congestion cost of e increases as well. The formulation of congestion cost in this stage can be expressed as follows:

$$congCost(e) = p_e * (1 + (h_e)^2) \quad (5)$$

, where p_e denotes congestion penalty, and h_e represents history cost. The congestion penalty is defined as follows:

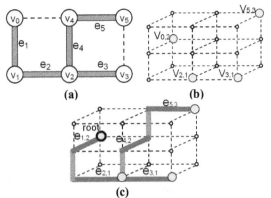

(a)   **(b)**

(c)

Fig. 5. Example of a single net layer assignment. (a) a 2D routing result of a net $N(V_N, E_N)$; (b) a set of N's pin locations in G^k; (c) layer assignment solution.

$$p_e = 1 + \frac{\alpha}{1 + exp^{\beta * (cap(e) - dem(e))}} \qquad (6)$$

, where $cap(e)$ and $dem(e)$ refer to the capacity and demand of e, respectively. Notably, α and β are user defined constants. The history cost h_e increases by one as the grid edge e overflows; in addition, h_e remains unchanged if e does not overflow. The value of h_e in the k-th iteration can be expressed as,

$$h_e^{k+1} = \begin{cases} h_e^k + 1, & \text{if } e \text{ overflows} \\ h_e^k, & \text{otherwise} \end{cases} \qquad (7)$$

During the post optimization stage, each net is re-assigned to greedily improve the via count under wire congestion constraints. The following congestion cost formula is used in the post optimization stage,

$$congCost(e) = \begin{cases} \sigma * (1 + dem(e) - cap(e)), & \text{if } dem(e) \geq cap(e) \\ 0, & \text{otherwise} \end{cases} \qquad (8)$$

, where σ denotes an extremely large constant.

3.2. Antenna avoidance Layer Assignment

In current antenna avoidance layer assignment works [1][4], each single net layer assignment comprises two steps. In the first step, the separators location is determined using a tree-partitioning algorithm, in which the separators decompose a net into several sub-nets, subsequently causing the antenna ratio of each sub-net to be less than A_{max}. In the second step, the wire segments of each sub-net are assigned to the layers lower than the surrounding separators. In the second step, no solution or an inferior quality solution with many vias may be found since the wire congestion and via count are not considered in the first step. In contrast to [1][4], this work presents a single net layer assignment algorithm called NALAR, capable of determining the separator locations and assigning wire segments to the corresponding layers in a single step. The via count, wire congestion and antenna rule can be optimized simultaneously as well. Additionally, current works [1][4] lack the ability to control the number of separators, while this work formulates the separator cost into the objective function of NALAR in order to control the number of separators.

4. PROPOSD ANTENNA AVOIDANCE LAYER ASSIGNMENT (ANLA)

To avoid antenna violations, this work presents a negotiation-based single-net layer assignment considering antenna rule (NALAR). If a net has at least an antenna-violation-safe assignment solution, NALAR can identify the minimum-cost antenna-violation-safe solution for this net. Sub-section 4.1 details NALAR. Sub-section 4.2 presents the flow of ANLA which adopts NALAR to assign each single net.

4.1. Negotiation-Based Single-Net Layer Assignment Considering Antenna Rule (NALAR)

The single-net layer assignment problem is formulated as follows. Given a 2D routing result of a net $N(V_N, E_N)$ where V_N and E_N denote the sets of 2D grid nodes and 2D grid edges passed by N, respectively; and given a set of pin locations of N in G^k, which is denoted by P_N^k. The net N can be regarded as a 2D tree, and the root and the leaves of the tree must contain a pin. The single-net layer assignment problem identifies a 3D tree in G^k to connect all pins of P_N^k. As the layer assignment result of N, this 3D tree consists of a set of 3D grid edges, denoted as E_N^k, the edges of which E_N^k are the corresponding edges of E_N. For instance, a single net layer assignment reads a 2D routing result $N(V_N, E_N)$ in Fig. 5(a) and a set of pins P_N^k in Fig. 5(b), $V_N = \{v_0, v_1, v_2, v_3, v_4, v_5\}$, $E_N = \{e_1, e_2, e_3, e_4, e_5\}$ and $P_N^k = \{v_{0,2}, v_{2,1}, v_{3,1}, v_{5,3}\}$. The single net layer assignment then identifies a 3D tree, as shown in Fig. 5(c), $E_N^k = \{e_{1,2}, e_{2,1}, e_{3,1}, e_{4,2}, e_{5,3}\}$. NALAR is developed to solve this single-net layer assignment problem and avoid antenna violations. Before detailed NALAR, some notations are introduced first.

1. k: layer number of the 3D grid graph.
2. $ch(v_i)$: set of child nodes of v_i. In Fig. 5(a), $ch(v_2) = \{v_3, v_4\}$.
3. $ch\_e(v_i)$: set of grid edges connecting v_i to its child nodes. In Fig. 5(a), $ch\_e(v_2) = \{e_3, e_4\}$.
4. $pinL(v_i, P_N^k)$: if a pin is located at a corresponding node of v_i, $pinL(v_i, P_N^k)$ represents the layer of the corresponding node. In Fig. 5(b), $pinL(v_5, P_N^k) = 3$.
5. $T(v_i)$: a 2D tree rooted at v_i.
6. $t_{i,z}$: a 3D tree rooted at $v_{i,z}$. The $t_{i,z}$ is an assignment solution of $T(v_i)$. A 3D tree consists of a set of 3D grid edges and a set of required vias for connecting these grid edges and pins. Figure 5(c) illustrates a 3D tree $t_{0,2}$.
7. $S(v_{i,z})$: set of 3D trees rooted at $v_{i,z}$.
8. $S(v_i)$: set of assignment solutions of $T(v_i)$.
9. $via(\Delta)$: Δ denotes a set of layers, and $via(\Delta)$ represents a set of required vias connecting the layers of Δ.

4.1.1. Overview of NALAR

NALAR is a 2-phase algorithm based on dynamic-programming method. Figure 6 presents the pseudo code of NALAR. Lines 2 to 13 are the bottom-up phase; and the procedure **InitSol** initializes a 3D tree rooted at the leaf node. The procedure **EnumSol** enumerates all antenna-violation-safe 3D trees rooted at the internal node $v_{i,z}$ and identifies a set of separators for each enumerated 3D tree. Additionally,

Algorithm NALAR
Input: net $N(V_N, E_N)$, pin locations P_N^k, 3D grid graph G^k;
1.   //Bottom-Up phase
2.   **foreach** node v_i in the order given by a postorder traversal of V_N, v_i is not the root.
3.   **foreach** layer z from 1 to k
4.   **set** $S(v_{i,z})$ to Φ
5.   **if** v_i is a leaf node
6. $S(v_{i,z}) \leftarrow$ InitSol$(v_{i,z}, P_N^k)$
7.   **else**
8. $S(v_{i,z}) \leftarrow$ EnumSol$(v_{i,z}, ch(v_i), ch\_e(v_i), P_N^k)$
9.   **end if**
10.   PruneSol$(S(v_{i,z}))$
11.   **end foreach**
12. $S(v_i)=S(v_{i,0}) \cup S(v_{i,1}) \cup ... \cup S(v_{i,k})$
13. **end foreach**
14. //Top-Down phase
15. $x=pinL(v_0, P_N^k)$
16. $S(v_{0,x}) \leftarrow$ EnumSol$(v_{0,x}, ch(v_0), ch\_e(v_0), P_N^k)$
17. $t_{0,x} \leftarrow$ Select_solution$(S(v_{0,x}))$
18. TopDown_Assignment$(N, G^k, t_{0,x})$
19. **end**

Fig. 6. The pseudo code of NALAR

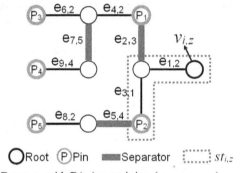

○Root Ⓟ Pin ▬Separator ⸽⸽⸽ $st_{i,z}$

Fig. 7. 3D tree $t_{i,z}$ with $B(t_{i,z})$ containing $\{e_{2,3}, e_{5,4}, e_{7,5}\}$

$B(t_{i,z})$ refers to the separator set of $t_{i,z}$. For instance, Fig. 7 shows a 3D tree $t_{i,z}$, in which $B(t_{i,z})$ contains $\{e_{2,3}, e_{5,4}, e_{7,5}\}$. Notably, a separator, states $e_{a,al}$, has the following properties:
Property 1. The layer of a separator must be higher than the layer of its neighboring antennas' grid edges. For instance, in Fig. 7, $\{e_{6,2}, e_{4,2}\}$ and $\{e_{9,4}\}$ are the neighboring antennas of the separator $e_{7,5}$. Therefore, the layer of $e_{7,5}$ is higher than that of $e_{6,2}, e_{4,2}$ and $e_{9,4}$.
Property 2. A separator $e_{b,bl}$ appears before $e_{a,al}$ in a path from a leaf to the root, the layer of $e_{a,al}$ must not be higher than $e_{b,bl}$, owing to that if the layer of $e_{a,al}$ is higher than $e_{b,bl}$, $e_{b,bl}$ cannot be regarded as a separator according to the separator's definition.
The 3D trees generated by **InitSol** and **EnumSol** are inserted into $S(v_{i,z})$. At line 10, the procedure **PruneSol** discards the redundant trees from $S(v_{i,z})$, and allows the size of $S(v_{i,z})$ to be limited in an acceptable range. Lines 15 to 18 are the top-down phase. Line 16 enumerates all antenna-violation-safe assignment solutions for the entire 3D tree into $S(v_{0,x})$. Line 17 extracts the minimum-cost assignment solutions from $S(v_{0,x})$. Finally, each wire segment of N is assigned to the corresponding layers according to the minimum-cost assignment solution. The cost of a 3D tree $t_{i,z}$ is

Procedure InitSol
Input: node $v_{i,z}$, 3D pins location P_N^k
1.   I= via(pinL(v_i, P_N^k), z)
2. $t_{i,z}$=I
3. $B(t_{i,z})=\Phi$
4. $cost(t_{i,z})=|I| * viaCost$
5. $AL(t_{i,z})=0$
6. $LS(t_{i,z})=\infty$
7. $TW(t_{i,z})=0$
8.   **inset** $t_{i,z}$ into $S(v_{i,z})$
9.   **return** $S(v_{i,z})$

Fig. 8. Procedure InitSol of NALAR

evaluated by the following equation,

$$cost(t_{i,z}) = sepCost \times numSP(t_{i,z}) + viaCost \times numVia(t_{i,z}) + \sum_{e \in t_{i,z}} congCost(e) \quad (9)$$

, where *sepCost* denotes a user defined constant for the cost of a single separator and *numSP*$(t_{i,z})$ represents the separator number of $t_{i,z}$. Adjusting the value of *sepCost* significantly affects the ability to control the number of separators. In Section 5, experimental results indicate the effectiveness of adjusting the value of *sepCost*.

Two issues in NALAR must be addressed: how to enumerate antenna-violation-safe assignment solutions and how to prune the surplus solutions in order to reduce the size of a solution set with the minimum-cost assignment solution still remaining in the solution set.

4.1.2. Enumerating Antenna-Violation-Safe LA Solutions

To address the first issue, in **InitSol** and **EnumSol**, the $AL(t_{i,z})$, $LS(t_{i,z})$ and $TW(t_{i,z})$ of each 3D tree $t_{i,z}$ are defined. These attributes are used to determine whether $t_{i,z}$ violates the antenna rule. Notably, $AL(t_{i,z})$ represents the antenna length of $st_{i,z}$ which is the maximum sub-tree of $t_{i,z}$ rooted at $v_{i,z}$ and is surrounded by separators. Also, $TW(t_{i,z})$ represents the highest layer containing at least one wire segment in $st_{i,z}$. Additionally, $LS(t_{i,z})$ represents the lowest layer containing at least one separator in $t_{i,z}$. For instance, in Fig. 7, $AL(t_{i,z})=2$, $TW(t_{i,z})=2$ and $LS(t_{i,z})=3$. Notably, if $t_{i,z}$ contains no separator, $LS(t_{i,z})$ is initialized to an extremely large constant. If $st_{i,z}$ contains no wire segment, $TW(t_{i,z})$ and $AL(t_{i,z})$ are initialized to zero.

Figure 8 shows the pseudo code of **InitSol**. Line 4 calculates the cost of $t_{i,z}$, where $|I|$ denotes the number of required vias connecting the pin's layer to layer z. Lines 5 to 7 initialize the attribute values of $AL(t_{i,z})$, $LS(t_{i,z})$ and $TW(t_{i,z})$. Figure 9 shows the pseudo code of **EnumSol**. For an easy explanation, assume that v_i has three child nodes in the pseudo code. The loop from lines 1 to 12 enumerates all combinations of the sub-trees of $t_{i,z}$. The $t_{a,la}$, $t_{b,lb}$ and $t_{c,lc}$, are the sub-trees of $t_{i,z}$, and root at $v_{a,la}$, $v_{b,lb}$ and $v_{c,lc}$, respectively. The $v_{a,la}$, $v_{b,lb}$ and $v_{c,lc}$ are the corresponding nodes of $ch(v_i)$. Let $e_{a,la}$, $e_{b,lb}$ and $e_{c,lc}$ denote the 3D grid edges connecting $v_{a,la}$, $v_{b,lb}$ and $v_{c,lc}$ to the 3D nodes $v_{i,la}$, $v_{i,lb}$ and $v_{i,lc}$, respectively. The loop from lines 2 to 11 enumerates all combinations of (f_a, f_b, f_c). Notably, the notations f_a, f_b and f_c are the boolean values implying whether $e_{a,la}$, $e_{b,lb}$, $e_{c,lc}$ are the separators, respectively. The two loops explore all 3D trees rooted at $v_{i,z}$ with different separator assignments. At line 4, a 3D tree $t_{i,z}$ is constructed by

Input: node $v_{i,z}$, $ch(v_i)=\{v_a, v_b, v_c\}$, $ch\_e(v_i)=\{e_a, e_b, e_c\}$, 3D pins location P_N^k

1.   **foreach** solution $t_{a,la}$, $t_{b,lb}$ and $t_{c,lc}$ of $S(v_a)$, $S(v_b)$ and $S(v_c)$, respectively. Let $e_{a,la}$, $e_{b,lb}$ and $e_{c,lc}$ are the 3D grid edges connecting the roots of $t_{a,la}$, $t_{b,lb}$ and $t_{c,lc}$ to the 3D nodes $v_{i,la}$, $v_{i,lb}$ and $v_{i,lc}$, respectively.
2.   **foreach** combination (f_a, f_b, f_c)
3.   I= $via(la, lb, lc, pinL(v_i, P_N^k), z)$
4. $t_{i,z}= t_{a,la}+t_{b,lb}+ t_{c,lc}+e_{a,la}+e_{b,lb}+ e_{c,lc}$+I
5. $B(t_{i,z})=B(t_{a,la})+B(t_{b,lb})+B(t_{c,lc})+$
                          $f_a*e_{a,la}+f_b*e_{b,lb}+f_c*e_{c,lc}$
6. $calculateCost(t_{i,z}, f_a, f_b, f_c)$
7. $setAttributes(t_{i,z}, f_a, f_b, f_c)$
8.   **if** $t_{i,z}$ **is an antenna-violation-safe tree**
9.   **insert** $t_{i,z}$ into $S(v_{i,z})$
10.   **endif**
11.   **end foreach**
12. **end foreach**
13. **return** $S(v_{i,z})$

Fig. 9. Procedure EnumSol of NALAR

composing $t_{a,la}$, $t_{b,lb}$, $t_{c,lc}$, $e_{a,la}$, $e_{b,lb}$, $e_{c,lc}$ and I. The vias in I connect $v_{i,la}$, $v_{i,lb}$, $v_{i,lc}$ and $v_{i,z}$. If a pin is located at a corresponding node of v_i, vias also connect to this node. Figure 10 illustrates an example of constructing a 3D tree $t_{i,3}$, which consists of $t_{a,2}$, $t_{b,3}$, $t_{c,4}$, $e_{a,2}$, $e_{b,3}$, $e_{c,4}$, and I. If $(f_a, f_b, f_c)=\{1, 0, 1\}$, $e_{a,2}$ and $e_{c,4}$ are regarded as the separators. Line 5 builds the separator set of $t_{i,z}$. Line 6 calculates the cost of $t_{i,z}$ by Eq. (10), and then line 7 sets the attribute values of $AL(t_{i,z})$, $LS(t_{i,z})$ and $TW(t_{i,z})$ by Eqs. (11), (12) and (13), respectively.

$$cost(t_{i,z})=|I|*viaCost + \sum_{v_s \in ch(v_i)} \frac{cost(t_{s,ls})+f_s*sepCost}{+congCost(e_{s,ls})} \quad (10)$$

$$AL(t_{i,z}) = \sum_{v_s \in ch(v_i)} (1-f_s)*(AL(t_{s,ls})+1) \quad (11)$$

$$LS(t_{i,z}) = \min_{v_s \in ch(v_i)} (LS(t_{s,ls}), (1-f_s)*k+ls) \quad (12)$$

$$TW(t_{i,z}) = \max_{v_s \in ch(v_i)} ((1-f_s)*TW(t_{s,ls})) \quad (13)$$

Line 8 verifies whether $t_{i,z}$ is antenna-violation-safe. Only the *antenna-violation-safe tree* can be inserted into the solution set $S(v_{i,z})$.

Definition 2. *antenna-violation-safe tree*: The $t_{i,z}$ is regarded as an *antenna-violation-safe tree* if v_i is the leaf node or v_i is the internal node such that the following four conditions are true. Assume that $v_s \in ch(v_i)$ and $t_{s,ls}$ is one of sub-trees of $t_{i,z}$.

a).   Each $t_{s,ls}$ is an antenna-violation-safe tree.
b). $AL(t_{i,z}) \leq A_{max}$.
c).   For each $t_{s,ls}$, $ls > TW(t_{s,ls})$ and $ls \leq LS(t_{s,ls})$ if $e_{s,ls} \in B(t_{i,z})$.
d).   For each $t_{s,ls}$, $ls < LS(t_{s,ls})$ if $e_{s,ls} \notin B(t_{i,z})$.

The first condition ensures that all sub-trees of $t_{i,z}$ are antenna-violation-safe trees. The second condition ensures that the antenna length of $st_{i,z}$ is shorter than A_{max}, which prevents the global-antenna-violation and local-antenna-violation. The third and fourth conditions make the separators of $t_{i,z}$ conform to Properties 1 and 2. According to the third and the fourth

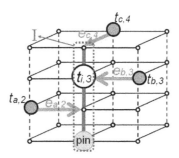

Fig. 10. Illustrative example for constructing a 3D tree $t_{i,3}$, which consists of $t_{a,2}$, $t_{b,3}$, $t_{c,4}$, $e_{a,2}$, $e_{b,3}$, $e_{c,4}$, and I

conditions, if $e_{s,ls}$ is a separator, $e_{s,ls}$ must be located at one of the layers between $(TW(t_{s,ls}), LS(t_{s,ls})]$; if $e_{s,ls}$ is not a separator, $e_{s,ls}$ must be located at one of the layers lower than $LS(t_{s,ls})$. Therefore, if the layer range of $(TW(t_{s,ls}), LS(t_{s,ls})]$ increases, the candidate layers of $e_{s,ls}$ also increase. A 3D tree $t_{s,ls}$ with a larger layer range of $(TW(t_{s,ls}), LS(t_{s,ls})]$ is regarded as having a higher flexibility.

4.1.3. Pruning Inferior Solutions

To limit the size of solution set in an acceptable range, **PruneSol** discards the *inferior trees* from $S(v_{i,z})$. An inferior tree is defined as follows:

Definition 3. *inferior tree*: A 3D tree $t_{i,z}^w$ is regarded as an *inferior tree* if another 3D tree $t_{i,z}^u$ holding the following equations exists.

$$AL(t_{i,z}^w) \geq AL(t_{i,z}^u), \quad cost(t_{i,z}^w) \geq cost(t_{i,z}^u),$$
$$TW(t_{i,z}^w) \geq TW(t_{i,z}^u) \text{ and } LS(t_{i,z}^w) \leq LS(t_{i,z}^u) \quad (14)$$

where the layer range of $(TW(t_{i,z}^w), LS(t_{i,z}^w)]$ is covered by the layer range of $(TW(t_{i,z}^u), LS(t_{i,z}^u)]$. Therefore, $t_{i,z}^w$ has lower flexibility than $t_{i,z}^u$. Moreover, $t_{i,z}^w$ has a higher cost and longer antenna than $t_{i,z}^u$, thus $t_{i,z}^w$ is totally worse than $t_{i,z}^u$. Accordingly, $t_{i,z}^w$ should be discarded from $S(v_{i,z})$. To identify the inferior trees in $S(v_{i,z})$, the most intuitive method compares all pairs of the solutions in $S(v_{i,z})$. As the original size of $S(v_{i,z})$ is n, the time complexity of the intuitive method is $O(n^2)$. The method of complexity $O(n^2)$ is unrealistic as n increases up to a very large number. Therefore, this work adopts a bin-based pruning method to discard inferior trees, in which the time complexity of the bin-based pruning method is $O(n+k^2A_{max})$. Since n is markedly larger than k^2A_{max}, $O(n+k^2A_{max})$ is much faster than $O(n^2)$. In the bin-based pruning method, each solution in $S(v_{i,z})$ is assigned to the corresponding bin according to its layer range and antenna length. For instance, a solution with the layer range of $(3, 5]$ and the antenna length of 15 is assigned to the bin $B[3,5,15]$. In each bin, only the minimum-cost solution is reserved because other solutions in the same bin are totally worse than the minimum-cost one. The time complexity of the bin-based pruning method is analyzed as follows. Assigning each solution to the corresponding bin takes the time of $O(n)$, and the time complexity of selecting the minimum-cost solution in each bin is also $O(n)$ because each solution must be scanned once. Finally, collecting the remained solution from each bin to rebuild $S(v_{i,z})$ requires the time of $O(k^2A_{max})$ since there are $(k(k-1)A_{max})/2$ bins. Thus, the time complexity of the bin-based pruning is $O(n)+O(n)+O(k^2A_{max})=O(n+k^2A_{max})$, which is much faster than the intuitive method. However, the bin-based pruning method can not prune all inferior trees since the solutions in different bins can not be compared with each other.

4.2. Algorithm Flow of ANLA

The algorithm flow of ANLA resembles the framework of [3]. The initial layer assignment stage initially does not consider the antenna rule and wire congestion constraints, and greedily identifies the minimal via assignment solution of each net by NANA. Next, the wire overflow reduction stage re-assigns the nets with overflows or antenna violations, in which NANA is used to assign the short nets; meanwhile, the long nets are assigned using NALAR. Notably, a net is regarded as a short net if the length of the net is less then A_{max}; otherwise, the net is regarded as a long net. Because the assignment result of the short net must conform to the antenna rule, performing NANA to assign the short net can always identify the minimum-cost antenna-violation-safe solution; it is also more efficient than performing NALAR. The wire overflow reduction stage iterates until all overflows are eliminated or the overflows do not decline for three consecutive rounds. Notably, the net ordering of assignment generally influences the quality of final results. But, this work simply assigns nets in increasing order of net number because ANLA adopting negotiation scheme can lower the quality variation of assignments, that is demonstrated in [3]. Finally, in the post optimization stage, NANA and NALAR are used to rip up and re-assign each short net and long net, respectively, to further reduce the via count. In each stage, NANA and NALAR adopts different congestion cost formulae for different objectives. Section 3.1 details the congestion cost formulae. In the post optimization stage, Eq. 8 enlarges the congestion penalty to avoid overflows. If NALAR identifies an antenna-violation-safe solution with wire congestion violation for a net, this net will be re-assigned again by NANA to meet wire congestion constraints even the number of antenna-violation increases. Accordingly, the final assignment result must conform to wire congestion constraints.

5. EXPERIMENTAL RESULTS

The proposed algorithms were implemented in C/C++ language on a 2.4 GHz Intel Xeon-based linux server with 48GB memory, and use the benchmarks from the ISPD'08 global routing contest [19]. Each benchmark has six or eight routing layers. The preferred direction in even layers is horizontal, while that in odd layers is vertical. An used net edge that does not follow the preferred direction produces a wire overflow. To compare ANLA with previous works, each algorithm reads the same 2D global routing results of NTHU-Route 2.0 [9] for ISPD'08 benchmarks. Then, each layer assignment algorithm transforms 2D routing results to 3D.

Table 1 compares ANLA with the existing works. Note that, sepCost and viaCost is set to 1 and 100 in ANLA, respectively. In Table 1, COLA [2] and NVM [3] focus on the via count minimization but does not consider the antenna effect, and LAVA [4] is an antenna avoidance layer assignment algorithm which addresses on eliminating global-antenna-violations yet was unaware of local-antenna-violation. NVM and ANLA perform on our machine. The routing results of COLA and LAVA are quoted from [4]. Because the performing machine of COLA and LAVA is different than ANLA and NVM, the runtimes of COLA and LAVA are normalized by the clock rate. The number of overflows is not listed since the results of all layer assignment algorithms have the same number of overflows. Table 1 reveals that the results of LAVA contain a lot of nets with antenna violations while ANLA can effectively reduce the number of nets with antenna violations (#vn). In 11 benchmarks, ANLA can yield 5 violation-free assignments while the other works yield no violation-free assignment. As for the total number of antenna violations in all benchmarks, this work, COLA, NVM and LAVA yield 21, 43506, 41261 and 29671 antenna violations, respectively. In addition, the via count of ANLA is less than COLA and LAVA by 4.9% and 4.6%, respectively. However, ANLA consumes more runtime than other algorithms. The cpu_1 denotes to the runtime of ANLA with the intuitive inferior tree pruning method, and cpu_2 denotes to the runtime of ANLA with the bin-based pruning method. The ANLA with the bin-based pruning method can improve the runtime by 10.8% than that with the intuitive pruning method. Note that, no matter if the intuitive pruning method or bin-based pruning method is used, NALAR can identify the minimum-cost antenna-violation-safe solution for a net. However, if the number of minimum-cost antenna-violation-safe solutions of a net exceeds one, these two pruning methods may identify different solutions. Experiments show that the quality variation is relatively small. In Table 1, the maximum variation of via count of two results obtained by ANLA with and without the bin-based pruning is less than 0.003% and two methods yield the same number of antenna violations.

To demonstrate that ANLA can easily control the separators number by adjusting sepCost, Table 2 shows the separators number (#sep) of ANLA with the different values of sepCost when viaCost is set to 100. This experiment reveals that the separators can be significantly reduced if the value of sepCost increases. ANLA with sepCost=500 can reduce 94% separators than that with sepCost=0 and only increases the via count by 9.4%. However, the runtime of ANLA increases as the value of sepCost increases.

6. CONCLUSIONS

This work proposes a negotiation-based single-net layer assignment called NALAR which can identify the minimum-cost antenna-violation-safe layer assignment solution for a net, and can simultaneously minimize the via count and separators. An antenna avoidance layer assignment algorithm (ANLA) adopting NALAR is presented in this work, ANLA not only avoids global-antenna-violations, but also eliminates local-antenna-violations. As compared to previous works, ANLA can reduce the number of antenna violations significantly.

REFERENCES

[1] D. Wu, J. Hu, and R. Mahapatra,"Antenna avoidance in layer assignment,"*IEEE Trans. Comput.-Aided Design Integr. Circuits Syst.*, vol. 25, no. 4, pp. 1643-1656, 2006.

[2] T.-H. Lee, T.-C. Wang, "Congestion-constrained layer assignment for via minimization in global routing," *IEEE Trans. On Computer-Aided Design of Integrated Circuits and Systems*, pages 1643-1656, 2008.

[3] Wen-Hao Liu and Yih-Lang Li, "Negotiation-Based Layer Assignment for Via Count and Via Overflow Minimization," in *The 16th Asia and South Pacific Design Automation Conference*, Jan. 2011, P. 539-544.

[4] T.-H. Lee and T.-C. Wang, "Simultaneous antenna avoidance and via optimization in layer assignment of multi-layer global routing," *Proc. Intl. Conf. on Computer-Aided Design*, pp. 312-318, 2010.

[5] J. A. Roy and I. L. Markov, "High-performance routing at the nanometer scale,", *IEEE Trans. on Computer-Aided Design*, vol. 27, no. 6, pp. 1066-1077, 2008.

[6] M. D. Moffitt, "MaizeRouter: Engineering an effective global router," in *Proc. Asia and South Pacific Design Automation Conf.*, pages 232-237, 2008.

[7] M. Cho, K. Lu, K. Yuan, and D. Z. Pan, "BoxRouter 2.0: Architecture and implementation of a hybrid and robust global router," in *Proc. Int. Conf. Comput.-Aided Des.*, pages 503-508, 2007

[8] M. M. Ozdal and M. D.F. Wong, "ARCHER: A history-driven global routing algorithm," in *Proc. Int. Conf. Comput.-Aided Des.*, pages 488-495, 2007.

[9] Y.-J. Chang, Y.-T. Lee, and T.-C. Wang, "NTHU-Route 2.0: a fast and stable global router," in *Proc. Int. Conf. Comput.-Aided Des.*, pages 338-343, 2008.

[10] H.-Y. Chen, C.-H. Hsu, and Y.-W. Chang. "High-performance global routing with fast overflow reduction." in *Proc. Asia and South Pacific Design Automation Conf.*, pages 582–587, 2009.

[11] Y. Xu, Y. Zhang and C. Chu, "FastRoute 4.0: Global Router with Efficient Via Minimization," In Proc. *Asia and South Pacific Design Automation Conf.*, pages 576-581, 2009.

[12] K.-R. Dai, W.-H. Liu, and Y.-L. Li, "Efficient Simulated Evolution Based Rerouting and Congestion-Relaxed Layer Assignment on 3-D Global Routing," in *Proc. Asia and South Pacific Design Automation Conf.*, pages 570-575, 2009.

[13] W.-H. Liu, W.-C. Kao, Y.-L. Li and K.-Y. Chao, "Multi-Threaded Collision-Aware Global Routing with Bounded-Length Maze Routing" in *Proc. Des. Autom. Conf.*, pages. 200-205, June 2010.

[14] N. J. Naclerio, S. Masuda, and K. Nakajima, "The via minimization problem is NP-complete," *IEEE Trans. Comput,* pages 1604-1608, Nov. 1989.

[15] T.-H. Lee, T.-C. Wang, "Robust layer assignment for via optimization in multi-layer global routing," In *Proc. Intl. Symp. On Physical Desgin*, pages 159-166, 2009.

[16] Z. Chen and I. Koren, "Layer reassignment for antenna effect minimization in 3-layer channel routing," in *Proc. IEEE Int. Symp. Defect Fault Tolerance*, Boston, MA, 1996, pp. 77–85.

[17] H. Shin, C.-C. King, and C. Hu, "Thin Oxide Damage by Plasma Etching and Ashing Process," *Proceedings of International Reliability Physics Symposium*, pp. 37–41, 1992.

[18] H. Watanabe, J. Komori, K. Higashitani, M. Sekine, and H. Koyama,"A Wafer Level Monitoring Method for Plasma-charging Damage Using Antenna PMOSFET Test Structure," *IEEE Transactions on Semiconductor Manufacturing*, Vol. 10, No. 2, pp. 228–232, 1997.

[19] http://archive.sigda.org/ispd2008/contests/ispd08rc.html

TABLE1. COMPARISON BETWEEN THE PROPOSED ANTENNA AVOIDANCE LAYER ASSIGNMENT AND PREVIOUS WORKS.

Bench-mark	COLA [2]			NVM [3]			LAVA [4]			ANLA			
	#vn	vias (10^5)	cpu* (min)	#vn	vias (10^5)	cpu (min)	#vn	vias (10^5)	cpu* (min)	#vn	vias (10^5)	cpu_1 (min)	cpu_2 (min)
adaptec1	911	17.69	0.46	709	16.69	0.80	602	17.51	0.73	4	16.72	14.82	13.95
adaptec2	879	19.30	0.43	712	18.31	0.70	568	19.07	0.64	0	18.33	12.86	10.57
adaptec3	2959	34.91	1.38	2919	32.90	2.11	2194	34.58	1.94	5	33.00	60.03	53.12
adaptec4	2009	32.15	1.25	1925	30.82	1.76	1931	31.93	1.61	4	30.90	41.11	37.44
adaptec5	4166	52.40	1.65	3744	49.30	2.28	2465	51.9	2.09	0	49.43	53.53	47.65
newblue1	328	22.22	0.38	460	21.42	0.58	273	24.95	0.53	6	21.43	4.93	4.37
newblue2	681	29.46	0.66	534	28.14	0.94	444	29.15	0.86	1	28.18	11.79	10.52
newblue3	466	30.23	0.99	429	29.00	1.58	251	29.42	1.44	1	29.08	29.48	27.3
newblue4	874	47.05	1.33	849	44.73	1.68	617	46.59	1.54	0	44.77	20.23	17.51
newblue5	3009	84.51	2.25	2766	80.16	3.52	2137	83.79	3.23	0	80.30	77.93	68.68
newblue6	3453	74.66	1.76	3280	71.01	2.39	2736	73.83	2.19	0	71.12	33.27	30.75
newblue7	10286	166.01	5.10	8628	157.21	6.47	5844	164.52	5.93	0	157.50	354.41	301.97
bigblue1	1841	18.73	0.64	1459	17.60	0.93	1423	18.57	0.85	0	17.65	19.71	18.51
bigblue2	392	42.11	0.87	389	40.32	1.24	264	41.72	1.13	0	40.34	16.25	15.31
bigblue3	3576	52.43	1.51	3631	50.55	2.49	2692	51.99	2.28	0	50.66	233.01	200.95
bigblue4	7676	109.14	2.78	8627	104.69	4.32	5230	108.28	3.96	0	104.93	301.91	256.57
sum	43506			41261			29671			21			
ratio		1.049	0.036		0.998	0.053		1.046	0.123		1	1	0.892

TABLE2. THE SEPARATOR NUMBER OF THE PROPOSED ANTENNA AVOIDANCE LAYER ASSIGNMENT WITH THE DIFFERENT VALUES OF *sepCost*

Benchmark	ANLA (*sepCost*=0)				ANLA (*sepCost*=1)				ANLA (*sepCost*=100)				ANLA (*sepCost*=500)			
	#vn	#sep (10^5)	vias (10^5)	cpu (min)	#vn	#sep (10^5)	vias (10^5)	cpu (min)	#vn	#sep (10^5)	vias (10^5)	cpu (min)	#vn	#sep (10^5)	vias (10^5)	cpu (min)
adaptec1	4	8.93	16.72	8.83	4	7.54	16.72	13.95	2	1.51	17.89	16.02	2	0.94	18.99	18.58
adaptec2	0	5.85	18.33	5.14	0	4.68	18.33	10.57	0	0.27	18.90	8.45	1	0.17	19.07	8.84
adaptec3	5	32.61	32.97	24.31	5	25.72	33.00	53.12	6	4.99	35.71	44.58	8	3.72	38.30	52.48
adaptec4	4	26.00	30.87	18.81	4	20.20	30.90	37.44	4	1.49	32.86	37.53	5	1.09	33.61	42.77
adaptec5	0	30.73	49.41	27.75	0	25.92	49.43	47.65	0	4.46	52.74	43.64	0	2.89	55.95	50.16
newblue1	6	2.53	21.43	2.68	6	1.93	21.43	4.37	6	0.21	21.82	4.36	5	0.07	22.03	4.50
newblue2	1	9.05	28.18	7.39	1	7.42	28.18	10.52	0	0.68	29.58	11.73	0	0.27	30.31	12.92
newblue3	1	18.08	29.06	13.30	1	15.35	29.08	27.30	0	2.02	31.27	21.57	0	1.28	32.78	23.41
newblue4	0	15.00	44.76	11.12	0	12.10	44.77	17.51	0	0.96	46.84	18.08	0	0.53	47.58	19.95
newblue5	0	34.39	80.28	44.38	0	28.09	80.31	68.68	0	2.93	84.30	67.82	0	1.34	87.25	78.43
newblue6	0	22.87	71.11	20.87	0	19.43	71.12	30.75	0	2.86	74.66	31.21	0	1.08	77.89	34.17
newblue7	0	45.51	157.47	187.66	0	36.30	157.50	301.97	0	2.88	162.76	281.94	0	1.84	164.55	286.05
bigblue1	0	11.30	17.65	12.34	0	9.79	17.65	18.51	0	2.87	19.19	17.45	0	1.54	22.26	19.29
bigblue2	0	7.27	40.33	11.86	0	5.98	40.34	15.31	0	1.29	41.72	19.57	0	0.62	42.93	23.31
bigblue3	0	24.45	50.65	61.38	0	19.18	50.66	200.95	0	0.73	52.40	97.65	0	0.50	52.80	102.09
bigblue4	0	36.69	104.90	148.10	0	28.88	104.93	256.57	0	2.06	108.84	205.14	0	1.28	110.23	217.54
sum	20				21				18				21			
ratio		1	1	1		0.813	1	1.79		0.106	1.052	1.618		0.06	1.094	1.79

Simultaneous Clock and Data Gate Sizing Algorithm with Common Global Objective

Gregory Shklover
Intel Corporation
MATAM, Haifa 31015, Israel
gregory.shklover@intel.com

Ben Emanuel
Intel Corporation
MATAM, Haifa 31015, Israel
ben.emanuel@intel.com

ABSTRACT

We present an algorithm that performs gate sizing circuit optimization for VLSI designs. Contrary to existing approaches that usually target either clock or data gate sizing as separate optimization steps, our algorithm performs simultaneous optimization of both data and clock gate sizes with common global optimization objective. To do so, we extend traditional gate sizing by Lagrangian Relaxation method with clock-related formulations and use Dynamic Programming to solve those optimally. We demonstrate that on a set of industrial blocks such simultaneous optimization achieves superior results when compared to classical separate optimization flow.

Categories and Subject Descriptors

J.6 [**Computer-aided Engineering**]: Computer-aided Design; B.5.2 [**Design Aids**]: Optimization

General Terms

Algorithms

Keywords

VLSI, gate sizing, clock optimization

1. INTRODUCTION

Gate sizing optimization is one of the main tools in achieving high quality VLSI circuit design. Given a gate-level circuit and a standard cell library, the goal of such optimization is to find gate sizes that would yield best combination of total circuit power, performance and area. Note that under the notion of "gate sizing" we generalize the process of selection from a discrete set of possible implementation variants with different device sizes or technology parameters.

In a synchronous gate-level design, the circuit is usually comprised of *data gates* that implement the logical function

of the block and *clock gates* responsible for distributing common synchronization signal to different state elements in the circuit.

Most gate sizing optimization algorithms separately target either the data gates or the clock gates with different optimization objectives [2, 3, 4, 11, 14]. There are a number of reasons for such separation including: *design methodology*, different *circuit structure*, and *optimization problem class*.

Design methodology. With traditional design methodologies logic part of the circuit is usually sized for best performance vs power or area, while clock gates are sized to meet minimum skew and skew variability [3, 11]. Modern design methods use more liberal skew constraints and also utilize the concept of *useful skew* [5] by either manual or tool-assisted flows.

Structure. Clock gates are usually structured as a grid or as a tree, suitable for efficient solution space exploration with methods such as Dynamic Programming. Data gates are structured as general directed graphs making efficient application of such methods very challenging [6, 7] and usually computationally-prohibitive for industrial size designs.

Problem classification. Another reason for separation is the convexity classification: whereas solving data gate sizing with setup timing constraints results in a convex optimization problem, clock gate sizing within the same formulation is not a convex optimization problem making some convex optimization methods inapplicable.

Previous publications in the area of simultaneous optimization focused on constructing clock routing tree (without buffers) with useful skew along with pre-computed data gate sizing combinations [15]. Complexity of the industrial designs grew significantly since, requiring consideration for buffered clock tree sizing and making the methods used to pre-compute combinatorial gate sizes computationally-prohibitive. The problem formulation also assumed that a solution without timing violations exists which is usually not the case throughout most of the design cycle in high performance industrial projects.

Newer publications relied on more efficient gate sizing algorithms based on Lagrangian Relaxation (LR) and suggested co-optimizing data sizing with clock skew assignment [10, 13]. These approaches benefit from the efficiency of convex optimization formulation, but neglect the cost of implementing a clock skew assignment by either gate sizing or clock tree synthesis.

The area of clock tree optimization is dominated by algorithms based on Dynamic Programming (DP) similar in style to that originally proposed by van Ginneken [12] for buffer placement in RC-tree networks. Tsai *et al* [11] used similar approach to obtain optimal power zero skew clock trees. Unlike the minimum delay objectives used in [12], zero skew objective used in [11] has non-convex nature resulting in weaker pruning condition and wider solution space.

Main contribution of this work is that we naturally combine clock and data sizing decisions to solve a common global objective. We use the efficient LR method with a practical objective formulation as the base for our algorithm. In addition to regular data-related formulations, we derive clock-related formulations within the same objective. To overcome the inapplicability of traditional convex sub-optimization methods to this extended formulation, we use Dynamic Programming algorithm to optimally solve the clock-related part of the relaxed objective.

Since the underlying optimization problem is non-convex, our LR-based method cannot guarantee globally optimal solution. Yet since at each step we either solve convex part of the objective with traditional LR optimization, or use systematic search DP algorithm for the non-convex part of the objective, we claim that overall the algorithm arrives at high quality global solutions superior to those obtained with separate optimization.

The rest of the paper is organized as follows. In Section 2.1 we give an outline of the Lagrangian Relaxation gate sizing algorithm. In Section 3 we derive the clock-related terms within the LR objective for simultaneous optimization. Section 4 presents the algorithm for solving these terms by DP clock sizing algorithm. In Section 5 we analyse complexity and optimality of the new algorithm. Section 6 presents comparative results of this algorithm versus separate clock and data optimization flows on a set of blocks from a production micro-processor project. Finally Section 7 concludes this paper.

2. GATE SIZING BY LAGRANGIAN RELAXATION

2.1 Algorithm Formulation

Proposed by Chen *et al* [2], gate sizing by Lagrangian Relaxation offers a method for efficiently optimizing large industrial circuits. Below is a formal definition of one of the variants of this approach where we consider gate sizing to minimize weighted sum of total circuit power and setup timing violations. Unlike the types of objectives discussed in [2], such an objective is more practical for sizing industrial circuits where timing constraints cannot necessarily be met during earlier stages of circuit design [8]. For simplicity of presentation we omit other practical constraints usually present in an industrial design such as maximum slew or max capacitive load limitations.

Let the circuit be represented by a directed graph $G = (V, E)$ with input nodes $Inputs \subseteq V$ and output nodes $Outputs \subseteq V$. Let $Gates$ be the set of gates in the circuit, where each gate $g \in Gates$ can be assigned different size from a library L_g.

The set $Seq \subseteq Gates$ is the set of sequential elements (flip-flops, latches) where each element $s \in Seq$ has an input data node s_d, output node s_q, and clock node s_{clk}. Let a_n denote the latest arrival time for a signal at node n.

The algorithm is required to find size assignment that min-

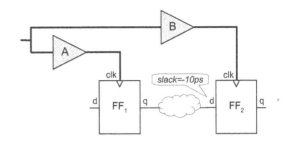

Figure 1: Combined data sizing with clock skew assignment

imizes the following objective:

$$Objective = \alpha \times \sum_{g \in Gates} power_g + \\ + \sum_{n \in Seq_d \cup Outputs} \max(0, -slack_n) \quad (1)$$

Subject to:

$$\begin{aligned} &\forall i \in Inputs : a_i = const_i \\ &\forall u \to v \in E : a_u + delay_{u \to v} \leq a_v \\ &\forall s \in Seq : slack_{s_d} = cycle + a_{s_{clk}} - a_{s_d} - setup_s \\ &\forall o \in Outputs : slack_o = cycle - a_o - const_o \end{aligned} \quad (2)$$

Detailed description of Lagrangian Relaxation for this objective is provided in [8] and will not be fully repeated here. In the process of relaxation, Lagrangian $Objective_\lambda$ formulation is obtained by integrating constraints in (2) into the objective (1) with Lagrange multipliers (λ). To eliminate a_n variables, Karush–Kuhn–Tucker condition [1] is applied to compute partial derivatives. Substituting back the results of the partial derivatives yields the following simplified formulation:

$$Objective_\lambda = \alpha \times \sum_{g \in Gates} power_g + \sum_{\forall u \to v} \lambda_{u,v} delay_{u \to v} + \\ + \sum_{\forall s \in Seq} \lambda_s setup_s + const \quad (3)$$

A tail of constant terms in the formulation originates from the constant terms used in original constraints such as *cycle*, a_i, and $a_{s_{clk}}$ assumed to be constant in data-only LR gate sizing. Such formulation can be efficiently solved by iterative algorithm with steps that either use subgradient method [2] or more sophisticated optimization algorithms [10, 13].

2.2 LR with skew optimization

Previous publications suggested combining LR-based data gate sizing with optimal clock skew ($a_{s_{clk}}$) solution [10, 13]. Such problem formulations remain convex, allowing LR algorithm to find theoretically optimal solution.

These formulations leave the question of clock implementation outside the scope of the optimization, disregarding the power cost associated with implementing a given clock skew and its contribution to the global objective. Such separation may potentially lead to sub-optimal results.

Figure 1 gives a simple example of such a case. Unaware of the power cost of different clock skew solutions, [10] - style algorithm could propose a clock skew assignment that speeds up a_{clk} at FF_1 to fix the timing violation, subsequently requiring up-size of gate A in the clock distribution

```
ALGORITHM LRDP:
    Initialize()
    for i := 1 to max_iteration do
        LRSize()
        DPSize()
        UpdateTiming()
        UpdateMultipliers()
    end for
```

Figure 2: Combined clock and data gate sizing algorithm

```
ALGORITHM DPSize:
    for all n ∈ TopoOrder(Nodes) do
        if isLeaf(n) then
            S_n ← {(c_n, 0)}
        else if isMerge(n) then
            S_n ← S_{left_n} ⊗ S_{right_n}
        else if isGate(n → o) then
            S_n ← L_g ⊕ S_o
        end if
        Prune(S_n)
        if isRoot(n) then
            Apply(MIN_{obj}(S_n))
        end if
    end for
```

Figure 3: Dynamic Programming clock gate sizing algorithm

tree. An alternative, and more power-efficient, solution for this example could be to delay a_{clk} at FF_2 by down-sizing gate B.

To address such issues we propose to use simultaneous clock and data optimization algorithm with common global objective that captures both timing and total power aspects. In this work we consider simultaneous data and clock gate sizing, although a natural extension of our work could also consider simultaneous data sizing and clock tree synthesis.

3. COMBINED CLOCK AND DATA SIZING

Let us consider the set of clock gates $Clock \subseteq Gates$ and the set of sequential elements Seq. There are two constraints associated with each sequential element $s \in Seq$:

$$a_{s_d} + setup_s - a_{s_{clk}} - cycle \leq -slack_{s_d}$$
$$a_{s_{clk}} + delay_{s_{clk} \to s_q} \leq a_{s_q} \quad (4)$$

Integrating these into $Objective_\lambda$ with Lagrange multipliers $\lambda_{s,d}$ and $\lambda_{s,q}$ respectively and performing simplifications yields the following clock-related terms:

$$Objective_\lambda = ... + \alpha \times \sum_{g \in Clock} power_g +$$
$$+ \sum_{s \in Seq} (\lambda_{s,q} - \lambda_{s,d}) \times a_{s_{clk}} \quad (5)$$

In data-only gate sizing optimization presented in Section 2.1, these are considered to be constants and are disregarded during the optimization process.

After expanding each $a_{s_{clk}}$ as a sum of delays from the root of the clock tree to the corresponding leaf and rearranging the terms we obtain the following formulation:

$$Objective_\lambda = ... + \alpha \times \sum_{g \in Clock} power_g +$$
$$+ \sum_{(u,v) \in ClockTree} delay_{u \to v} \times \sum_{s \in Leaves_v} (\lambda_{s,q} - \lambda_{s,d}) \quad (6)$$

Where $Leaves_v$ denotes the set of clock tree leaves in v's sub-tree.

Under convex gate delay model the formulation (6) is not convex since some of the multiplier aggregations are negative. Applying sub-gradient descent techniques for optimizing this part of the objective would hence be inappropriate. Instead we propose using Dynamic Programming (DP) algorithm which performs systematic search over the solution space and thus is immune to the non-convexity of the problem.

Top level clock and data optimization algorithm is listed in Figure 2. It has the same structure as traditional LR optimization algorithm: Lagrange multiplier initialization is followed by a number of iterations of gate sizing and timing-based multiplier update. In addition to traditional gate sizing procedure $LRSize()$ clock gates are sized in procedure $DPSize()$ explained later.

4. CLOCK SIZING BY DYNAMIC PROGRAMMING

4.1 Algorithm Formulation

We consider clock gates in the circuit structured as a buffered tree. In this section we assume binary tree with general tree structure discussed in Section 5.1.

For clarity of presentation first we only consider repeater gates without enable signals. Additional considerations required in the presence of clock gates with enable signals (*clock gating*) are described in Section 4.2. We will also assume that gate delay model is a function of output load without input slew dependency. Support for accurate delay model with input slew dependency is discussed in Section 4.2.

Although the algorithm is applicable to wire sizing, for simplicity of presentation we focus on gate sizing alone and omit wire delays from following formulations. Similarly, we omit other details required in practical implementation such as rise and fall transition separation or different sampling vs generating delay modeling.

At each LR iteration, the DP algorithm is required to find clock gates sizes that minimize the following objective:

$$Objective_{DP} = \alpha \times \sum_{g \in Clock} power_g +$$
$$+ \sum_{(u,v) \in ClockTree} delay_{u \to v} \times \sum_{s \in Leaves_v} (\lambda_{s,q} - \lambda_{s,d}) \quad (7)$$

The DP algorithm pseudo-code is listed in Figure 3.

For each node n in the clock tree, the algorithm computes a set of solutions S_n. Each solution $s_n \in S_n$ is a tuple $s_n = \langle c, obj \rangle$ where c is the associated downstream capacitance and obj is the corresponding objective value. Each solution s_n is also associated with an assignment of gate sizes in n's sub-tree. Solutions are propagated upwards from tree leaves to the root node.

For a leaf node n the algorithm generates a single solution:

$$S_n = \{\langle c_n, 0 \rangle\} \qquad (8)$$

New solutions are generated when considering different gate sizes. For an edge $u \to v$ corresponding to a gate g with a set of sizes L_g, the set of solutions S_u (prior to pruning) is obtained as follows (denoted by $\oplus$ in Figure 3):

$$S_u = \{\langle c_\tau, obj + \Delta obj_\tau(c) \rangle \mid \forall \langle c, obj \rangle \in S_v, \forall \tau \in L_g\}$$
$$where \qquad (9)$$
$$\Delta obj_\tau(c) = \alpha \times power_\tau + delay_\tau(c) \times \sum_{s \in Leaves_v} (\lambda_{s_q} - \lambda_{s_d})$$

Where $power_\tau$ denotes gate g power with size τ and $delay_\tau(c)$ denotes $u \to v$ delay with size τ and output load c.

At a branch node n with two sub-trees l and r, solutions from the sub-trees are merged as follows (denoted by $\otimes$ in Figure 3):

$$S_n = \{\langle c_l + c_r, obj_l + obj_r \rangle \mid$$
$$\forall \langle c_l, obj_l \rangle \in S_l, \langle c_r, obj_r \rangle \in S_r\} \qquad (10)$$

At the root node the algorithm chooses and applies a solution with lowest objective value.

Pruning of sub-optimal solutions is performed when new solutions are generated. A solution $s = \langle c_s, obj_s \rangle$ at n is said to be sub-optimal if there is a solution $q \in S_n \setminus \{s\}$ such that:

$$(c_q = c_s) \wedge (obj_q \le obj_s) \qquad (11)$$

Note that this pruning criterion is different from those applicable to shortest delay convex optimization formulations such as those in [12] $((c_q \le c_s) \wedge (obj_q \le obj_s))$. This is due to the presence of negative coefficients in the objective, where global optimal solution may be obtained from a sub-solution with higher local sub-tree capacitance and higher sub-tree objective. Note also that following these definitions the number of essential (not sub-optimal) solutions at an input to a gate g with set of possible sizes L_g is limited by $|L_g|$.

THEOREM 1. *Algorithm in Figure 3 solves objective in (7) optimally.*

PROOF. Let $s^* = \langle c^*, obj^* \rangle$ be the theoretical optimal solution to (7). Without loss of generality, we will assume that s^* is strictly optimal: $obj^* < obj'$ for any other solution $s' = \langle c', obj' \rangle$.

It is easy to see that without pruning the DP algorithm enumerates all possible combinations of gate sizes, subsequently choosing the lowest objective solution at the root. Hence to prove DP algorithm correctness we need to show that pruning criterion (11) will not eliminate any of the partial solutions comprising s^*. We will use s_n^* to denote the corresponding partial solution for the sub-tree of a node n.

Let us assume that there exists a node $n \in ClockTree$ where optimal partial solution $s_n^* = \langle c_n^*, obj_n^* \rangle$ was pruned. From our pruning criterion it follows that there exist another solution $s_n^+ = \langle c_n^+, obj_n^+ \rangle$ such that $c_n^+ = c_n^*$ and $obj_n^+ \le obj_n^*$.

Let s^{**} be the a solution obtained from s^* by substituting sizing decisions in s_n^* by those in s_n^+. Since $c_n^* = c_n^+$ this substitution does not affect any of the objective terms (delay or power) that are not in n's sub-tree. The effect on objective

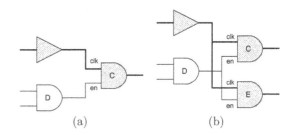

Figure 4: **Additional sizing effects: (a) accurate delay modeling for single receiver, (b) approximate delay modeling for multiple receivers**

terms that belong to n's sub-tree is captured by $obj_n^+ - obj_n^*$. Thus $obj^{**} = obj^* + (obj_n^+ - obj_n^*) \le obj^*$ which is a contradiction to our earlier assumption of s^* (strict) optimality. Hence pruning criterion (11) is correct in preserving optimal solution s^*. $\square$

4.2 Additional Considerations

Side-load effects. Consider a clock gate with enable signal controlled by the logic part of the circuit (gate C in Figure 4(a)). Sizing the gate affects both clock- and data-related terms in $Objective_\lambda$, where data-related terms include the setup time of the signal en and delays of the gate D affected by the change in load capacitance on the enable signal.

The DP objective formulation in Section 4.1 can be extended to account for these effects with sufficient level of accuracy. To do so, we add affected setup time and delays with corresponding multipliers to the Δobj of the generated partial solutions:

$$\Delta obj_\tau(c) = \alpha \times power_\tau + delay_\tau(c) \times \sum_{s \in Leaves_v} (\lambda_{s_q} - \lambda_{s_d}) +$$
$$+ \lambda_{en} \times setup_{en,\tau} + \sum_{(u \to v) \in Fanin_{en}} \lambda_{u,v} \times delay_{u \to v}(c_\tau)$$
$$(12)$$

This computation can accurately capture the side-load effect for single receiver cases such as that in Figure 4(a), but bares some inaccuracy in general case with multiple receivers and resistive interconnect. This is because gate and wire delays have non-linear dependency on load capacitance in presence of resistive interconnect [9].

Accurate delay calculation for a multiple receiver case such as the one presented in Figure 4(b) would require coordination between different sizing decisions, subsequently significantly complicating the DP algorithm. To avoid this complexity we rely on iterative nature of our algorithm: as the algorithm progresses we expect gate sizes to stop changing significantly (discussed in more detail in Section 5.2), so that we can rely on first order approximation for delay with multi-receiver load changes:

$$delay_{u \to v}(c_{ref} + \Delta c_1 + \Delta c_2) \simeq delay_{u \to v}(c_{ref} + \Delta c_1) +$$
$$+ delay_{u \to v}(c_{ref} + \Delta c_2) -$$
$$- delay_{u \to v}(c_{ref}) \qquad (13)$$

Where c_{ref} is the load capacitance corresponding to gate sizes selected in the previous iteration. $-delay_{u \to v}(c_{ref})$ is a constant term for a given iteration and can be omitted from the Δobj calculations with no effect on the correctness of pruning or combining steps of the DP algorithm. This approximation allows for use of the decoupled formulation in (12).

Input slews. Accurate gate delay models also require one to account for accurate input signal slews. Being dependent on yet undetermined gate sizes, slew values are not available when computing solutions bottom-up – a challenge common to many DP bottom-up algorithms. To address this issue systematically one could extend the definition of a DP solution to account for different input slews: $s_n = \langle c, slew, obj \rangle$. The algorithm would then be modified to explore solutions with different possible input slews.

Alternatively, we can rely on convergence of the gate sizes: slew values at a given stage have stronger dependence on local gate sizes, but their dependence on previous stages quickly diminishes with distance. With decreasing degree of gate size variations between iterations, slews are expected to stop changing significantly so that we can rely on slew values of the previous iteration when computing (9).

5. ALGORITHM ANALYSIS

5.1 Complexity

Traditional LR gate sizing complexity is discussed in detail in [2]. We will limit our discussion to the new DP clock sizing part of the optimization algorithm described in Section 4.1.

Number of distinct solutions that the algorithm may hold for a given node n is determined by the number of distinct capacitive loads that may be observed at n. For a gate output node this number is exponential in the n's fanout degree. Whereas for binary trees this does not pose a problem, this dependency might be computationally-prohibitive in general case.

To address this issue we use sampling technique suggested in [11]. Instead of holding a complete set of solutions at n we limit the number of solutions to p samples. The samples are chosen by mapping different c values to a geometrically-spaced grid and selecting lowest obj solution from each grid sub-range. N-way merge can then be implemented by $(n-1)$ pairwise $p \times p$ merges.

Using p as the bound for number of different solutions per node, we can derive the complexity for the DP algorithm: $O(max(p,|L|)pN)$. At each of the N nodes, the algorithm either generates new solutions with buffers from the library L or combines solutions from node's sub-trees.

5.2 Convergence

Applying LR algorithm to non-convex optimization problem does not guarantee convergence. To address this challenge we borrow the *cooling* concept from simulated annealing optimization method and introduce additional control parameter R^i as follows:

$$\forall i \in [2..max\_iteration]:$$
$$\forall (u,v) \in Clock : |delay_{u \to v}^i - delay_{u \to v}^{i-1}| \le R^i \quad (14)$$
$$R^i < R^{i-1}$$

As the algorithm progresses, the radius of the clock optimization changes can be gradually decreased subsequently reducing the global optimization problem to a convex formulation and forcing algorithm convergence.

Implementing such control mechanism into DP algorithm does not require significant changes and has no impact on the DP algorithm complexity bounds.

5.3 Optimality

For simultaneous clock and data gate sizing both the primal formulation *Objective* (1) and the LR formulation *Objective*$_\lambda$ (6) are non-convex. These conditions do not guarantee zero duality gap [1]. Subsequently, global optimality of the LR algorithm is not theoretically guaranteed.

Whereas separate optimization techniques explore solutions along orthogonal cut planes (with either data or clock sizes fixed), the method proposed in this work allows deeper exploration of the solution space. We use LR relaxation method to guide the exploration and step size control to ensure convergence. Section 6 presents empirical data in support of the statement that the algorithm produces high quality solutions superior to comparable separate optimization methods.

Use of sampling technique presented in Section 5.1 is another source of possible sub-optimality. The latter can be shown to behave ϵ-*optimally*: for arbitrarily small ϵ there exist finite p such that the error between optimal solution and the solution obtained with p-sampling is less than ϵ. We will omit formal proof of this claim due to page limitations.

6. EXPERIMENTAL RESULTS

In this section we compare results of running separate vs simultaneous clock and data optimization algorithms on a set of blocks from an industrial VLSI design project. An existing industrial flow with separate clock and data optimization steps was used for reference. Below is a detailed description of the algorithms/flows compared.

ref Industrial reference flow. The flow performs separate clock and data optimization. First, data gate sizes are optimized by LR algorithm with timing violations vs leakage power as objective. Then the clock gates are sized for minimal total power (leakage + dynamic) while allowing deviation from existing clock skew schedule at leaves with positive timing slacks.

new Simultaneous clock and data gate sizing presented in this work. For data part of the circuit only leakage power was considered. For clock part of the circuit the optimization targeted both leakage and dynamic power. The objective coefficients were defined to favour reducing timing violations over power reduction.

Table 1 presents the results of running these algorithms. Note that due to IP reasons all values in the table are scaled by process-dependent units.

Due to implicit useful skew optimization, simultaneous clock and data gate sizing algorithm achieves significant reduction in timing violations. This useful skew is also translated by the LR algorithm into overall leakage reduction. Due to separation of optimization steps, the reference algorithm defines conservative limitations on the clock sizing step to avoid degradations on critical paths. These result in reduced potential for dynamic power optimization in clock. The new algorithm is free of this conservatism. By allowing the algorithm to explicitly balance between the timing and power optimization we obtain not only better timing

Table 1: Separate vs simultaneous clock and data optimization

Block	Negative Slack		Leakage		Clk Dyn. Power		Total Power	
	ref	new	ref	new	ref	new	ref	new
block1	-0.038	-0.044	2.26	2.10	2.07	1.77	4.33	3.87
block2	-0.051	-0.015	1.80	1.77	1.38	1.36	3.19	3.14
block3	-2.387	-1.902	6.59	6.22	5.51	5.18	12.10	11.40
block4	-0.032	-0.030	1.42	1.39	1.46	1.44	2.88	2.84
block5	-0.275	-0.206	3.86	3.77	4.44	4.20	8.30	7.97
block6	-0.087	-0.056	6.05	5.95	0.25	0.27	6.31	6.22
block7	-0.207	-0.158	3.61	3.57	3.42	3.33	7.03	6.90
block8	-0.407	-0.179	5.61	5.09	2.30	2.26	7.92	7.35
block9	-1.075	-0.537	6.49	6.24	0.96	0.89	7.44	7.12
block10	-0.108	-0.066	3.31	3.08	1.65	1.55	4.96	4.63
block11	-0.794	-0.529	7.73	7.42	2.84	2.70	10.57	10.12
block12	-0.154	-0.121	3.47	2.98	2.44	2.39	5.91	5.37
block13	-0.171	-0.058	3.00	2.93	0.50	0.52	3.50	3.44
block14	-0.168	-0.072	2.57	2.51	1.78	1.70	4.35	4.20
block15	-0.062	-0.063	3.10	3.02	2.33	1.97	5.43	4.99
Total	-6.02	-4.03	60.88	58.03	33.32	31.52	94.20	89.55

results, but also lower clock dynamic power. Besides timing improvements, the new algorithm also achieves 5% lower total power ($94.20 \rightarrow 89.55$).

Table 2 demonstrates the necessity of the *cooling* technique presented in Section 5.2. $\alpha = 0$ was used for the power term when sizing clock tree, subsequently targeting timing optimization only. The algorithm was run twice: with and without cooling method applied.

Without cooling applied, small changes in the LR multipliers resulted in significant changes in the clock skew schedule, subsequently causing *overshoot* and lack of convergence to a better solution. This effect was significantly reduced when cooling was applied resulting in overall better solution obtained by the algorithm.

7. SUMMARY

In this work we presented a novel approach to simultaneous clock and data gate sizing optimization. Contrary to previously proposed solutions that target either clock or data separately, our method exploits implicit useful skew in the circuit while balancing performance versus total power of the circuit.

The algorithm behaviour was analysed and methods for addressing known challenges such as performance and convergence were suggested.

Although this work concentrated on gate sizing only, our method is also applicable to wire sizing and buffer insertion. We also believe further research could extend this method to handle simultaneous gate sizing and clock tree synthesis.

8. REFERENCES

[1] S. Boyd and L. Vandenberghe. *Convex optimization.* Cambridge Univ. Press, 2004.

[2] C.-P. Chen, C. Chu, and D. Wong. Fast and exact simultaneous gate and wire sizing by lagrangian relaxation. *IEEE Transactions on Computer-Aided Design of Integrated Circuits and Systems,* 18(7):1014–1025, jul 1999.

[3] J. Chung and C.-K. Cheng. Skew sensitivity minimization of buffered clock tree. In *Proceedings of*

Table 2: Results of applying *cooling* method

Block	Negative Slack	
	cooling off	cooling on
block1	-0.023	-0.023
block2	-0.019	-0.019
block3	-2.649	-1.885
block4	-0.036	-0.013
block5	-0.166	-0.160
block6	-0.153	-0.064
block7	-0.126	-0.118
block8	-0.224	-0.211
block9	-0.693	-0.535
block10	-0.185	-0.083
block11	-0.662	-0.553
block12	-0.102	-0.118
block13	-0.073	-0.032
block14	-0.055	-0.053
block15	-0.130	-0.052
Total	-5.29	-3.92

the *1994 IEEE/ACM International Conference on Computer-Aided Design,* ICCAD '94, pages 280–283, Los Alamitos, CA, USA, 1994. IEEE Computer Society Press.

[4] O. Coudert. Gate sizing: A general purpose optimization approach. In *Proceedings of the 1996 European Conference on Design and Test,* EDTC '96, page 214, Washington, DC, USA, 1996. IEEE Computer Society.

[5] J. Fishburn. Clock skew optimization. *IEEE Transactions on Computers,* 39(7):945–951, jul 1990.

[6] S. Hu, M. Ketkar, and J. Hu. Gate sizing for cell library-based designs. In *Proceedings of the 44th Annual Design Automation Conference,* DAC '07, pages 847–852, New York, NY, USA, 2007. ACM.

[7] Y. Liu and J. Hu. A new algorithm for simultaneous gate sizing and threshold voltage assignment. *IEEE*

Transactions on Computer-Aided Design of Integrated Circuits and Systems, 29(2):223–234, feb. 2010.

[8] M. Ozdal, S. Burns, and J. Hu. Gate sizing and device technology selection algorithms for high-performance industrial designs. In *Proceedings of the International Conference on Computer-Added Design*, ICCAD '11, nov, 2011.

[9] J. Qian, S. Pullela, and L. Pillage. Modeling the "effective capacitance" for the rc interconnect of cmos gates. *IEEE Transactions on Computer-Aided Design of Integrated Circuits and Systems*, 13(12):1526–1535, dec 1994.

[10] S. Roy, Y. H. Hu, C. C.-P. Chen, S.-P. Hung, T.-Y. Chiang, and J.-G. Tseng. An optimal algorithm for sizing sequential circuits for industrial library based designs. In *Proceedings of the 2008 Asia and South Pacific Design Automation Conference*, ASP-DAC '08, pages 148–151, Los Alamitos, CA, USA, 2008. IEEE Computer Society Press.

[11] J.-L. Tsai, T.-H. Chen, and C.-P. Chen. Zero skew clock-tree optimization with buffer insertion/sizing

and wire sizing. *IEEE Transactions on Computer-Aided Design of Integrated Circuits and Systems*, 23(4):565 – 572, april 2004.

[12] L. van Ginneken. Buffer placement in distributed rc-tree networks for minimal elmore delay. In *Proceedings of the IEEE International Symposium on Circuits and Systems*, pages 865–868 vol.2, may 1990.

[13] J. Wang, D. Das, and H. Zhou. Gate sizing by lagrangian relaxation revisited. In *Proceedings of the IEEE/ACM International Conference on Computer-Aided Design*, ICCAD 2007, pages 111 –118, nov. 2007.

[14] K. Wang and M. Marek-Sadowska. Buffer sizing for clock power minimization subject to general skew constraints. In *Proceedings of the 41st annual Design Automation Conference*, DAC '04, pages 159–164, New York, NY, USA, 2004. ACM.

[15] J. Xi and W.-M. Dai. Useful-skew clock routing with gate sizing for low power design. In *Proceedings of the 33rd Design Automation Conference*, DAC '96, pages 383–388, jun. 1996.

Construction of Realistic Gate Sizing Benchmarks With Known Optimal Solutions

Andrew B. Kahng
UC San Diego
La Jolla, CA 92093
abk@ucsd.edu

Seokhyeong Kang
UC San Diego
La Jolla, CA 92093
shkang@vlsicad.ucsd.edu

ABSTRACT

Gate sizing in VLSI design is a widely-used method for power or area recovery subject to timing constraints. Several previous works have proposed gate sizing heuristics for power and area optimization. However, finding the optimal gate sizing solution is NP-hard [1], and the suboptimality of sizing solutions has not been sufficiently quantified for each heuristic. Thus, the need for further research has been unclear.

In this work, we describe a new benchmark generation approach for leakage power-driven gate sizing (the subject of the forthcoming ISPD-2012 contest) which constructs realistic circuit netlists with known optimal solutions. The generated netlists resemble real designs in terms of gate count, maximum path depth, interconnect complexity (Rent parameter), and net degree distributions. Using these benchmark circuits with known optimal gate size, we have studied the suboptimality of several leakage-driven gate sizing heuristics, including two commercial tools, with respect to key circuit topology parameters. Our study shows that common sizing methods are suboptimal for realistic benchmark circuits by up to 52.2% and 43.7% for V_t-assignment and gate sizing formulations, respectively. The results also suggest that (1) commercial tools may still suffer from significant suboptimality, and/or (2) existing methods have "similar" degrees of suboptimality.

Categories and Subject Descriptors

B.7.2 [**Hardware**]: INTEGRATED CIRCUITS—*Design Aids*

General Terms

Algorithms, Design.

Keywords

Benchmarks, Gate Sizing, Dynamic Programming, Leakage Power, Static Power, Optimization.

1. INTRODUCTION

The *sizing problem* in VLSI design seeks to assign design parameters (width and/or threshold voltage) to each gate, so as to optimize timing, area and/or power of the design subject to constraints. Gate sizing is widely applied during optimization and design closure phases of the implementation flow, since it helps to meet design constraints with minimal overall disruption. The problem has been extensively studied, and a number of heuristics have been proposed. Greedy heuristics for gate sizing to optimize power/area/delay subject to delay/area constraints are found in [2] and [3]. Sensitivity-based gate downsizing and V_t-assignment techniques are given in [4] and [5]. The sizing problem has been formulated as a linear program (LP) in [6] and [7]. Lagrangian relaxation (LR) based optimization is proposed in [8].

However, finding an optimal gate sizing solution is NP-hard [1], and the suboptimality of sizing solutions has not been quantified and analyzed sufficiently for available heuristics. Real circuits have unknown optimal solution quality, and thus do not shed much light on heuristic suboptimality. On the other hand, artificial circuits with known optimal solution quality – along with any implications they might have for suboptimality of heuristics – are viewed as unrealistic. Thus, the need for further research and development on gate sizing methods has been unclear. In this work, we focus on sizing for leakage reduction, and propose a new method for generating *realistic sizing benchmark circuits with known optimal sizing solutions*, which enables systematic and quantitative comparisons of available gate-sizing heuristics.

For evaluation of CAD heuristics, several methods of generating synthetic benchmarks that match real designs have been proposed. Darnauer and Dai [9] generate random benchmark circuits based on Rent's rule. Their code generates random circuits with a specified number of inputs, outputs, blocks, terminals per cell, and Rent parameter. Hutton et al. [10] define properties such as size, delay, physical shape, edge-length distribution and fanout distribution, and generate combinational circuits to match a given parameterization. Stroobandt et al. [11] provide parameterized (by Rent exponent and net degree distribution) benchmarks with user-selected library cells. With these synthetic benchmarks, various CAD heuristics can be compared to each other, but the suboptimality of the heuristics cannot be measured.

Suboptimality of existing heuristics has been studied for VLSI problems such as synthesis, placement, partitioning, and buffer insertion. Hagen et al. [12] show how to quantify the suboptimality of heuristic algorithms for NP-hard placement and partitioning problems arising in VLSI layout. They construct scaled instances from the original problem and execute the heuristic. If the heuristic solution cost increases at a faster rate than the scaling of the heuristic instance itself, this establishes a lower bound on the heuristic's suboptimality. PEKO (placement examples with known optimal solutions) [13] and its extension PEKU (placement examples with known upper bounds) [14] enable estimation of suboptimality of

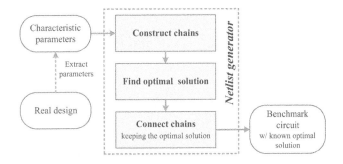

Figure 1: Generation of benchmark circuits with known optimal solutions.

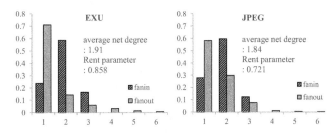

Figure 2: Circuit characteristics (fanin distribution, fanout distribution, average net degree and Rent parameter) for two real designs (EXU: OpenSPARC T1 execution unit; JPEG: JPEG encoder).

several timing-driven placement algorithms; the core approach involves perturbing an original design to obtain a new design with similar topological properties and a known optimal solution.

Our present work builds on the recent work of Gupta et al. [15], which to our knowledge is the only work in the literature to address suboptimality of (leakage-driven) gate sizing heuristics. The authors of [15] (1) propose *eyechart* benchmark circuits which can be optimally sized using dynamic programming methods, and (2) use eyecharts to evaluate the suboptimalities of several gate sizing algorithms. However, [15] does not address the difference or similarity between real designs and eyechart circuits. In [15], the *eyechart* circuits are built from three basic topologies – chain, mesh and star – and the resulting topologies differ substantially from those of real designs in terms of Rent parameter, path length and other parameters.[1] Thus, the eyecharts may be helpful in measuring suboptimality of heuristics, but do not have clear implications for heuristic performance on real designs. Furthermore, [15] does not provide any automated flow for eyechart circuit generation.

In this paper, we provide more realistic benchmarks with known optimal solutions for gate sizing problems. Figure 1 shows the flow of our benchmark circuit generation. (1) To create a circuit with known optimal gate sizing solution, we construct multiple chains (for which optimal sizing solutions can be found by dynamic programming), then connect the chains with inter-chain nets *without affecting the property of having a known leakage-optimal sizing solution*. (2) During the circuit construction, circuit topology is constrained according to user-specified parameters (path depth, and fanin / fanout distributions) so that the constructed benchmarks show similar characteristics to real designs. (3) The inter-chain connections can be added in many possible ways, which gives the potential for greater topological diversity than the previous construction of [15].

Our main contributions are summarized as follows.

- We propose benchmark circuits with known optimal solutions for gate (width and/or V_t) sizing, specifically, for leakage minimization subject to a (setup) delay constraint.

- The proposed benchmarks resemble real designs in terms of size, path depth (number of logic stages), interconnect complexity (Rent parameter), and net degree distribution. These parameters are extracted from real designs. The property of known optimal solution quality is maintained.

- We assess the suboptimality of standard gate sizing approaches, including two commercial tools, with respect to

the above circuit parameters. Our results suggest that (1) commercial tools may suffer from significant suboptimality, and/or (2) existing methods have "similar" degrees of suboptimality for real designs.

The rest of this paper is organized as follows. Section 2 discusses how we address the two main considerations for gate sizing benchmarks – realism in circuit topology and tractability to optimal solution. Section 3 presents details of our benchmark generation procedure. Section 4 provides experimental results and analysis, including suboptimality studies of several heuristics and comparisons between real and artificial circuits. Section 5 summarizes and concludes the paper.

2. BENCHMARK CONSIDERATIONS

For benchmark circuit generation, *realism* and *tractability to analysis* are opposing goals since (1) determining the optimum solution is usually intractable in real designs, and (2) constructions for which optimum solution costs are known are often considered "artificial" [12]. We begin by considering this tension between realism and tractability in benchmark circuits.

First, to construct a *realistic benchmark*, we must use characteristic design parameters in the benchmark generation. Many works in the literature classify or parameterize circuits according to an empirical power-law scaling phenomenon that governs statistics of interconnects among and within subcircuits (cf. the well-known Rent parameter or Rent exponent [16]). Distributions of net degrees, or of the numbers of fanins and fanouts per cell instance, are additional important circuit characteristics. Figure 2 shows circuit characteristics of two real design blocks; each shows different characteristic parameters. In our work, to construct realistic benchmarks we use four design characteristic parameters: (1) number of primary (PIs and POs), (2) (maximum) path depth, (3) fanin distribution, and (4) fanout distribution. These four parameters can be configured in advance, and our benchmark generator makes net connections according to the given parameters subject to a given (setup) timing constraint.

Second, for generated benchmarks to permit *known optimal gate sizing solutions*, some simplifications are required. The eyechart work of [15] achieved tractable optimal solutions by simplifying the cell timing library to eliminate slew dependency. With such a simplified library, it is possible to find an optimal sizing solution for simple (chain) topologies using dynamic programming (DP). Star and mesh topologies can be reduced to equivalent chain topologies so that they, too, can be optimally sized using DP. In our present work, we use the same library simplification approach as [15]. However, we would also like to consider all possible topologies in order to satisfy our goal of realistic benchmark topologies; un-

[1]Eyecharts used in [15] have large depth (650 stages) and small Rent parameter (0.17). Table 4 below shows that real designs have path depths of $20 \sim 70$, and Rent parameter values of $0.72 \sim 0.86$.

fortunately, this makes optimal sizing intractable to DP even with the simplified timing library.

Our key insight is that instead of separating the netlist generation and optimization stages as in the eyechart approach, we can find optimal cell sizes *during* the benchmark netlist generation. We then augment the benchmark circuit without disturbing the existing, known optimal solution. More precisely: (1) we construct gate-chains to realize a specified number of primary input/output ports and a specified path depth; (2) we add fanins and fanouts to cells on the chains to match given fanin and fanout distributions; (3) we find optimal sizing solutions for cells in each chain using DP; and (4) finally, we connect the chains using *connection cells* while preserving the optimal gate sizing solution of each chain.[2]

3. BENCHMARK GENERATION DETAILS

Table 1 shows input parameters to our benchmark generation process. To simplify the procedure, we assume that the numbers of primary inputs and primary outputs are both equal to N. I and O respectively indicate the maximum numbers of fanins and fanouts to any given cell instance. Given the five input parameters, our flow generates N chains, each of which consists of K cells. We connect the chains using *connection cells* according to the prescribed fanin and fanout distributions. The result is a netlist with $K \cdot N + C$ cells, where C is the number of connection cells.

Table 1: Input parameters for benchmark generation.

parameter	description
T	timing path delay upper bound
N	number of primary inputs/outputs
K	(maximum) data path depth
$fid(i)$	fanin distribution (#cells with $i = 1, ..., I$ fanins)
$fod(j)$	fanout distribution (#cells with $j = 1, ..., O$ fanouts)

To generate the circuit properly, the input parameters must satisfy three constraints.

1. The timing budget T should be larger than minimum delay of a chain of K cells.

2. The total numbers of fanins and fanouts in the circuit should satisfy the equality of Equation (1).

3. The prescribed proportion of single-fanout cells, $fod(1)$, should be larger than the proportion of connection cells since connection cells have only one fanout.

We note that in real circuit designs (such as shown in Figure 2), fanout distribution tends to follow a power law, with $fod(1)$ typically greater than 0.6. Thus the third constraint above can be easily satisfied in realistic benchmarks.

$$\sum_{i=1}^{I} i \cdot fid(i) = \sum_{o=1}^{O} o \cdot fod(o) \qquad (1)$$

[2]As discussed in [15], a leakage-optimal gate sizing solution can be known when the nonlinear delay model (NLDM) timing library (e.g., Synopsys .lib format) for the standard cells in eyecharts is modified to eliminate slew-dependence. Also, interconnect delays are omitted for simplicity. This departure from real performance libraries incurs the risk of misleading conclusions; thus, below we show comparisons made using real performance libraries (for which optimal solutions cannot be known).

Algorithm 1 describes the procedure of benchmark generation. In the pseudocode, $gate(i, j)$ represents a gate at the jth stage of the ith chain. G_{co} is the set of connection cells. G_{fi} is the set of gate cells with open fanin ports. $DP(G_{chain}, T)$ is a dynamic programming procedure which finds an optimal cell sizing to minimize leakage power subject to the timing constraint T. We consider arrival times at the output side of any given gate, e.g., the arrival time at the output of gate g is denoted by a_g. Cell delay along the timing arc of cell g from the input that is connected to cell c is denoted by d_g^c. Finally, net delay along the net connecting c and g is denoted by $w_{c,g}$.

Algorithm 1 Netlist generation flow.

Procedure $NetlistGen(T, K, N)$
1. Initialize $gate(i, j)$, where $i = 1, ..., N$ and $j = 1, ..., K$;
2. $G_{co} \leftarrow \emptyset$, $G_{fi} \leftarrow \emptyset$;
3. **for** $j = 1$; $j \leq K$; $j \leftarrow j + 1$ **do**
4.   **for** $i = 1$; $i \leq N$; $i \leftarrow i + 1$ **do**
5.   Assign fanin number to $gate(i, j).fanin$;
6.   Assign fanout number to $gate(i, j).fanout$;
7.   **for** $k = 2$; $k \leq gate(i, j).fanout$; $k \leftarrow k + 1$ **do**
8.   Attach connection gate c to $gate(i, j)$;
9. $G_{co} \leftarrow G_{co} \cup \{c\}$;
10.   **end for**
11.   **if** $gate(i, j).fanin > 1$ **then**
12. $G_{fi} \leftarrow G_{fi} \cup \{gate(i, j)\}$;
13.   **end if**
14.   **end for**
15. **end for**
16. **for** $i = 1$; $i \leq N$; $i \leftarrow i + 1$ **do**
17. $G_{chain} \leftarrow gate(i, j)$, where $j = 1, ..., K$;
18. $DP(G_{chain}, T)$; // find optimal gate size under T
19. **end for**
20. Update timing for all gates ($gate(*)$ and G_{co});
21. **while** $G_{co} \neq \emptyset$ **do**
22.   Select gate c from G_{co} with maximum arrival time;
23.   **for each** gate $g \in G_{fi}$ **do**
24.   Select gate g with minimum arrival time;
25.   **if** $a_c + w_{c,g} + d_g^c \leq a_g$ **then**
26.   Connect c and g;
27. $G_{fi} \leftarrow G_{fi} - \{g\}$;
28.   **break**
29.   **end if**
30.   **end for**
31. $G_{co} \leftarrow G_{co} - \{c\}$;
32. **end while**
33. Assign logic high or low to open input ports of $g \in G_{fi}$;

First, we generate N chains, each with depth K (Lines 1 $\sim$ 15), as shown in Figure 3(a). For each of the $K \cdot N$ cells, we assign (i.e., instantiate) a gate according to the fanin distribution fid (Line 5). Cells in the first stage ($stage_1$) should be assigned one-input gates. Then, we assign the number of fanouts to the output of each cell (Line 6). Cells in the last stage ($stage_k$) have a single fanout. For remaining cells, the number of fanouts is assigned according to the fod. We have explored two alternative strategies for the fanin and fanout assignments: (1) *arranged assignment*, which assigns larger fanins to later stages and larger fanouts to earlier stages, and (2) *random assignment*, which assigns fanins and fanouts in arbitrary order. The arranged assignment improves connectability among the chains, while the random assignment improves diversity of the resulting topology.

Second, we attach connection cells to open fanouts (Lines 7 $\sim$ 10), as illustrated by the red lines in Figure 3(a). The number of connection cells, C, is the same as the number of open fanin ports,

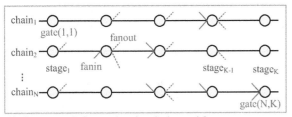

(a) Construct chains, and assign fanins and fanouts.

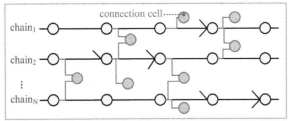

(b) Attach connection cells, and execute DP to determine optimal sizing.

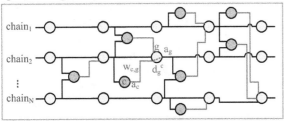

(c) Connect chains subject to the (setup) timing constraint.

Figure 3: Netlist generation flow.

as expressed by Equation (2).

$$C = K \cdot N \sum_{i=1}^{I} (i-1) \cdot fid(i) \qquad (2)$$

The fanin number of connection cells follows the fanin distribution (fid). For connection cells which have more than one fanin, open fanouts in the same stages are connected to the connection cell (Lines 11 ~ 13), as illustrated in Figure 3(b).

After attaching all connection cells, we perform the dynamic programing (DP) with timing budget T for each chain (Line 18). The DP finds the optimal gate sizing which minimizes the leakage power for the chain. After the gate sizing, the sizes of attached connection cells will be set to minimum possible values since they do not have a timing constraint.

Finally, we connect all connection cells to any cells having open fanin ports (Lines 21 ~ 32). Before connecting them, the arrival time for each cell is computed with static timing analysis (STA) with the timing budget T (Line 20). Connections between connection cells and open fanin cells are made only if the timing constraints are satisfied. In Figure 3(c), cell c and cell g can be connected when the arrival time of g via c ($a_c + w_{c,g} + d_g^c$) is less than the arrival time of g through the chain path. If timing slack of the connection cell is large, sizing heuristics can recognize them easily and the problem complexity will be the same as with a chain topology. To prevent this situation, we minimize timing slack of connection cells when making connections. Connection cells and open fanin ports are sorted according to their arrival time. Then, a connection is tried first between a connection cell with large arrival time and an open fanin port with small arrival time (Lines 22, 24).

The connection cells do not change the optimal chain solution since they have minimum gate size. If we upsize them, there is no benefit to the timing slack of the main chain, and the optimal gate sizing of the chain does not change. Without timing constraints, our algorithm guarantees complete connection between open fanins and fanouts by virtue of Equation (1). With timing constraints, some ports can remain unconnected, which we address in Section 4.2 below. The open input ports are assigned with logic high (VDD) or low (VSS) according to the logic type. This assignment does not change the optimal solution.

After completing all the connections, we end up with a benchmark circuit of $K \cdot N + C$ cells with known optimal gate sizing for minimum leakage. (A small detail: when we use the generated circuit as a sizing benchmark, we initially assign maximum cell size (with highest leakage and fastest timing) to each instance, so as to avoid giving the leakage optimization tool any information about the optimal solution.)

4. EXPERIMENTAL SETUP AND RESULTS

4.1 Experimental Setup

Our netlist generator is implemented in C++ and produces a benchmark netlist in *Verilog HDL* (.v) with the corresponding delay models (.lib). Two types of delay and power models are used from the previous *eyechart* work [15][3] – (1) LP: linear increase in power with size for gate sizing context, and (2) EP: exponential increase in power with size for V_t or gate-length bias. The LP and EP power models have eight and three gate sizes (i.e., cell variants per master), respectively. To analyze the problem complexity of generated netlists, and suboptimality of standard sizing tools, we perform experiments on a 2.8 GHz Linux workstation with 24 GB RAM, using three different gate sizing methods – (1) two commercial gate sizing and leakage optimization tools (*BlazeMO v2008* [21] and *Cadence Encounter v9.1* [22]),[4] (2) a web-available *UCLA sizing tool* [25] (*Greedy*) which greedily swaps cells according to a $\Delta power / \Delta delay$ sensitivity function, and (3) a web-available UCSD sensitivity-based leakage optimizer [26] (*SensOpt*) with $\Delta power \times slack$ sensitivity function.[5] To generate realistic benchmark circuits, we use six open-source designs – *SASC* (asynchronous serial controller), *SPI* (serial peripheral interface), *AES* (data encryption), *JPEG* (image processing) and *MPEG* (video processing) from the *OpenCores* site [19], and *EXU* (execution unit) from *OpenSPARC T1* [20]. In our experiments, we measure the suboptimality of the various gate sizing heuristics, as is defined in Equation (3).

$$Suboptimality = \frac{power_{heur} - power_{opt}}{power_{opt}} \qquad (3)$$

We use the same timing and power analysis tool (*Synopsys PrimeTime C2009.6* [24] to evaluate results.

---

[3]According to the authors of [15], EP corresponds to the multi-V_t context, and LP corresponds to the gate-length biasing context.
[4]These are referred to as *Comm1* and *Comm2* below. We do not give the mapping – i.e., which tool is *Comm1* and which is *Comm2* – in order to maintain anonymity as required by the tools' licenses.
[5]At the website [26], details of the UCSD *SensOpt* tool are given. The tool performs post-layout cell swapping using the *Tcl* socket interface to a golden STA tool, *Synopsys PrimeTime*.

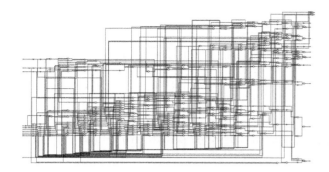

Figure 4: Schematic of generated netlist ($N = 10$, $K = 20$).

4.2 Generated Benchmarks

In this subsection, we present the results of generated benchmarks and their complexity in a power (leakage) optimization. Then, we compare the benchmarks and real designs in terms of characteristic parameters. Figure 4 shows the schematic of a generated netlist with 10 chains and path depth of 20. In the netlist, chains are connected to each other in arbitrary order, and various topologies can be found.

A connection between chains can be made when the newly generated path has positive (or zero) slack with respect to the timing constraint. As a result, some cells in the chain will have open ports and some connection cells will remain unconnected. If the number of unconnected cells is large, the generated netlist will deviate from the specified fanin and fanout distributions. As noted above, to improve the connectability we can assign the larger fanins to later stages, and larger fanouts to earlier stages. However, such an *arranged assignment* can reduce the difficulty of the sizing optimization for the benchmark: many connection cells will have loosely constrained timing (i.e., large slack), and this makes it easy to find the optimal solution. To consider both connectability and optimization difficulty, we mix the two alternative strategies – arranged and random assignments – in Algorithm 1, Lines 5 and 6. Table 2 shows the failure rate of connections among chains and problem complexity (suboptimality) according to the different mixtures of the arranged and random assignments. In the experiment, N and K values are fixed (40), and the EP power model is used. Suboptimality and runtime are obtained for the commercial tool (*Comm1*). The results show that 25% of arranged assignment in practice results in over 99% of connectivity, while also affording a sufficient problem complexity (11.2% suboptimality). The 100% random assignment shows smaller suboptimality (7.7%) for gate sizing because it results in many unconnected gates (17%). In all experiments reported below, we use the 75% random / 25% arranged assignment.

Since our benchmark generator makes chains first, then connects the chains to each other, we have assessed the problem complexity of benchmarks before and after the chain connection. Table 3 shows the suboptimality of leakage reduction for the commercial tool and the greedy method. The results show that the complexity (difficulty) of gate sizing increases with the number of chain connections. The chain-only structures are easy to solve, and heuristics show small suboptimalities ($\sim$3%). However, with added chain connections, the observed suboptimality (and inferred instance difficulty) increase significantly.

Table 4 shows the characteristic parameters of (a) real designs and (b) generated benchmarks. In the table, the Rent parameter has been evaluated using [18]. The real circuits do not follow the second constraint of our netlist generator (Equation (1)) since the numbers of primary inputs and primary outputs differ. For this reason, we select fanin and fanout distribution numbers that are only similar (not identical) to those of the real design when we perform the benchmark generation. From the results, generated circuits show similar design size, path depth, Rent parameter and average fanin (fanout); this offers hope that our benchmark generation approach can provide realistic benchmark circuits for gate sizing.

Table 2: Connectability and complexity (suboptimality) of generated netlists according to different proportions of arranged and random assignments.

arranged	random	unconnected	subopt.	runtime
100%	0%	0.00%	2.6%	108 sec.
75%	25%	0.00%	6.8%	97 sec.
50%	50%	0.25%	10.3%	120 sec.
25%	75%	0.75%	11.2%	225 sec.
0%	100%	17.0%	7.7%	311 sec.

Table 3: Instance complexities (difficulties) of chain-only and connected-chain topologies.

# of chain	# of stage	chain-only		connected	
		Comm1	*Greedy*	*Comm1*	*Greedy*
(a) EP library					
40	20	2.4%	0.3%	10.4%	8.7%
40	40	2.1%	1.3%	10.3%	11.1%
80	20	2.0%	0.5%	10.3%	10.9%
80	40	2.1%	1.3%	9.9%	10.9%
(b) LP library					
40	20	1.7 %	3.1%	7.7%	17.9%
40	40	2.4 %	3.5%	12.0%	18.5%
80	20	1.9 %	3.3%	12.3%	19.1%
80	40	2.5 %	3.5%	15.9%	19.6%

4.3 Suboptimality of Heuristics

In this section, we show suboptimality of standard heuristic solutions with our benchmarks. Figure 5 (respectively, Figure 6) shows suboptimality and runtime of heuristics (including *Comm1*) when the number of chains (respectively, number of stages) increases in the benchmark circuits.[6] From Figure 5, we see that the suboptimality increases slightly according to the design size, but runtime increases exponentially with the number of chains since total number of paths increases significantly with respect to the chain number. When the number of stages increases (Figure 6), the suboptimality increases especially for the *Comm1* and *SensOpt* solvers. We recall that the LP library model has a larger number (eight) of sizing candidates than the EP library model (three). We believe that the likely cause of the LP model showing a larger suboptimality and runtime for the greedy and sensitivity-based optimizations.

Figure 7 shows suboptimality and runtime results when the testcases have different topological complexities. To estimate the effect of netlist complexity, we change fanin and fanout distributions in the benchmark generation, such that each benchmark has a different average net degree. From the results, we see that subop-

---

[6]The same fanin and fanout distributions have been used for the experiments in Figure 5 and Figure 6 – fid: 0.3, 0.6, 0.1, fod: 0.6, 0.1, 0.2, 0.1 .

Table 4: Characteristic parameters of real designs and generated benchmarks.

testcase	path depth	#instances	Rent parameter	average fanin	fanin distribution			fanout distribution					
					1	2	3	1	2	3	4	5	6
(a) characteristic parameters of real designs													
SASC	20	624	0.858	2.06	0.169	0.606	0.225	0.663	0.177	0.045	0.029	0.015	0.071
SPI	33	1092	0.880	1.813	0.345	0.497	0.158	0.735	0.077	0.090	0.038	0.020	0.039
EXU	31	25560	0.858	1.91	0.237	0.587	0.165	0.711	0.142	0.059	0.032	0.014	0.007
AES	23	23622	0.810	1.89	0.237	0.637	0.126	0.694	0.120	0.062	0.039	0.026	0.015
JPEG	72	141165	0.721	1.84	0.280	0.597	0.123	0.582	0.299	0.077	0.011	0.005	0.004
MPEG	33	578034	0.848	1.59	0.334	0.567	0.04	0.681	0.244	0.021	0.007	0.003	0.002
(b) characteristic parameters of generated benchmarks													
ng_SASC	20	631	0.865	2.06	0.17	0.60	0.23	0.66	0.18	0.05	0.05	0.02	0.04
ng_SPI	33	1079	0.877	1.80	0.35	0.50	0.15	0.73	0.10	0.08	0.03	0.02	0.04
ng_EXU	31	24733	0.814	1.90	0.25	0.60	0.15	0.7	0.05	0.05	0.1	0.05	0.05
ng_AES	23	23780	0.820	1.88	0.24	0.64	0.12	0.7	0.05	0.05	0.12	0.03	0.05
ng_JPEG	72	132479	0.831	1.84	0.28	0.6	0.12	0.58	0.25	0.05	0.04	0.03	0.05
ng_MPEG	33	527995	0.848	1.60	0.42	0.56	0.02	0.68	0.2	0.04	0.03	0.02	0.03

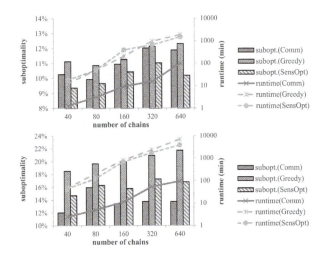
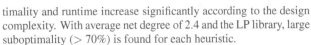

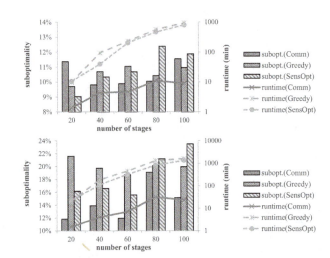

Figure 5: Suboptimality and runtime for different number of chains N **(stage** $K = 40$**) with EP library (above) and LP library (below).**

Figure 6: Suboptimality and runtime for different number of stages K **(#chain** $N = 100$**) with EP library (above) and LP library (below).**

timality and runtime increase significantly according to the design complexity. With average net degree of 2.4 and the LP library, large suboptimality ($> 70\%$) is found for each heuristic.

In addition, we study the effect of delay constraints on suboptimality and complexity. Figure 8 shows suboptimality and runtime results when the testcases are generated with different delay constraints. The testcases have the same topology (number of chains, number of stages and net degree). However, the suboptimalities achieved by each heuristic differ widely according to the timing constraint. From the results, netlists with tight delay constraint lead to greater heuristic suboptimality, especially with the LP library.

Table 5 shows suboptimality and runtime results for the generated netlists in Table 4 and one example *eyechart*[7] circuit [15]. The results[8] show that common sizing methods, including two commer-

---

[7]Specifically, we use the EP_NLD and LP_NLD eyechart circuits with 5ns timing budget from [27].

[8]Results for the *ng_MPEG* testcase are missing for the *UCLA* greedy sizing tool, which cannot handle the large ($\sim$500K instance) instance size.

cial tools (*Comm1, Comm2*), are suboptimal for realistic benchmark circuits by up to 16.7%, 52.2%, 29.0% and 26.9% for the commercial tools, greedy method and sensitivity-based method, respectively. Among the testcases, *ng_JPEG* shows the largest suboptimality; we believe that this a consequence of having larger path depth than the other testcases.

Finally, because new realistic benchmarks may not induce the same relative performance across heuristics as real designs, we have also compared the same leakage optimizers using real circuits and real timing/leakage libraries. Table 6 shows the leakage optimization results with the real circuits and libraries. The designs are implemented with a TSMC 65GP library (65nm), and are synthesized, placed and routed with *Synopsys Design Compiler C-2009* [23] and *Cadence Encounter v9.1* [22]. For the leakage optimization, both multi-V_t (NVT, HVT, LVT) library (part (a) of the table) and multi-L_{gate} NVT library (part (b) of the table) have been used. Reported suboptimality is calculated from the best result in the four result columns. In the table, each optimizer shows different suboptimality according to each design. The tools suffer from

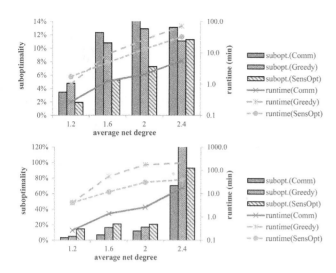

Figure 7: Suboptimality and runtime for different average net degrees (chain $N = 40$, stage $K = 40$) with EP library (above) and LP library (below).

significant suboptimality, e.g., *EXU* circuits for *Comm1* (40.8%) and *Greedy* (28.7%), and *AES* circuits for *Comm2* (18.2%). Since suboptimalities are calculated relative to the best heuristic result, there is further suboptimality from the actual optimal solution. It is clear from Tables 5 and 6 that – somewhat unfortunately – the artificial and real netlists and performance libraries suggest different relative and absolute suboptimalities for the sizing heuristics. We believe that this is for several reasons, including (1) our enhanced eyechart-like benchmarks consist of netlists without wire capacitance, and (2) we use only one kind of library cell for each number of fanin ports. We continue to explore ways to improve the matching to results on real designs and real libraries, while maintaining the important property of having a known optimal sizing solution.

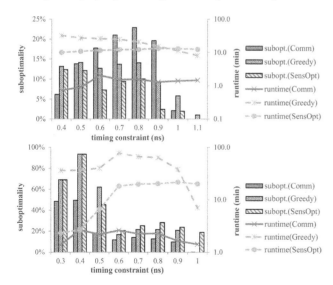

Figure 8: Suboptimality and runtime for different timing constraints (chain $N = 40$, stage $K = 40$, average net degree = 2.0) with EP library (above) and LP library (below).

Table 5: Suboptimality with respect to known optimal solution for generated netlists in Table 4.

testcase	optimal leakage (W)	Comm1	Comm2	Greedy	SensOpt
(a) EP library					
eyechart	3.46E-6	18.9%	25.4%	24.0%	20.1%
ng_SASC	9.35E-8	10.9%	24.9%	4.9%	2.8%
ng_SPI	2.62E-7	11.4%	21.0%	13.5%	10.9%
ng_AES	5.97E-6	14.2%	13.7%	11.7%	8.8%
ng_EXU	6.62E-6	15.2%	20.9%	10.8%	8.9%
ng_JPEG	3.29E-5	16.7%	33.1%	11.9%	12.0%
ng_MPEG	9.54E-5	10.9%	52.2%	-	10.1%
(b) LP library					
eyechart	3.38E-6	39.3%	44.3%	29.0%	29.1%
ng_SASC	1.55E-7	13.0%	31.4%	22.5%	16.0%
ng_SPI	3.91E-7	6.9%	27.2%	24.3%	24.0%
ng_AES	8.62E-6	11.3%	31.9%	22.4%	23.5%
ng_EXU	9.33E-6	13.9%	40.5%	24.6%	23.4%
ng_JPEG	4.74E-5	16.3%	38.3%	29.0%	26.9%
ng_MPEG	1.65E-4	15.5%	43.7%	-	21.6%

5. CONCLUSIONS

In this work, we have proposed a new benchmark generation technique for gate sizing, which constructs *realistic* circuits with known optimal solutions. Our generated netlists closely resemble real designs in terms of instance count, path depth, interconnect complexity, and net degree / fanin / fanout distributions; all of these attributes are parameters of the netlist generation. When we compare our generated benchmarks with real designs, we also see similarities with respect to other circuit characteristics such as average net degree and Rent parameter.

Our benchmarks with known optimal solutions enable systematic and quantitative study of the suboptimality of common sizing heuristics, with respect to key parameters of the circuit topology. In particular, our experimental results with web-available academic tools and commercial tools show that common leakage-driven sizing methods are suboptimal for realistic benchmark circuits by up to 52.2% and 43.7% for V_t-assignment and CD-biasing formulations, respectively. At the same time, our results also show discrepancies between inferences obtained using our generated circuits and those obtained from real circuits (and libraries). However, all of our results suggest that (1) commercial tools may still suffer from significant suboptimality, and/or (2) existing methods have "similar" degrees of suboptimality, especially as instance size increases (cf. results for JPEG and MPEG in Table 6). Our ongoing work seeks to address the above-mentioned discrepancies. In addition, we are working to handle more realistic delay models, possibly in the context of realistic benchmarks with tight upper bounds on optimal gate leakage.

6. REFERENCES

[1] W. N. Li, "Strongly NP-Hard Discrete Gate Sizing Problems", *Proc. ACM/IEEE International Conference on Computer-Aided Design*, 1993, pp. 468–471.

[2] J. P. Fishburn and A. E. Dunlop, "Tilos: A Polynomial Programming Approach to Transistor Sizing", *Proc. ACM/IEEE International Conference on Computer-Aided Design*, 1985, pp. 326–328.

Table 6: Suboptimality (with respect to best heuristic solution) for real performance libraries and netlists.

testcase	minimum leakage (W)	Comm1	Comm2	Greedy	SensOpt
(a) TSMC65 GPLUS Multi-V_t (NVT, HVT, LVT) library					
SASC	2.46E-5	15.3%	5.9%	13.0%	0.0%
SPI	6.26E-5	33.0%	24.3%	10.8%	0.0%
AES	6.29E-4	0.0%	18.2%	13.8%	9.8%
EXU	9.27E-4	17.3%	26.1%	21.0%	0.0%
JPEG	5.76E-3	1.7%	2.7%	3.5%	0.0%
MPEG	1.38E-2	22.3%	18.7%	-	0.0
(b) TSMC65 GPLUS L_{gate} biased NVT library					
SASC	2.85E-5	23.6%	27.7%	4.4%	0.0%
SPI	6.55E-5	1.9%	0.0%	1.4%	0.5%
AES	5.63E-4	30.2%	12.6%	9.1%	0.0%
EXU	9.41E-4	40.8%	3.4%	28.7%	0.0%
JPEG	5.24E-3	13.7%	12.4%	5.7%	0.0%
MPEG	1.81E-2	0.0%	5.3%	-	2.9%

[3] P. Pant, R. K. Roy and A. Chatterjee, "Dual-Threshold Voltage Assignment with Transistor Sizing for Low Power CMOS Circuits", *IEEE Trans. on VLSI Systems* 9(2) (2001), pp. 390–394.

[4] S. Sirichotiyakul, T. Edwards, C. Oh, R. Panda and D. Blaauw, "Duet: An Accurate Leakage Estimation and Optimization Tool for Dual-Vt Circuits", *IEEE Trans. on VLSI Systems* 10(2) (2002), pp. 79–90.

[5] P. Gupta, A. B. Kahng, P. Sharma and D. Sylvester, "Gate-length Biasing for Runtime-leakage Control", *IEEE Trans. on Computer-Aided Design* 25(8) (2006), pp. 1475–1485.

[6] M. R. C. M. Berkelaar and J. A. G. Jess, "Gate Sizing in MOS Digital Circuits with Linear Programming", *Proc. EURO-DAC*, 1990, pp. 217–221.

[7] K. Jeong, A. B. Kahng and H. Yao, "Revisiting the Linear Programming Framework for Leakage Power vs. Performance Optimization", *Proc. International Symposium on Quality Electronic Design*, 2009, pp. 127–134.

[8] Y. Liu and J. Hu, "A New Algorithm for Simultaneous Gate Sizing and Threshold Voltage Assignment", *IEEE Trans. on Computer-Aided Design* 29(2) (2010), pp. 223–234.

[9] J. Darnauer and W. W. Dai, "A Method for Generating Random Circuits and Its Application to Routability Measurement", *Proc. ACM/SIGDA International Symposium on Field Programmable Gate Arrays*, 1996, pp. 66–72.

[10] M. D. Hutton, J. Rose, J. P. Grossman and D. Corneil, "Characterization and Parameterized Generation of Synthetic Combinational Circuits", *IEEE Trans. on Computer-Aided Design* 17(10) (1998), pp. 985–996.

[11] D. Stroobandt, P. Verplaetse and J. V. Campenhout, "Generating Synthetic Benchmark Circuits for Evaluating CAD Tools", *IEEE Trans. on Computer-Aided Design* 19(9) (2000), pp. 1011–1022.

[12] L. Hagen, J. H. Huang and A. B. Kahng, "Quantified Suboptimality of VLSI Layout Heuristics", *Proc. ACM/IEEE Design Automation Conference*, 1995, pp. 216–221.

[13] C. Chang, J. Cong and M. Xie, "Optimality and Scalability Study of Existing Placement Algorithms", *Proc. Asia South-Pacific Design Automation Conference*, 2003, pp. 621–627.

[14] J. Cong, M. Romesis and M. Xie, "Optimality and Stability Study of Timing-Driven Placement Algorithms", *Proc. ACM/IEEE International Conference on Computer-Aided Design*, 2003, pp. 472–478.

[15] P. Gupta, A. B. Kahng, A. Kasibhatla and P. Sharma, "Eyecharts: Constructive Benchmarking of Gate Sizing Heuristics", *Proc. ACM/IEEE Design Automation Conference*, 2010, pp. 597–602.

[16] B. S. Landman and R. L. Russo, "On a Pin versus Block Relationship for Partitions of Logic Graphs", *IEEE Trans. on Computers* C-20(12) (1971), pp. 1469–1479.

[17] C. Chen, C. Chu and D. Wong, "Fast and Exact Simultaneous Gate and Wire Sizing by Lagrangian Relaxation", *IEEE Trans. on Computer-Aided Design* 18(7) (1999), pp. 1014–1025.

[18] K. Jeong, A. B. Kahng and H. Yao, "Rent Parameter Evaluation Using Different Methods", http://vlsicad.ucsd.edu/WLD/RentCon.pdf.

[19] *OpenCores: Open Source IP-Cores*, http://www.opencores.org .

[20] *Sun OpenSPARC Project*, http://www.sun.com/processors/opensparc .

[21] *Blaze MO User's Manual*, http://www.tela-inc.com .

[22] *Cadence Encounter User's Manual*, http://www.cadence.com .

[23] *Synopsys Design Compiler User's Manual*, http://www.synopsys.com .

[24] *Synopsys PrimeTime User's Manual*, http://www.synopsys.com .

[25] *Enhanced OAGear-Static-Timer with Heuristics for Gate Sizing*, http://nanocad.ee.ucla.edu/Main/DownloadForm .

[26] *Sensitivity-Based Leakage Optimizer*, http://vlsicad.ucsd.edu/SIZING/optimizer.html#SensOpt .

[27] *UCLA Eyecharts*, http://nanocad.ee.ucla.edu/Main/DownloadForm .

The ISPD-2012 Discrete Cell Sizing Contest and Benchmark Suite .

Muhammet Mustafa Ozdal[1], Chirayu Amin[1], Andrey Ayupov[1], Steven Burns[1], Gustavo Wilke[2], and Cheng Zhuo[2]

[1] Strategic CAD Labs, Intel Corporation, Hillsboro, OR 97124
[2] Core CAD Technologies, Intel Corporation, Hillsboro, OR 97124
{mustafa.ozdal, chirayu.s.amin, andrey.ayupov, steven.m.burns, gustavo.r.wilke, cheng.zhuo}@intel.com

ABSTRACT

Circuit optimization is essential to minimize power consumption of designs while satisfying timing constraints. The CAD problem focused on in the ISPD-2012 Contest is simultaneous gate sizing and threshold voltage assignment. In this paper, we describe an overview of the contest objectives and the provided benchmark suite. Furthermore, some details are provided in terms of the standard cell library, timing models, and the evaluation metrics of the ISPD-2012 Contest.

Categories and Subject Descriptors

B.7.2 [**Hardware, Integrated Circuits**]: Design Aids

General Terms

Algorithms, Design, Experimentation, Performance

Keywords

Gate sizing, Circuit optimization, Benchmarks, Physical Design

1. INTRODUCTION

As the complexity of mobile devices are increasing, it is becoming more important to have designs with both high performance and low power. Circuit optimization is an important stage in VLSI design to minimize power consumption while satisfying performance constraints. In the ISPD-2012 Contest, we focus on the problem of gate sizing and device parameter selection. Although this problem has been studied extensively in the literature, there are still various challenges that make it difficult to apply the existing academic algorithms to modern industrial designs, as outlined in [5]:

• Discrete cell sizes: There are many existing sizing algorithms that are based on continuous optimization, such as linear programming [1, 3], convex programming [4, 8], iterative analytical solving [2, 6, 9], and fractional network flow [7, 10]. Since industrial libraries contain discrete cells, these algorithms require rounding to the nearest cell sizes available in the library, which may lead to suboptimalities. Furthermore, different technology parameters introduce discrete cell families, with very few elements in between. For example, only a few levels of threshold voltages exist for a typical library.

Assuming continuous sizes during optimization, and rounding to discrete choices at the end may lead to suboptimalities.

• Cell timing models: Modern cell libraries are characterized based on complex timing models. Many academic algorithms assume simple delay models such as piecewise linear model [1] and RC switch gate model [2, 7, 9, 10]. In reality, even higher-order convex models may not be able to capture the delay characteristics of real cell libraries, because delay is not a convex function of cell size and output load, due to various reasons such as transistor folding [3]. This effect was illustrated for a cell family from a 32nm industrial library in [5].

• Complex timing models constraints: High-performance designs typically have multiple clock domains, multi-cycle overrides, transparent latches, false paths, etc. Furthermore, interconnect timing models are significantly more complex than the simple Elmore delays assumed by many existing works. A realistic sizing algorithm should not rely on the simplicity of timing models, and should be able to be extended to capture complex timing constraints.

• Slew effects: Cell delays depend on both output load and input slew rate. For example, downsizing a cell can lead to worse slew rate at its output, which in turn affects the delays of the successor cells. Ignoring slew effects can lead to significant inaccuracies during optimization.

• Large design sizes: Some designs contain millions of cells, and a large portion of the cells may be on critical or near-critical paths. Realistic sizing algorithms should be scalable enough to handle large designs.

One of the objectives of the ISPD-2012 Discrete Cell Sizing Contest is to expose some of these challenges to the academic community, while keeping the problem complexity still manageable. Specifically, we provide cell libraries with discrete sizes and relatively realistic delay and slew models, as well as a benchmark suite with a range of small, medium, and large designs. On the other hand, some simplifications have been done in the problem formulation to avoid overwhelming the algorithmic-CAD community with complexities of timing analysis. In particular, only a single clock domain exists for each benchmark design, and there are no extra timing overrides such as false paths. Furthermore, to simplify interconnect timing computations, each net is defined as a lumped capacitance with zero resistance. The purpose here is to allow the contestants to focus on the algorithmic aspects without having to implement a full-scale timing engine. Nevertheless, the cell timing models are still kept complex; however, this complexity is hidden in the pre-characterized lookup tables that define the delay and slew functions of the timing arcs.

1.1 Contest Overview

The objective of the ISPD-2012 Contest is to perform simultaneous gate sizing and threshold voltage assignment to optimize total leakage power under timing constraints. A standard cell library and a set of benchmarks are provided for the contest. For each benchmark, the

Prefix	Description	Function
in01	Inverter	!A
na02	2-input NAND	!(A & B)
na03	3-input NAND	!(A & B & C)
na04	4-input NAND	!(A & B & C & D)
no02	2-input NOR	!(A | B)
no03	3-input NOR	!(A | B | C)
no04	4-input NOR	!(A | B | C | D)
ao12	AND-OR-INV	!(A | (B & C))
ao22	AND-OR-INV	!((A & B) | (C & D))
oa12	OR-AND-INV	!(A & (B | C))
oa22	OR-AND-INV	!((A | B) & (C | D))
ms00	Flip Flop	sequential

Figure 1: Cell families in the ISPD-2012 Benchmark Suite

following set of files exist: 1) verilog netlist, 2) interconnect parasitics, and 3) timing constraints. In addition, C++ helper classes are provided to parse the contest specific data from each of these files, as well as the cell library.

The ranking of the submitted solutions will be done based on the total leakage power and timing violations. The timing violations (if any) will be computed by running Synopsys PrimeTime®.

The rest of the paper is organized as follows. In Section 2, we describe the standard cell library provided. In Section 3, we summarize the timing models and the timing infrastructure of the contest. The benchmarks provided are outlined in Section 4. After that, the evaluation metrics and the ranking criteria are described in Section 5.

2. STANDARD CELL LIBRARY

A cell library in Synopsys Liberty™ format is provided as part of the *ISPD-2012 Benchmark Suite*. There are cells with 12 different logic functions (11 combinational and 1 sequential) included in this library, as listed in Figure 1.

Each combinational library cell has 3 threshold voltage (Vt) levels and 10 sizes for each Vt level. In other words, there are 30 different library cells that implement the same combinational logic function. As will be explained in Section 3, sequential sizing is not allowed in this contest. Hence, there is only a single Vt level and a single size for the sequential library cell.

For each library cell, timing arcs are defined from rising/falling edges of the signal at the input pin(s) to the falling/rising edges at the output pin. Each timing arc is associated with lookup tables corresponding to the pre-characterized delay and transition time (i.e. slew) functions. Specifically, the delay of a timing arc is a function of the input transition time (at the input pin) and the total output load (at the output pin). Similarly, the output transition time of a timing arc is a function of the input transition time and the output load. These functions are defined in the form of lookup tables in the library, and the intermediate values (not provided in the table) are computed based on 2-D linear interpolation.

A simple current source model was used to characterize cell delays and transition times. This model comprehends the input transition time, the output load, the threshold voltage of the cell, and the p/n transistor ratio. Figure 2 illustrates the delay values of a timing arc through an inverter for different cell sizes and threshold voltages. In this graph, the ratio of cell size to the output load is kept constant. For example, if the inverter with $size = S$ drives $load = L$, then the inverter with $size = 2S$ drives $load = 2L$. If a simple delay model was used, we would expect delay to be constant when the size/load ratio is kept constant. However, as shown in this figure, delay changes non-monotonically, especially for cells with small sizes. The main reason for this behavior is due to transistor folding in the layout, and

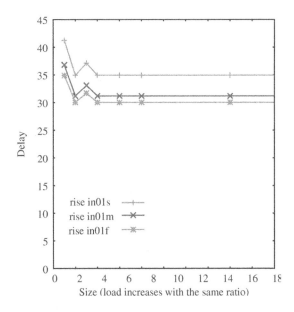

Figure 2: Delay as a function of size and load for 3 inverter cell families with different threshold voltages. In the x axis, the size and load values are changed such that size/load ratio is kept constant.

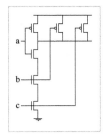

Figure 3: The transistor diagram of a 3-input NAND gate.

the relatively more internal loading on the smaller cells. A similar behavior was illustrated in [5] for a 32nm industrial library.

In Figure 2, the delay functions for 3 cell families are demonstrated: *in01s*, *in01m*, and *in01f*, corresponding to high, medium, and low threshold voltage (Vt) levels, respectively. The cell family with high Vt has the largest delay, but the lowest leakage power (not shown in this figure), and vice versa. Proper tradeoff needs to be done in the sizer to choose the right threshold voltages as well as cell sizes to minimize the total leakage power, while satisfying all timing constraints.

The p/n transistor size ratios were chosen in such a way to balance the drive strengths of the rising and falling transitions. For each inverter cell, the p:n size ratio was chosen to be 1:1 because of symmetric pmos and nmos transistors. However, for a 3-input NAND gate, the p:n size ratio was chosen to be 1:3, because the three pmos transistors are connected in parallel, while the 3 nmos transistors are connected in series, as shown in Figure 3. However, due to discreteness, the exact 1:3 ratio cannot always be achieved, especially for small cell sizes, as listed in Figure 4. For this reason, the rising and falling drive strengths are not always symmetric, and further non-linearities exist when delay of the 3-input NAND gate is plotted against cell size (while keeping size/load ratio constant as before) in Figure 5. This is another realistic effect captured in the contest library.

In addition to delay and slew functions, the following data is provided for each library cell:

na03m01			na03m02			na03m03			na03m04			na03m06			na03m08			na03m10			na03m20			na03m40			na03m80		
p	n	tot	p	n	tot	p	n	tot	p	n	tot	p	n	tot	p	n	tot	p	n	tot	p	n	tot	p	n	tot	p	n	tot
1	1	6	1	2	9	1	3	12	2	4	18	2	6	24	3	8	33	6	16	66	11	32	129	22	64	258	43	128	513

Figure 4: The p/n ratios for different sizes of the 3-input NAND gate cell family "na03m" in the contest library. Here, "p" and "n" denote the size of one pmos and nmos transistor, respectively. The total size (denoted as "tot") is the sum of 3 pmos and 3 nmos transistors.

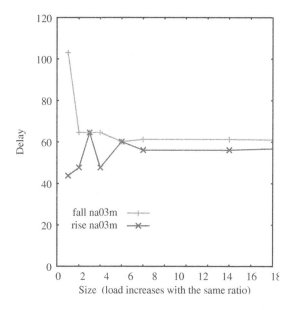

Figure 5: Rising and falling signal delay as a function of size and load for cell family "na03m" of the contest library. In the x axis, the size and load values are changed such that size/load ratio is kept constant.

- Cell leakage power value

- Input pin capacitances

- Maximum capacitance limit for the output pin (i.e. the maximum load that the output pin can drive)

A C++ class is also provided with the cell library to extract the relevant data (for this contest) from the library file.

3. TIMING INFRASTRUCTURE

The cell delay and slew functions have been provided as lookup tables in the standard cell library, as described in Section 2. For simplicity, the interconnects are modeled as lumped capacitances with zero resistances in this contest. All sequential cells in the benchmarks are rising-edge triggered flip flops. Sequential sizing is not allowed, and there is only a single size available in the contest library for sequential cells, for which the set-up time is 0, and there is no hold constraint. For contest purposes, the following ideal clock network assumptions hold:

- Clock port at the top level is directly connected to all sequential pins without clock buffers.

- Arrival time of the clock signal at all sequential inputs is the same (i.e. zero skew).

- Sequential delay is independent of the clock slew.

- Clock input pin capacitances are zero for all sequential cells.

- Clock net has zero lumped capacitance.

Name	# Inputs	# Outputs	# Comb. Cells	# Seq. Cells	# Total Cells
usb_phy	15	19	514	98	612
DMA	683	276	23109	2192	25301
pci_bridge32	160	201	29844	3359	33203
des_perf	234	140	102427	8802	111229
vga_lcd	85	99	147812	17079	164891
b19	22	25	212674	6594	219268
leon3mp	254	79	540352	108839	649191
leon2	615	85	644905	149381	794286
netcard	1836	10	860949	97831	958780

Figure 6: Statistics of the ISPD-2012 Contest Benchmarks

Only a single clock domain is defined, and there are no complex timing constraints such as false paths, transparent latches, multi-cycle overrides, etc. The purpose here is to simplify the timing model so that the contestants can focus their efforts on the combinational sizing optimization. On the other hand, the cell timing and slew models are kept realistic, and this complexity needs to be taken into account during sizing optimization.

In terms of the timing engine, the contestants have the following two options: 1) Implement a timing engine to be used during optimization. 2) Use a stand-alone industrial timing engine as a black-box during optimization. If the contestants choose the second option, we have provided a timing infrastructure (with C++ and TCL APIs) for the contestants to be able to call the Synopsys PrimeTime® timing engine as a black-box from their sizing engines.

4. BENCHMARKS

A set of designs are provided as the contest benchmarks, the statistics of which are as shown in Figure 6. Each design consists of the following files:

- Design netlist: The netlists provided in this contest are based on IWLS 2005 benchmark circuits. These netlists have been mapped to the contest standard cell library (Section 2), and minor changes were made for simplification purposes. The file format is a subset of the structured Verilog format, restricted to enable easy parsing. The following properties hold for each netlist:

    - No hierarchy, no buses, no behavioral keywords.

    - Single clock domain.

    - No unconnected pins, no escape characters in names.

    - No power or ground nets.

- Interconnect Parasitics: As described in Section 3, the interconnect parasitics are modeled as lumped capacitances with zero resistances. The lumped capacitance values are provided in IEEE SPEF format. In this file, a single lumped capacitance value is provided for each net in the design.

- Timing Constraints: Timing constraints for each design are provided in Synopsys Design Constraints (SDC) format. The following data is provided in this file:

    - The clock period.

    - The load independent arrival time at each input port.

- The combinational delay from each output port to the next presumed sequential cell (outside this design).

- The presumed driving cell connected to each input port.

- The presumed load capacitance connected to each output port.

Note that these timing constraints define the environment in which this design is instantiated. For example, a specific driving cell (outside this design) is *presumed* to be connected to each input port. In other words, this presumed driver defines the drive strength at the corresponding input port. Similarly, a certain load is presumed to be connected at each output port. This is to make sure that the cell driving this output port is strong enough to handle this *presumed* load outside.

We have provided C++ parser classes to extract the contest-specific data from these benchmarks.

5. EVALUATION METRICS

A subset of the benchmarks listed in Figure 6 will be used for contest evaluations. The basic objective of the contest is to obtain results with minimum total leakage power, while satisfying all timing constraints. There will be two separate rankings:

- Primary ranking: Solution quality will be the main metric. Runtime will be used for tie-breaking.

- Secondary ranking: Both solution quality and runtime will be important.

The timing results of the optimized designs will be evaluated by running Synopsys PrimeTime® on the sizing results. There are 3 different violation types defined in the evaluation metrics:

1. Negative timing slacks

2. Slew violations

3. Maximum capacitance violations

The total violation value for each benchmark result will be computed by summing up all violations. The timing constraints are set such that it is possible to obtain results with zero violations. For this reason, the total violation value is the most important ranking metric for both primary and secondary criteria mentioned above. In other words, a solution with smaller violations will always be ranked higher than another solution with larger violations. We expect that the top-ranked solutions will have zero violations, and the actual ranking will be based on the other metrics outlined below.

If the violation values are identical, then the solution quality will be determined by the total cell leakage power. The submitted binaries will be run on a system with 12 Xeon cores with 3GHz frequency, and multi-core implementations are encouraged. There will be a hard-limit on the runtime of each submission. This limit is determined as a function of the number of cells in each benchmark. Specifically, the runtime limit will be 5 hours plus 1 hour for every 35K cells in the design.

In summary, the primary ranking criteria is defined in lexicographic order as follows: 1) total violations, 2) total leakage power, and 3) runtime. In other words, total leakage power will be used to rank results with identical total violation values. When the total violation and leakage power values are identical, then runtime will be used for tie-breaking.

The secondary ranking will be done based on a tradeoff between solution quality and runtime. Specifically, the secondary metric is defined in lexicographic order as follows: 1) total violations, and 2)

combined cost metric as a function of total leakage power and runtime. The following equation defines this combined cost metric:

$$combined\_cost = \frac{Power}{Power_{REF}} + \gamma \frac{Runtime}{Runtime_{REF}} \quad (1)$$

In this equation, the leakage power and runtime values are normalized with respect to the reference values. The reference values will be set based on the best quality solution (first-ranked solution in the primary ranking criteria). The value of γ is set to 0.05. This means that 1% degradation in solution quality can be compensated by a 20% runtime reduction with respect to the reference values.

6. CONCLUSIONS

The focus of ISPD-2012 Contest is the problem of simultaneous gate sizing and threshold voltage assignment to optimize total leakage power under timing constraints. We have tried to expose some of the industrial challenges for this problem to the academic community, while keeping the complexity still manageable. A standard cell library with realistic delay and slew functions is provided, as well as benchmarks with various sizes. We hope that the provided benchmark suite helps the academic community to advance the state-of-the-art for this problem.

7. ACKNOWLEDGEMENTS

The contest organizers would like to thank Troy Wood and Robert Hoogenstryd (Synopsys), Noel Menezes, Jason Xu, Nanda Kuruganti, Shishpal Rawat, and Robert Nguyen (Intel) for helping with the contest organization.

8. REFERENCES

[1] M. R. C. M. Berkelaar and J. A. G. Jess. Gate sizing in MOS digital circuits with linear programming. In *Proc. of DATE*, pages 217–221, 1990.

[2] C. P. Chen, C. C.-N. Chu, and D. F. Wong. Fast and exact simultaneous gate and wire sizing by Lagrangian relaxation. *IEEE Trans. on Computer-Aided Design*, 18(7):1014–1025, July 1999.

[3] D. Chinnery and K. Keutzer. Linear programming for sizing, vth and vdd assignment. In *Proc. of ISLPED*, pages 149–154, 2005.

[4] H. Chou, Y.-H. Wang, and C. C.-P. Chen. Fast and effective gate sizing with multiple-Vt assignment using generalized Lagrangian relaxation. In *Proc. of ASPDAC*, pages 381–386, 2005.

[5] M. M. Ozdal, S. Burns, and J. Hu. Gate sizing and device technology selection algorithms for high-performance industrial designs. In *Proc. of ICCAD*, Nov. 2011.

[6] M. Rahman, H. Tennakoon, and C. Sechen. Power reduction via near-optimal library-based cell-size selection. In *Proc. of DATE*, 2011.

[7] H. Ren and S. Dutt. A network-flow based cell sizing algorithm. In *Workshop Notes, Int'l Workshop on Logic Synthesis*, 2008.

[8] S. Roy, W. Chen, C. C.-P. Chen, and Y. H. Hu. Numerically convex forms and their application in gate sizing. *IEEE Trans. on Computer-Aided Design*, 26(9):1637–1647, Sept. 2007.

[9] H. Tennakoon and C. Sechen. Gate sizing using Lagrangian relaxation combined with a fast gradient-based pre-processing step. In *Proc. of ICCAD*, pages 395–402, 2002.

[10] J. Wang, D. Das, and H. Zhou. Gate sizing by Lagrangian relaxation revisited. *IEEE Trans. on Computer-Aided Design*, 28(7):1071–1084, July 2009.

Towards Layout-Friendly High-Level Synthesis

Jason Cong[1,2,3]   Bin Liu[1]   Guojie Luo[2,3]   Raghu Prabhakar[1]

[1] Computer Science Department, University of California, Los Angeles
[2] Center for Energy-Efficient Computing and Applications (CECA), Peking University
[3] UCLA/PKU Joint Research Institute in Science and Engineering
{cong,bliu,raghu}@cs.ucla.edu   gluo@pku.edu.cn

ABSTRACT

There are two prominent problems with technology scaling: increasing design complexity and more challenges with interconnect design, including routability. High-level synthesis has been proposed to solve the complexity problem by raising the abstraction level. In this paper, we share our vision that high-level synthesis can potentially help the routability problem as well. We show that many interconnect problems that occur in layout can be avoided or mitigated by adopting a layout-friendly RTL architecture generated from high-level synthesis. We also evaluate some structural metrics that can be used to estimate the routability impact of design decisions in high-level synthesis. Experimental results have demonstrated correlations between the metrics and the routability of the resulting design.

Categories and Subject Descriptors

B.6.3 [**Logic Design**]: Design Aids—*automatic synthesis, optimization*; G.3 [**Mathematics of Computing**]: Probability and Statistics—*correlation and regression analysis*

General Terms

Design, Experimentation

Keywords

High-Level Synthesis, Routability, Interconnect Estimation

1. INTRODUCTION

Technology scaling has led to the increasing difficulty in resolving interconnect problems. This is due to relatively scarce routing resources and large interconnect delays. Various efforts to solve these problems have reshaped almost every aspect of the IC industry in the past twenty years. Advanced process technologies have offered new materials and more metal layers. 3-D integration provides a further opportunity to reduce the length of interconnects. In the EDA community, early work mainly focused on congestion optimization during placement and routing [8, 26, 28, 30]. As interconnect problems become worse, it is insufficient to repair routability failures only during physical design. With the prevalence of RTL-based design flows, early interconnect estimation and optimization techniques during logic synthesis have been proposed, [24, 25, 27, 31]. On the designer side, optimization of global interconnects is the central task of design planning and architecture definition. New design methodologies and styles have been established to address interconnect challenges [6, 9, 11, 33].

Along with technology scaling, another trend is the rapid increase of complexity in system-on-chip designs. This has encouraged the design community to seek better productivity. Electronic system-level design automation has been widely identified as the next productivity boost for the semiconductor industry, where high-level synthesis (HLS) plays a central role, by enabling the automatic synthesis of high-level, untimed or partially timed specifications (in languages such as C/C++, SystemC, Matlab) to cycle-accurate RTL models. These RTL models can then be accepted by the downstream RTL synthesis flow for implementation.

HLS has been an active research topic for more than 30 years. Early attempts to deploy HLS tools began when RTL-based flows were well adopted. In 1995, Synopsys announced Behavioral Compiler, which accepts behavioral HDL code and connects to downstream flows. Similar tools include Monet from Mentor Graphics and Visual Architect from Cadence. This wave of tools received wide attention, but failed to widely replace RTL design. This is partly ascribed to the use of behavioral HDLs, which are not popular among algorithm and system designers and require steep learning curves. Since 2000, a new generation of HLS tools has been developed in both academia and industry. Unlike their predecessors, many of them use C-based languages for design capture. This makes them more accessible to algorithm and system designers. It also enables hardware and software to be specified in the same language, facilitating software/hardware co-design and co-verification. The use of C-based languages also makes it easy to leverage new techniques in software compilers for parallelization and optimization. As of 2012, notable commercial C-based tools include Cadence C-to-Silicon Compiler, Calypto Catapult C (formerly a product of Mentor Graphics), NEC CyberWork-Bench, Synopsys Synphony C (formerly a product of Synfora, and originating from the HP PICO project), and Xilinx AutoESL (originating from the UCLA xPilot project [10]). More detailed surveys on the history and progress of HLS are available from [15, 21].

In our experimental HLS system based on xPilot, compiler transformations are first performed on the behavioral specification to obtain an optimized intermediate code represented as a control-data flow graph (CDFG). Operation scheduling then assigns operations in the CDFG to control states. The result of scheduling is a finite-state machine with datapath (FSMD), on which binding is applied to allocate resources in the datapath. After that, an RTL netlist can be generated. The basic flow is illustrated in Figure 1.

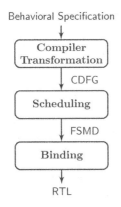

Figure 1: A typical high-level synthesis flow.

Numerous academic research and industry practices have demonstrated the gain in performance, area, and power of implementations obtained by improving the HLS algorithm, or by exploring different options/directives in HLS. This is primarily because decisions at a higher abstraction level often have bigger influences. Thus, we believe that huge opportunity for interconnect optimization exists in HLS. Instead of expanding time and effort to fix problems and making various compromises for a given netlist, the designer could use a HLS tool to generate a *layout-friendly* RTL netlist, which is easier for downstream tools to implement.

In this paper we show the impact of HLS on routability in Section 2. We then describe some structural metrics that can be used in the synthesis engine (i.e., scheduling and binding) in Section 3 and experimentally evaluate their effectiveness in Section 4. Discussions on future directions are presented in Section 5.

2. IMPACT OF HIGH-LEVEL SYNTHESIS

In this section we experimentally study the impact of high-level synthesis on routability. Our study is based on the xPilot HLS tool [10]. Separate studies are performed for the synthesis engine and for compiler transformations.

2.1 Routability Evaluation Flow

In order to evaluate routability, we feed the RTL netlist under evaluation into an implementation flow, which includes the stages of (i) RTL elaboration, (ii) logic synthesis, (iii) placement, and (iv) routing.

For a given netlist, its routability mainly depends on the amount of routing resources on the target platform. Generally, the target platform is either ASIC-style or FPGA-style. The routing resources of an ASIC-style platform are determined by the number of metal layers and the wire pitches. The routing resources of an FPGA-style platform, assuming the architecture model in [4], are determined by the number of tracks between the configurable logic blocks (CLBs).

In this paper we mainly focus on an FPGA-style routability evaluation flow, illustrated in Figure 2. We use Altera Quartus II version 9.1 [1] as an RTL frontend, and then use ABC version 70731 [2] to synthesize and map the netlist into 4-input lookup tables (4-LUTs). The mapped netlist is packed into CLBs by T-VPACK 5.0.2 [3], where each CLB contains ten 4-LUTs and ten flip-flops. We adopt a simplified routing architecture from [4]. The CLBs form a regular array: the space between two neighboring CLB rows or columns is called a channel, and the space between two neighboring CLBs is called a segment. There are multiple routing tracks in the segments, where we assume the span of a track is one CLB, and the number of tracks (channel width) can be different for different segments. The packed netlist is then placed by VPR 5.0.2 [3] with the total bounding-box wire length as the objective. Routing is also done by VPR, which minimizes the maximum channel width using a binary search.

It is obvious that a design will not be routable if the channel width is too small. Thus we consider maximum channel width (CW_max) and average channel width (CW_avg) as indicators for routability. In addition, total wire length (WL_tot) and average wire length (WL_avg) are also used in the evaluation.

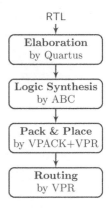

Figure 2: The RTL implementation flow.

2.2 Impact of the Synthesis Engine

To demonstrate how decisions in the synthesis engine impact routability, we generate multiple RTL models from the same CDFG by varying strategies in scheduling and binding. For scheduling, we adjust the optimization objective as well as resource constraints to obtain different FSMDs. The strategies in scheduling are listed in Table 1. Similarly, different objectives and constraints listed in Table 2 are specified for binding to generate different RTL netlists.

By combining each strategy for scheduling with each strategy in binding, we can obtain 60 different strategies in the synthesis engine. In practice, different strategies can sometimes lead to equivalent or identical RTL netlists. We extract several computation-intensive kernels as test cases from DSP applications. These cases generally perform many multiplications and addition/subtractions. For each test case, we measure the routability for all RTL models generated using different synthesis strategies, and report the minimum and maximum values for each metric in Table 3. The results indicate that different RTL models generated from the same CDFG can have drastically different routability.

Table 1: Scheduling strategies.

	objective	constraint
1	ASAP	None
2	ALAP	None
3	MINREG	None
4	ALAP	$M = \lceil 0.25 \times m \rceil$
5	ALAP	$M = \lceil 0.25 \times m \rceil$, $A = \lceil 0.4 \times a \rceil$
6	MINREG	$M = \lceil 0.1 \times m \rceil$, $A = \lceil 0.2 \times a \rceil$

The ASAP/ALAP objective tries to schedule operations as soon/late as possible, subject to optional resource constraints [17]; the MINREG objective tries to minimize the total lifetime of variables in order to reduce registers. M and A are the number of available multipliers and adders in the resource constraints, respectively; m and a are the number of multiplication and addition operations in the CDFG, respectively.

Table 2: Binding strategies.

	objective	constraint
1	total area	None
2	total area	$mux\_input \leq 4$
3	register	$mux\_input \leq 4$
4	multiplier	None
5	multiplier	$mux\_input \leq 4$
6	multiplier and register	None
7	multiplier and register	$mux\_input \leq 4$
8	multiplier and adder	None
9	multiplier and adder	$mux\_input \leq 4$
10	multiplier and adder and register	$mux\_input \leq 4$

Multiplier/adder/register in the objective means minimizing the number of corresponding components. mux_input is the maximum number of inputs for each multiplexer; here we try to avoid large multiplexers by setting an upper bound on mux_input.

2.3 Impact of Compiler Transformations

In fact, before CDFG is generated for scheduling and binding, the compiler front-end often applies a sequence of transformations on the intermediate representation. Some of the transformations are the same as ones found in conventional compilers, like dead code elimination; yet others are specifically designed for HLS. The performance implications of many transformations are well understood. Here we make a few observations about their impact on routability in the context of HLS.

We note that the following scalar transformations tend to reduce interconnect complexity.

- Expression simplification. For example, constant propagation and strength reduction can replace complex functional units with simpler ones.

- Expression structure optimization. For example, common sub-expression extraction, redundancy elimination and re-association can reduce the number of complex functional units and change the local structure of the datapath.

- Bitwidth optimization. Reducing the bitwidths of operands and results can reduce component sizes and the num-

Table 3: Routability measurements.

design	CW_max	CW_avg	WL_tot	WL_avg
test1	46/86	30/44	36K/166K	9/12
test2	60/98	38/47	61K/251K	11/13
test3	40/70	25/33	25K/103K	7/9
test4	52/86	31/40	47K/196K	10/12
test5	62/122	39/55	94K/562K	12/17

a/b indicates that a is the minimum value and b is the maximum value, among all results generated for the test case.

ber of wires. This optimization is particularly useful for HLS.

The following transformations often have more global influences.

- Transformations that change memory organization. In the implementation of xPilot, each array is mapped to a dedicated memory block. Transformations like array partitioning, array mapping, array reshaping can change the number of memory blocks, the number of ports on each memory block, as well as the way memory blocks connect to other components. For example, partitioning an array can increase throughput, but it often leads to more decoding/multiplexing logic and more interconnects [13].

- Loop transformations. Unrolling a loop creates opportunities for code optimization and parallelization; yet it often destroys the regular structure of the data transfers between different loop iterations and thus leads to more components and interconnects.

- Function-level transformations. In xPilot, each function is implemented as a hardware module. When a function is inlined, further optimization may be performed at call sites; on the other hand, when the function calls are optimized differently at different call sites, it becomes difficult for them to share the same set of components and interconnects.

Here we perform a case study on a simple design that multiplies two 8×8 32-bit integer matrices. The matrix multiplication is implemented straightforwardly as a three-level nested loop: the inner loop computes an element in the resulting matrix, the middle loop computes a row, and the outer loop computes the entire result. We generate three solutions.

- Solution 1: no loop unrolling.

- Solution 2: unroll the inner loop completely, and partition the two input matrices into row/column vectors.

- Solution 3: unroll the middle loop and the inner loop completely, partition the two input matrices completely (into scalars), and partition the output matrix into column vectors.

The three solutions are synthesized using default synthesis options and the resulting routability result is reported in Table 4. It is evident from the result that unrolling increases interconnect complexity drastically for this case. Note that a solution with higher interconnect complexity also has higher performance; thus a tradeoff between performance and routability needs to be considered when making decisions on loop unrolling.

Table 4: Routability for matrix multiplication.

solution	CW_max	CW_avg	WL_tot	WL_avg
1	30	18	4543	5
2	44	24	43610	7
3	80	36	223042	10

3. ROUTABILITY ESTIMATION

A frequent problem of optimizing at a higher level is the lack of good estimators: while there are many alternative RTL structures to explore in HLS, it is hard to decide whether one is superior to another. This is particularly true for interconnect optimization, because unlike other metrics (such as latency, throughput, area) which can be estimated to a reasonable accuracy for a given RTL model, the length and density of interconnects can hardly be decided without a layout. While different downstream tools can generate different layouts, the global structure of the block-level netlist plays an important role in deciding what a good layout looks like. The first step toward layout-friendly high-level synthesis is to be able to evaluate the layout-friendliness of an RTL netlist. The trivial approach of running through all downstream steps would give the most accurate result; yet it is often impractical when exploring a large number of alternatives. Most of the existing ways to predict routability fall in one of the following categories.

(1) Incorporate a rough layout in high-level synthesis. There are numerous efforts that combine high-level synthesis with floorplanning to help interconnect estimation and optimization [18, 20, 37, 39, 40]. This is quite a reasonable approach. However, since layout itself is a nontrivial problem, implementation of a stable and fast layout engine itself is a challenge.

(2) Use structural metrics to evaluate interconnect complexity. Such metrics are usually derived from a graph representation of the netlist without performing layout. Widely used structural metrics include the total multiplexer inputs [7,22,23,29], the number of global interconnects [12,32], the total cut size [24], etc. These metrics are often easier to obtain and are more stable (i.e., not dependent on the layout algorithm), but their accuracy is often a concern.

Given a scheduling/binding solution on a CDFG, the RTL netlist can be constructed. The netlist often consists of components (including functional units, registers, memories, multiplexers, pre-synthesized blocks, etc.) and wires which connect components through ports. A directed graph $G = (V, E)$ can be constructed to represent the netlist, where $V = \{1, 2, \ldots, n\}$ is the set of vertices each representing a component; and $E \subseteq V \times V$ is the set of directed edges each representing a net from the source component to the sink component. Note that an edge is present only when there are data transfers between the two components; if two components are connected in the netlist only because they are both sinks of a net, no edge is be created between the corresponding vertex pair.

The following metrics are considered in our evaluation.

- Total number of datapath nets (#nets).
- Total number of multiplexer inputs (#mux_input).
- The average cut size between components in the netlist (AMC). Here the bitwidth of each net is considered. More details are available in Section 3.1.

- The *spreading score* proposed in [14]. More details are available in Section 3.2.

3.1 A Metric Based on Cut Size

Cut size has been recognized as an indicator of interconnect complexity in a number of important synthesis steps, such as partitioning, clustering, and placement. Intuitively, smaller cut size implies less global connections, and thus better routability.

The cut-size minimization process is explicitly applied in every partitioning-based placer (e.g., Capo [34]). In fact, if we divide the placement region into unit squares, the total cut size of a placed netlist on the edges of these squares is approximately equal to the total wire length. Thus, the recursive cut-size minimization process in partitioning-based placers can be viewed as an approximate wire length minimization process. Moreover, the cut size across a local cut line directly captures the local congestion, and maintains better congestion information than the total wire length metric [35].

Kudva, Sullivan and Dougherty propose the *sum of all-pairs min-cut* (SAPMC) to evaluate the *adhesion* of a gate-level netlist, and use it in logic synthesis [24]. In a netlist represented by graph $G = (V, E)$, the cut size of the an *s-t* min-cut between two distinct vertices s and t is the minimum number of nets whose removal disconnects s and t. The SAPMC is the sum of min-cut sizes for all pairs of distinct vertices in V. Experimental results reported in [24] show a positive correlation between congestion and SAPMC; in addition, the results indicate that the correlation between congestion and the circuit size (as measured by the number of nodes in the netlist) is even stronger, and thus SAPMC is only used to break ties in optimization. Clearly, SAPMC is also positively correlated with the circuit size, because larger circuits naturally have more distinct node pairs, which lead to larger SAPMC. In an effort to obtain orthogonal metrics, we try to exclude the node count factor from SAPMC, and instead compute the *average min-cut* (AMC), by dividing SAPMC with $n(n-1)/2$, where n is the number of nodes.

3.2 A Metric Based on Graph Embedding

This section describes spreading score, a structural metric proposed in [14]. The metric is based on graph embedding, and can be computed efficiently using convex programming. Here we give a brief review, more details about the metric can be find in [14].

A layout of the netlist can be regarded as an embedding of G in the 2-dimensional Euclidean space $\mathbb{R}^2$. We associated each vertex i with a column vector $p_i = (x_i, y_i)^T$ to represent its position in the embedding. The length of the connection $(i, j) \in E$ can be measured as the distance in $\mathbb{R}^2$, i.e., $\|p_i - p_j\| = \sqrt{(x_i - x_j)^2 + (y_i - y_j)^2}$.

We consider the following optimization problem.

$$
\begin{aligned}
\text{maximize} \quad & \sum_{i=1}^{n} w_i \|p_i\|^2 \\
\text{subject to} \quad & \sum_{i=1}^{n} w_i p_i = 0 \\
& \|p_i - p_j\| \leq l_{ij} \quad \forall (i,j) \in E
\end{aligned}
\tag{1}
$$

Here $w = (w_1, w_2, \ldots, w_n)^T$ is the non-negative weight vector with w_i being the area of component i; l_{ij} is the maximum allowed length for the wire connecting i and j. The objective function measures how far components are spread from their weighted center of gravity, using a weighted 2-norm of the distance vector. Thus the problem in Equation

1 is to maximize component spreading, under the constraint that the length every connection $(i, j) \in E$ does not exceed l_{ij}.

With proper selection of l_{ij}, we expect that the optimal value of the problem in Equation 1 can be used to evaluate the layout-friendliness of a netlist. This is based on the following observation: if components in a netlist can spread over the chip region without introducing long wires, it will be easy to remove overlaps between components and obtain a layout with small wire length, and thus less congestion. This can be empirically verified using popular hand-designed interconnect topologies. For example, mesh and ring can all spread apart without long interconnects, and they are regarded as scalable layout-friendly topologies; on the other hand, spreading the full crossbar or hypercube on the 2D plane inevitably introduces long interconnects, and these topologies are generally much more expensive in interconnect cost. In our experiments, we set the distance l_{ij} based on estimated sizes of components as $l_{ij} = \sqrt{w_i} + \sqrt{w_j}$.

It is difficult to solve the problem in Equation 1 directly, because maximizing a convex function is generally NP-hard (note that minimizing a convex function is easy). We hereby propose a tractable relaxation.

Consider the graph G with n vertices. We use a $2 \times n$ matrix $P = (p_1, p_2, \ldots, p_n)$ to represent its embedding in $\mathcal{R}^2$, i.e.,

$$P = \begin{pmatrix} x_1 & x_2 & \cdots & x_n \\ y_1 & y_2 & \cdots & y_n \end{pmatrix}. \qquad (2)$$

Let $Q = P^T P$. Then Q is a symmetric semidefinite matrix with a rank of at most 2, and

$$Q_{ij} = p_i^T p_j = x_i x_j + y_i y_j. \qquad (3)$$

We can use Q as variables in the formulation in Equation 1 without losing any useful information, because p can be reconstructed from Q with Cholesky decomposition.

Using Equation 3, we can rewrite the objective and constraint functions in Equation 1 as follows.

$$\sum_{i=1}^{n} w_i \|p_i\|^2 = \sum_{i=1}^{n} w_i Q_{ii} = \langle \operatorname{diag}(w), Q \rangle \qquad (4)$$

$$\left\| \sum_{i=1}^{n} w_i p_i \right\|^2 = \sum_{i=1}^{n} \sum_{j=1}^{n} w_i w_j Q_{ij} = \langle ww^T, Q \rangle \qquad (5)$$

$$\|p_i - p_j\|^2 = Q_{ii} + Q_{jj} - 2Q_{ij} = \langle K^{ij}, Q \rangle \qquad (6)$$

Here $\operatorname{diag}(w)$ is the $n \times n$ diagonal matrix with w on its diagonal. e_i is the ith standard basis vector in $\mathcal{R}^n$ and $K^{ij} = (e_i - e_j)(e_i - e_j)^T$. $\langle X, Y \rangle$ is the Frobenius inner product of matrices X and Y, i.e.,

$$\langle X, Y \rangle = \sum_i \sum_j X_{ij} Y_{ij} = \operatorname{tr}(X^T Y) = \operatorname{tr}(Y^T X). \qquad (7)$$

Then we can rewrite the problem in Equation 1 to use Q as variables, and relax the rank constraint on Q.

$$\begin{aligned} \text{maximize} \quad & \langle \operatorname{diag}(w), Q \rangle \\ \text{subject to} \quad & \langle ww^T, Q \rangle = 0 \\ & \langle K^{ij}, Q \rangle \leq l_{ij}^2 \quad \forall (i, j) \in E \\ & Q \succeq 0 \end{aligned} \qquad (8)$$

This problem is convex. In fact, it is a semidefinite programming (SDP) problem. Like linear programs, SDP problems can be solved optimally in polynomial time, and efficient solvers have been developed in recent years. We will not discuss background on convex programming and SDP here. Interested readers may refer to books and survey papers on these topics [5, 38].

The problem in Equation 8 essentially asks for an embedding in $\mathcal{R}^n$ instead of $\mathcal{R}^2$, and thus its optimal value is the lower bound of the problem in Equation 1. In our implementation, we adopt this approximation and divide it by n^2 to normalize it. The result is referred to as spreading score. The expectation is that a larger spreading score implies a more layout-friendly datapath structure.

4. RESULT AND ANALYSIS

Results on the test cases in Table 3 are collected. We extract the pre-layout metrics described in Section 3 after HLS, and try to correlate them with post-layout routability data described in Section 2.1.

Single-variable linear regressions between the pre-layout characteristics and maximum channel width are illustrated in Figure 3. From the results, we can see that spreading score has a strong positive correlation with CW_max for several cases (e.g., test1), but shows a weak correlation on test2; #net shows strong positive correlations with CW_max as well; however, total number of multiplexer inputs performs poorly. Surprisingly, AMC tends to have a negative correlation with CW_max, i.e., larger average cut-size leads to smaller channel width. This seems to indicate that AMC is not a primal factor in deciding congestion in the context of HLS. Similar single-variable regressions can be performed for other post-layout metrics. Figure 4 illustrates the results for average wire length. It can be observed that spreading score has a consistently negative correlation with WL_avg, although the correlation is weak for test4 and test5; other metrics seem weaker.

To examine the usability of these metrics, we perform a two-variable polynomial regression with spreading score and AMC as independent variables, and CW_max as the dependent variable. We randomly select 70% of the data to fit a linear and a quadratic function, respectively, and then use the fitted function to "predict" the CW_max for the remaining 30% of cases. The random selections are performed ten times and we collect the average absolute error and the relative error for all the designs, as listed in Table 5. In addition, we obtain regressions for CW_max with spreading score, AMC and #net as independent variables in Table 7, and we obtain regressions for WL_avg with spreading score and AMC as independent variables in Table 6.

On average, the relative error of CW_max fitted with a linear function is less than 10%, while increasing the order of the fitting function would not help to increase the accuracy. For test2, increasing the order even produces a much less accurate fitted function. The data in Table 7 suggest that including more vaiables in the regression slightly reduces the error when we use a linear or a quadratic function.

Table 5: Errors of the polynomial regression for CW_max using spreading score and AMC.

design	linear		quadratic	
	error	%	error	%
test1	3.3	5%	7.6	11%
test2	5.8	8%	10.1	16%
test3	3.3	7%	4.3	9%
test4	4.4	7%	4.6	7%
test5	7.0	11%	5.7	9%

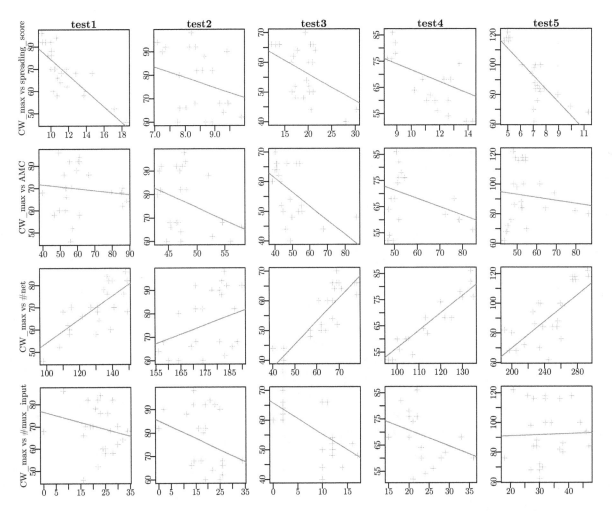

Figure 3: Linear regression of maximum channel width on four high-level metrics.

Table 6: Errors of the polynomial regression for WL_avg using spreading score and AMC.

design	linear		quadratic	
	error	%	error	%
test1	0.35	3%	0.41	4%
test2	0.27	2%	0.22	2%
test3	0.23	3%	0.56	7%
test4	0.43	4%	0.34	3%
test5	0.61	4%	1.1	7%

Table 7: Errors of the polynomial regression for CW_max using spreading score, AMC and #net.

design	linear		quadratic	
	error	%	error	%
test1	3.2	5%	3.7	6%
test2	1.8	2%	1.6	2%
test3	2.2	5%	3.5	8%
test4	3.5	6%	3.4	5%
test5	3.2	4%	2.8	4%

5. CONCLUDING REMARKS AND FUTURE DIRECTIONS

In this paper, we have demonstrated and quantified the opportunities for interconnect optimization in high-level synthesis, and evaluated several metrics in predicting wire length and congestion. This is a preliminary study. We see several directions for fruitful future research.

1. Although some metrics evaluated in this paper, such as spreading score and net number, show reasonably good correlations to results after layout on some designs, none of them can consistently predict routability with a high accuracy at this point. The problem of getting a good high-level routability estimator is still interesting and challenging. Any improvement in this direction can potentially improve the microarchitecture of the RTL design and reduce time and effort in layout. One possibility is to consider a combination of several metrics, together with some modeling of interconnects inside components, for more accurate routability prediction.

2. After obtaining a reasonably accurate metric, another challenge is how to use it to guide the optimization. One possible approach is through iterative refinement by restructuring a section of the HLS solution to gradually improve the routability (under timing constraints). The challenge is to ensure convergence (i.e. not to introduce new hotspots dur-

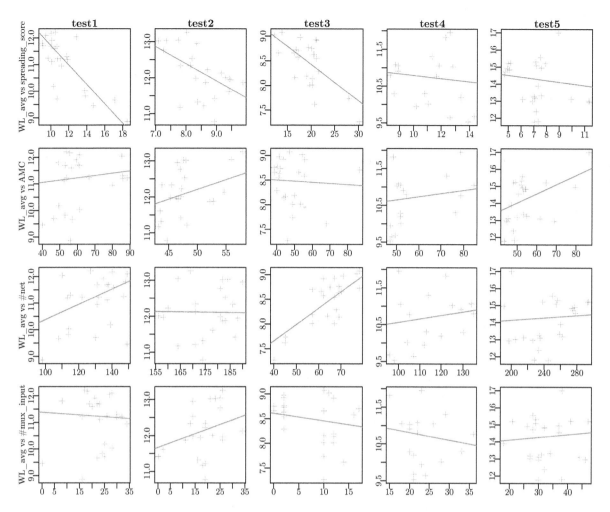

Figure 4: Linear regression of average wire length on four high-level metrics.

ing refinement). We used such an iterative approach to co-ordinate scheduling and resource binding [16] and achieved encouraging results.

3. As we have demonstrated, proper organization of compiler transformations also has a considerable influence on interconnect optimization. This is even more challenging as the existing metrics will not apply in the absence of RTL netlists. Some recent research efforts from the compiler community use statistical methods, like machine learning, for automated compiler optimization [19, 36]. We are in the process of exploring this direction.

6. ACKNOWLEDGEMENTS

This work was supported in part by the Semiconductor Research Corporation under Contract 2009-TJ-1879 and the UC Discovery program. Guojie Luo would like to thank the Center for Energy-Efficient Computing and Applications at Peking University for providing start-up research fund. The authors thank Janice Martin-Wheeler for her help in editing this paper.

7. REFERENCES

[1] Altera Quartus II. Available: http://www.altera.com.

[2] Berkeley Logic Synthesis and Verification Group, ABC: A System for Sequential Synthesis and Verification, Release 70731. Available: http://www.eecs.berkeley.edu/~alanmi/abc/.

[3] VPR and T-VPack 5.0.2 . Available: http://www.eecg.toronto.edu/vpr/.

[4] V. Betz, J. Rose, and A. Marquardt. *Architecture and CAD for Deep-Submicron FPGAs*. Kluwer Academic Publishers, 1999.

[5] S. Boyd and L. Vandenberghe. *Convex Optimization*. Cambridge University Press, New York, NY, 2004.

[6] L. Carloni, K. McMillan, and A. Sangiovanni-Vincentelli. Theory of latency-insensitive design. *IEEE Trans. on Computer-Aided Design of Integrated Circuits and Systems*, 20(9):1059–1076, Sept. 2001.

[7] D. Chen and J. Cong. Register binding and port assignment for multiplexer optimization. In *Proc. Asia and South Pacific Design Automation Conf.*, pages 68–73, 2004.

[8] C.-L. E. Cheng. Risa: accurate end efficient placement routability modeling. In *Proc. Int. Conf. on Computer-Aided Design*, pages 690–695, Nov. 1994.

[9] J. Cong. An interconnect-centric design flow for

nanometer technologies. *Proc. IEEE*, 89(4):505–528, 2001.

[10] J. Cong, Y. Fan, G. Han, W. Jiang, and Z. Zhang. Platform-based behavior-level and system-level synthesis. In *Proc. IEEE Int. SOC Conf.*, pages 199–202, 2006.

[11] J. Cong, Y. Fan, G. Han, X. Yang, and Z. Zhang. Architecture and synthesis for on-chip multicycle communication. *IEEE Trans. on Computer-Aided Design of Integrated Circuits and Systems*, 23(4):550–564, April 2004.

[12] J. Cong, Y. Fan, and W. Jiang. Platform-based resource binding using a distributed register-file microarchitecture. In *Proc. Int. Conf. on Computer-Aided Design*, pages 709–715, 2006.

[13] J. Cong, W. Jiang, B. Liu, and Y. Zou. Automatic memory partitioning and scheduling for throughput and power optimization. In *Proc. Int. Conf. on Computer-Aided Design*, pages 697–704, 2009.

[14] J. Cong and B. Liu. A metric for layout-friendly microarchitecture optimization in high-level synthesis. In *Proc. Design Automation Conf.*, 2012. in press.

[15] J. Cong, B. Liu, S. Neuendorffer, J. Noguera, K. Vissers, and Z. Zhang. High-level synthesis for FPGAs: From prototyping to deployment. *IEEE Trans. on Computer-Aided Design of Integrated Circuits and Systems*, 30(4):473–491, Apr. 2011.

[16] J. Cong, B. Liu, and J. Xu. Coordinated resource optimization in behavioral synthesis. In *Proc. Design, Automation and Test in Europe*, pages 1267–1272, 2010.

[17] J. Cong and Z. Zhang. An efficient and versatile scheduling algorithm based on SDC formulation. In *Proc. Design Automation Conf.*, pages 433–438, 2006.

[18] W. Dougherty and D. Thomas. Unifying behavioral synthesis and physical design. In *Proc. Design Automation Conf.*, pages 756–761, 2000.

[19] C. Dubach, T. M. Jones, E. V. Bonilla, G. Fursin, and M. F. P. O'Boyle. Portable compiler optimisation across embedded programs and microarchitectures using machine learning. In *Proc. Int. Symp. on Microarchitecture*, pages 78–88, 2009.

[20] Y.-M. Fang and D. F. Wong. Simultaneous functional-unit binding and floorplanning. In *Proc. Int. Conf. on Computer-Aided Design*, pages 317–321, 1994.

[21] R. Gupta and F. Brewer. *High-Level Synthesis: A Retrospective*, pages 13–28. Springer, 2008.

[22] C.-Y. Huang, Y.-S. Chen, Y.-L. Lin, and Y.-C. Hsu. Data path allocation based on bipartite weighted matching. In *Proc. Design Automation Conf.*, pages 499–504, 1990.

[23] T. Kim and X. Liu. Compatibility path based binding algorithm for interconnect reduction in high level synthesis. In *Proc. Int. Conf. on Computer-Aided Design*, pages 435–441, 2007.

[24] P. Kudva, A. Sullivan, and W. Dougherty. Measurements for structural logic synthesis optimizations. *IEEE Trans. on Computer-Aided Design of Integrated Circuits and Systems*, 22(6):665–674, June 2003.

[25] T. Kutzschebauch and L. Stok. Congestion aware layout driven logic synthesis. In *Proc. Int. Conf. on Computer-Aided Design*, pages 216–223, 2001.

[26] C. Li, M. Xie, C.-K. Koh, J. Cong, and P. Madden. Routability-driven placement and white space allocation. *IEEE Trans. on Computer-Aided Design of Integrated Circuits and Systems*, 26(5):858–871, May 2007.

[27] S. Liu, K.-R. Pan, M. Pedram, and A. Despain. Alleviating routing congestion by combining logic resynthesis and linear placement. In *Proc. European Conf. on Design Automation*, pages 578–582, Feb. 1993.

[28] S. Mayrhofer and U. Lauther. Congestion-driven placement using a new multi-partitioning heuristic. In *Proc. Int. Conf. on Computer-Aided Design*, pages 332–335, Nov. 1990.

[29] M. C. McFarland. Reevaluating the design space for register transfer hardware synthesis. In *Proc. Int. Conf. on Computer-Aided Design*, pages 262–265, 1987.

[30] L. McMurchie and C. Ebeling. PathFinder: a negotiation-based performance-driven router for FPGAs. In *Proc. Int. Symp. on FPGA*, pages 111–117, 1995.

[31] D. Pandini, L. Pileggi, and A. Strojwas. Congestion-aware logic synthesis. In *Proc. Design, Automation and Test in Europe*, pages 664–671, 2002.

[32] B. M. Pangre. Splicer: a heuristic approach to connectivity binding. In *Proc. Design Automation Conf.*, pages 536–541, 1988.

[33] S. Posluszny, N. Aoki, D. Boerstler, P. Coulman, S. Dhong, B. Flachs, P. Hofstee, N. Kojima, O. Kwon, K. Lee, D. Meltzer, K. Nowka, J. Park, J. Peter, J. Silberman, O. Takahashi, and P. Villarrubial. "Timing closure by design," a high frequency microprocessor design methodology. In *Proc. Design Automation Conf.*, pages 712–717, 2000.

[34] J. Roy, S. Adya, D. Papa, and I. Markov. Min-cut floorplacement. *IEEE Trans. on Computer-Aided Design of Integrated Circuits and Systems*, 25(7):1313–1326, July 2006.

[35] P. Saxena, R. S. Shelar, and S. S. Sapatnekar. *Routing Congestion in VLSI Circuits: Estimation and Optimization*. Springer, 2007.

[36] M. Stephenson, S. Amarasinghe, M. Martin, and U.-M. O'Reilly. Meta optimization: improving compiler heuristics with machine learning. In *Proc. ACM SIGPLAN Conf. on Programming Language Design and Implementation*, pages 77–90, 2003.

[37] S. Tarafdar, M. Leeser, and Z. Yin. Integrating floorplanning in data-transfer based high-level synthesis. In *Proc. Int. Conf. on Computer-Aided Design*, pages 412–417, 1998.

[38] L. Vandenberghe and S. Boyd. Semidefinite programming. *SIAM Review*, 38(1):49–95, 1996.

[39] J.-P. Weng and A. C. Parker. 3D scheduling: high-level synthesis with floorplanning. In *Proc. Design Automation Conf.*, pages 668–673, 1991.

[40] M. Xu and F. J. Kurdahi. Layout-driven RTL binding techniques for high-level synthesis using accurate estimators. *ACM Trans. on Design Automation of Electronics Systems*, 2:312–343, Oct. 1997.

Synthesis for Advanced Nodes – An Industry Perspective

Janet Olson
Synopsys
700 East Middlefield Rd
Mt View, CA 94043
650-584-1765
janeto@synopsys.com

ABSTRACT

As the industry moves to smaller geometries, physically-aware logic synthesis is required in the flow. With the complex designs under development, customers need to achieve high target frequencies with faster time-to-market. We are increasingly seeing the need to accurately model and optimize for physical effects early in the synthesis flow.

This talk will address the challenges for RTL synthesis today with small geometries. There is an increasing amount of physical effects that must be taken into account during synthesis – from congestion to various RC effects of smaller geometries to ensure convergence.

Categories and Subject Descriptors

B.6.3 [**Hardware**]: Design aids: *Automatic synthesis; Optimization*

General Terms

Algorithms, design

Keywords

Physical Synthesis

Reality-Driven Physical Synthesis

Patrick Groeneveld

Magma Design Automation, Inc.

1650 Technology Drive

San Jose, CA 95110, USA

+1 408 565 7654

patrick@magma-da.com

Abstract

A 'Physical Synthesis' design flow combines logical synthesis techniques with layout synthesis algorithms. It is constantly in flux, most recently due to extreme uncertainty of the delays in the latest process nodes. The flow is an intricate combination of a design steps and analysis runs, the tuning of which requires careful attention. Rather than addressing the individual algorithms, this presentation addresses the general structure of a successful physical synthesis design flow.

General Terms

Algorithms, Performance, Design, Verification.

Keywords

Placement, Routing, Logic Synthesis, Static Timing Analysis.

1. Introduction

For over a century, the 'net list' abstraction level has been a mainstay in electrical engineering. The net list is the 'blue print' of the design. It describes the circuit as a collection of components that represent transistors, capacitors, but also more complex objects such as logical gates or hierarchical blocks. 'Nets' represent the wires between the components. The net list abstraction allows designers to reason about function, speed, efficiency and correctness without the need for a physical implementation of the circuit. While much of analog circuit design is still performed at this level, digital design entry has moved on to the higher RTL abstraction level over the past decades.

With billions of transistors, ICs are arguably the most complex pieces of machinery that mankind has ever built. The combination of the RTL abstraction level and standard cell methodology has enabled IC designer productivity to follow Moore's law for exponential growth in integrated circuit density. It would have been impossible to design ICs without the innovations in electronic design automation, especially the fully automated physical synthesis.

Though EDA tools contain some sophisticated algorithms with good theoretical underpinnings, the core of a successful physical synthesis flow is based on common sense engineering trade-offs that receive little attention in the published literature. In this

presentation we address the underlying engineering concepts that shape a modern physical synthesis design flow.

2. Standard Cell Design Flow

Though often overlooked, the main enabler of successful digital design automation is the 'standard cell' design style. Building a logic circuit as net list of pre-designed and pre-characterized standard gates dramatically simplifies the design process. Many intricate transistor-level design details can be hidden inside of a standard cell. What remains on the outside of the standard cell is just a box of a certain size with connection points that has a pre-characterized delay.

The front-end of an automated standard cell design flow consists of logic synthesis steps that generate a net list of standard cells as output. The objective is to generate a functionally correct net list of logical gates, while simultaneously maximizing performance (circuit speed) and minimizing cost (component count and area). The back-end of the flow converts this standard cell net list into a correct layout mask pattern. This layout abstraction level is for historical reasons generally referred to as 'GDS2'. The back-end flow places the standard cells and then wires them up using between 6 and 10 levels of metal interconnect.

State of the art physical synthesis tools can handle up to 10 million standard cells. That translates into 2-5 miles of wire that are built from several hundreds of millions of layout rectangles. The margin of error for any of these wires is zero: an error in a single layout rectangle creates an open or a short that breaks the entire circuit.

3. Decline of the Net List Abstraction Level

The logic net list of standard cells has been the abstraction level that separated the distinct worlds of the logic and layout for decades. The standard cell abstraction provided the essential delay and area information that enabled logic optimization to trade off various design variants. In the previous century it was possible to sign off on the logic and speed of the circuit at the net list level. On the other side of this 'Berlin wall'-style net list abstraction the automatic placement and routing tools focused on generating a correct mask. Until the turn of the century the logical and layout worlds barely communicated even though cracks were forming well before that date.

There are several causes for the complete disintegration of net list as sign-off abstraction level, but the main one is the shift of the delay from the gates into the wires. The physics is simple: the capacitive load on the output mainly determines the delay of a gate. This output load is the parasitic capacitance of the wire plus the input capacitance of the gates it drives into. That input capacitance of a typical standard cell gate in an advanced 28nm node is in the order of 1fF. The unit capacitance for a wire in the same technology is 0.2fF/μm, which means that just 5 micrometer of wire length is equivalent to a gate load. With wire lengths ranging anywhere between 2μm and 400μm it is clear that this dominates the circuit delay.

The gate input capacitance is fully predictable at the net list abstraction level because it follows from the net list structure. The wire capacitance, unfortunately, is *not* because that depends on the detailed wire length and topology of each net. The actual wire is only known at the very end of the flow. Wire capacitance is responsible for a larger percentage of the stage delay with each process generation.

To make matters worse, crosstalk delay due to the parasitic coupling capacitance with neighboring nets adds dramatically to the delay uncertainty. For weaker gate drive strengths the crosstalk delay component could account for a 70% variation in delay for the same wire length, depending on the configuration and slew of the neighboring aggressors. The actual configuration is unknown until the very latest detail routing steps in the flow. This means that estimating wire length based on an intermediate placement and global routing has become inadequate to assess wire delay.

In 2012 - with 28nm and 22nm technology nodes - we have reached a point where is delay optimization at the net list level has become a very questionable exercise. Empirical correlation results on actual designs indicate a standard deviation of well over 150% between the net delay estimation at the net list level compared to the actual delay at the layout abstraction level. Strangely enough, significant effort is still spent on net list level optimization.

The collapse of the net list abstraction results in a chicken-and-egg problem: logical synthesis needs information on parasitics from the layout, while the router needs a net list of placed gates as its net list input. The Physical Synthesis flow attempts to bridge the net list divide, mainly by using stepwise refinement flow. The net list is modified during the placement and routing process to account for the latest data on the parasitics and local congestion. Typical actions to improve delay include buffer insertion and removal, gate sizing and gate cloning.

4. The Yin and Yang of Physical Synthesis

A physical synthesis flow breaks up the design task into many dozens of algorithms that are run in order. The goal is to gradually transform the design state, each step improving on the previous. Most often, it is a back-and-forth between:

1. **Synthesis tools** that generate or modify the design state (net list, layout). Placers, logic synthesis algorithms, routers or manual design fall into this category. Synthesis tools are wildly unpredictable, fast and hard to parallelize.

2. **Analysis tools** that measure or verify the current design state. The main workhorse is generally the static timing analyzer. Also SPICE or a DRC checking tool are analysis tools. They are generally accurate within 1%, very CPU intensive but they can take advantage of parallel hardware much better.

Ideally, these complementary categories work together harmoniously toward a common goal. In practice, however

entirely different groups of people with different mindsets and surprisingly little interaction develop them.

There is a general desire to make synthesis tools generate solutions that are closer to the requirements, and to make analysis tools faster. There is room. For instance, given the massive amount of unpredictability of any synthesis tool it makes sense to sacrifice analysis accuracy for speed.

5. Lining up multiple algorithms to do the job

The synthesis algorithms that are deployed in the flow are far from ideal and generally share a number of unpleasant characteristics:

1. They can only perform a rather small step in the transformation process towards a layout pattern. As a result we need to chain together several (dozens) of such algorithms. For example this could be a tool sequence: Placement, global routing, buffering, gate sizing, incremental placement, incremental global routing, gate sizing, detailed placement, detailed routing, extraction, timing, sizing, placement, routing, timing, etc, etc.

2. The long chain of algorithms makes it near impossible to reason about global quality, and especially optimality. It is exquisitely clear that optimal solutions at each step to not yield the optimal layout at the end of the flow.

3. Each algorithm needs a highly simplified model of the physical reality. Therefore, some relevant effect cannot be accounted for. Most routing algorithms, for instance, operate on a graph data structure in which the edges represent wire segments. This does not model congestion properly, nor does it capture the many intricate 28nm design rules fully. These omissions will need to be addressed using some 'hacks' or they 'patched up' at a later stage.

4. Related to this, algorithms generally cannot handle multiple objectives simultaneously in a satisfactory way. For example, most placement algorithms are geared towards minimizing total wire length. To handle congestion and timing objectives the wire length handle is mutilated to include an additional 'cost'. The result is often a compromise that is far from acceptable. The fragility of these concepts gainfully employs many people at EDA companies.

The flexibility of the TCL language gives the user the necessary control to modify the flow to work around the inevitable outliers. If design constraints are tight, the physical synthesis process is a multi-month process of repeated runs and tweaks. The process speeds up dramatically when design constraints such as area and speed are relaxed.

6. The ABCs of a design flow

The above imperfect nature of the algorithmic tools, combined with the many (and often conflicting) design objectives makes the architecture of the physical synthesis flow non-trivial. Upon closer examination, it is generally a combination of several interwoven incarnations of the following general 3-step synthesis procedure:

- Avoid potential issues by over-designing.

- Build, using a synthesis algorithm. This brings the design 'in the ballpark'.

- Correct the errors made by running an analysis tool, and then make small incremental changes.

To illustrate this, let's investigate how to apply the above ABC procedure is applied to delay optimization of a gate and the net it drives. Lets assume that the build step is detailed routing that generates the mask-level layout wires of the net.

There are several 'Avoid' ways to reduce the probability that eventual layout delay is worse than required after routing: a) Use a bigger drive-strength gate, b) force a double spacing NDR on the net wire to keep aggressors at a safe distance or c) force a high priority on the net that will cause the router to avoid detouring it. Notice that any of these avoidance methods generally comes at a significant cost of power, area and congestion. For this reason the temptation is to overdesign as little as possible. Therefore it the stronger high-impact avoidance is performed only on the more critical nets. Unfortunately, the large expected delay deviation described in section 3 makes any such criticality prediction quiet shaky

Next comes the router 'Build' step that finds a path for this wire. The router needs to trade off the requirements for this net against those of millions of other nets, while producing a DRC correct wire pattern. Congestion control and various obstacles may require detours on certain net. Also, there is no strong practical procedure yet that can control crosstalk delay during routing.

The 'Correct' step starts off with an extraction of the layout pattern that produces the actual parasitic capacitances and resistances. The delay calculator in the Static Timing Analyzer uses this to report the actual delays of all wires. The delays on failing paths could, for example, be improved incrementally in several ways:

- By up-sizing or buffering the driver gate. This will require a 'placement legalizer' tool to find room for the larger gate(s). In dense designs the nearest available slot might be at a larger distance that forces a significant wire detour. In some cases the extra parasitic capacitance and congestion of this action can neutralize the gain of the stronger gate.

- By re-routing the wire such that it is 'straightened'. In dense areas, this might require other nets to detour away, possibly resulting in a non-converging flow.

- In case the problem is due to crosstalk, by re-routing the net with a wider spacing NDR. This also requires other nets to move out of the way, which might not succeed or create new issues on those nets.

The above corrections are generally painfully slow due to the extensive use of analysis tools and the much lower efficiency and completion rate of incremental routing. Success depends on the density of the design and the tightness of the timing constraints. Correction quickly becomes infeasible with larger numbers of delay failures. In practice the limit is in the low single-digit percentage of the nets.

The key aspect of a physical synthesis flow is to find the proper balance between Avoidance and Correction. Too much avoidance results in overdesign and underperformance, while too much correction is just not feasible. With each technology node, and also between the designs, the balance between them needs to be re-adjusted.

7. Conclusions

The physical synthesis flow is built from of a large number of smaller steps to adapt to the imperfect nature of the algorithms and the wide variety of design objectives. The optimality of each individual algorithm is not very relevant. Instead, the quality of the final layout is determined by the careful tuning of the interactions between the steps. Balancing the effects and side effects of avoidance and correction steps is the key for a successful flow.

Optimal Slack-Driven Block Shaping Algorithm in Fixed-Outline Floorplanning *

Jackey Z. Yan
Placement Technology Group
Cadence Design Systems
San Jose, CA 95134 USA
zyan@cadence.com

Chris Chu
Department of ECE
Iowa State University
Ames, IA 50010 USA
cnchu@iastate.edu

ABSTRACT

This paper presents an efficient, scalable and optimal slack-driven shaping algorithm for soft blocks in non-slicing floorplan. The proposed algorithm is called *SDS*. Different from all previous approaches, *SDS* is specifically formulated for fixed-outline floorplanning. Given a fixed upper bound on the layout width, *SDS* minimizes the layout height by only shaping the soft blocks in the design. Iteratively, *SDS* shapes some soft blocks to minimize the layout height, with the guarantee that the layout width would not exceed the given upper bound. Rather than using some simple heuristic as in previous work, the amount of change on each block is determined by systematically distributing the global total amount of available slack to individual block. During the whole shaping process, the layout height is monotonically reducing, and eventually converges to an optimal solution. We also propose two optimality conditions to check the optimality of a shaping solution. To validate the efficiency and effectiveness of *SDS*, comprehensive experiments are conducted on MCNC and HB benchmarks. Compared with previous work, *SDS* is able to achieve the best experimental result with significantly faster runtime.

Categories and Subject Descriptors

B.7.2 [**Hardware, Integrated Circuits, Design Aids**]: Layout

General Terms

Algorithms, Design, Performance

Keywords

Block Shaping, Fixed-Outline Floorplan, Physical Design

1. INTRODUCTION

Floorplanning is a very crucial step in modern VLSI designs. A good floorplan solution has a positive impact on the placement, routing and even manufacturing. In floorplanning step, a design contains two types of blocks, hard and soft. A hard block is a circuit block

with both area and aspect ratio [1] fixed, while a soft one has fixed area, yet flexible aspect ratio. Shaping such soft blocks plays an important role in determining the top-level spatial structure of a chip, because the shapes of blocks directly affect the packing quality and the area of a floorplan. However, due to the ever-increasing complexity of ICs, the problem of shaping soft blocks is not trivial.

1.1 Previous Work

In slicing floorplan, researchers proposed various soft-block shaping algorithms. Stockmeyer [1] proposed the shape curve representation used to capture different shapes of a subfloorplan. Based on the shape curve, it is straightforward to choose the floorplan solution with the minimum cost, e.g., minimum floorplan area. In [2], Zimmermann extended the shape curve representation by considering both slicing line directions when combining two blocks. Yan *et al.* [3] generalized the notion of slicing tree [4] and extended the shape curve operations. Consequently, one shape curve captures significantly more shaping and floorplan solutions.

Different from slicing floorplan, the problem of shaping soft blocks to optimize the floorplan area in non-slicing floorplan is much more complicated. Both Pan *et al.* [5] and Wang *et al.* [6] tried to extend the slicing tree and shape curve representations to handle non-slicing floorplan. But their extensions are limited to some specific non-slicing structures. Instead of using the shape curve, Kang *et al.* [7] adopted the bounded sliceline grid structure [8] and proposed a greedy heuristic algorithm to select different shapes for each soft block, so that total floorplan area was minimized. Moh *et al.* [9] formulated the shaping problem as a geometric programming and searched for the optimal floorplan area using standard convex optimization. Following the same framework as in [9], Murata *et al.* [10] improved the algorithm efficiency via reducing the number of variables and functions. But the algorithm still took a long time to find a good solution. In [11], Young *et al.* showed that the shaping problem for minimum floorplan area can be solved optimally by Lagrangian relaxation technique. Lin *et al.* [12] changed the problem objective to minimizing the half perimeter of a floorplan, and solved it optimally by the min-cost flow and trust region method.

All of the above shaping algorithms for non-slicing floorplan were targeting at classical floorplanning, i.e., minimizing the floorplan area. But, in the nanometer scale era classical floorplanning cannot satisfy the requirements of hierarchical design. In contrast, fixed-outline floorplanning [13] enabling the hierarchical framework is preferred by modern ASIC designs. In [14], Adya *et al.* introduced the notion of *slack* in floorplanning, and proposed a slack-based algorithm to shape the soft blocks. Such shaping algorithm was applied inside an annealing-based fixed-outline floorplanner. There are two problems with this shaping algorithm: 1) It is a simple greedy heuristic, in which each time every soft block is shaped to use up all its slack

---

*This work was partially supported by IBM Faculty Award and NSF under grant CCF-0540998.

---

[1] The *aspect ratio* is defined as the ratio of the block height to the block width.

in one direction. Thus, the resulting solution has no optimality guarantee; 2) It is not formulated for fixed-outline floorplanning. The fixed-outline constraint is simply considered as a penalty term in the cost function of annealing. Therefore, in non-slicing floorplan it is necessary to design an efficient and optimal shaping algorithm that is specifically formulated for fixed-outline floorplanning.

1.2 Our Contributions

This work presents an efficient, scalable and optimal slack-driven shaping (*SDS*) algorithm for soft blocks in non-slicing floorplan. *SDS* is specifically formulated for fixed-outline floorplanning. Given a fixed upper bound on the layout width, *SDS* minimizes the layout height by only shaping the soft blocks in the design. If such upper bound is set as the width of a predefined fixed outline, *SDS* is capable of optimizing the area for fixed-outline floorplanning. As far as we know, none of previous work in non-slicing floorplan considers the fixed-outline constraint in the problem formulation. In *SDS*, soft blocks are shaped iteratively. At each iteration, we only shape some of the soft blocks to minimize the layout height, with the guarantee that the layout width would not exceed the given upper bound. The amount of change on each block is determined by systematically distributing the global total amount of available slack to individual block. During the whole shaping process, the layout height is monotonically reducing, and eventually converges to an optimal solution. Note that in [14] without a global slack distribution, all soft blocks are shaped greedily and independently by some simple heuristic. In their work, both the layout height and width are reduced in one shot (i.e., not iteratively) and the solution is stuck at a local minimum.

Essentially, we have three main contributions.

- **Basic Slack-Driven Shaping:** The basic slack-driven shaping algorithm is a very simple shaping technique. Iteratively, it identifies some soft blocks, and shapes them by a slack-based shaping scheme. The algorithm stops when there is no identified soft block. The runtime complexity in each iteration is linear time. The basic *SDS* can achieve an optimal layout height for most cases.

- **Optimality Conditions:** To check the optimality of the shaping solution returned by the basic *SDS*, two optimality conditions are proposed. We prove that if either one of the two conditions is satisfied, the solution returned by the basic *SDS* is optimal.

- **Slack-Driven Shaping (SDS):** Based on the basic *SDS* and the optimality conditions, we propose the slack-driven shaping algorithm. In *SDS*, a geometric programming method is applied to improve the non-optimal solution produced by the basic *SDS*. *SDS* always returns an optimal shaping solution.

To show the efficiency of *SDS*, we compare it with the two shaping algorithms in [11] and [12] on MCNC benchmarks. Even though both of them claim their algorithms can achieve the optimal solution, experimental results show that *SDS* consistently generates better solution on each circuit with significantly faster runtime. On average *SDS* is 253× and 33× faster than [11] and [12] respectively, to produce solutions of similar quality. We also run *SDS* on HB benchmarks. Experimental results show that on average after 6%, 10%, 22% and 47% of the total iterations, the layout height is within 10%, 5%, 1% and 0.1% difference from the optimal solution, respectively.

The rest of this paper is organized as follows. Section 2 describes the problem formulation. Section 3 introduces the basic slack-driven shaping algorithm. Section 4 discusses the optimality of a shaping solution and presents two optimality conditions. Section 5 describes the algorithm flow of *SDS*. Experimental results are presented in Section 6. Finally, this paper ends with a conclusion and the direction of future work.

2. PROBLEM FORMULATION

In the design, suppose we are given n blocks. Each block i ($1 \leq i \leq n$) has fixed area A_i. Let w_i and h_i denote the width and height of block i respectively. The range of w_i and h_i are given as $W_i^{min} \leq w_i \leq W_i^{max}$ and $H_i^{min} \leq h_i \leq H_i^{max}$. If block i is a hard block, then $W_i^{min} = W_i^{max}$ and $H_i^{min} = H_i^{max}$. Let x_i and y_i denote the x and y coordinates of the bottom-left corner of block i respectively. To model the geometric relationship among the blocks, we use the horizontal and vertical constraint graphs G_h and G_v, where the vertices represent the blocks and the edges between two vertices represent the non-overlapping constraints between the two corresponding blocks. In G_h, we add two dummy vertices 0 and $n+1$ that represent the left-most and right-most boundary of the layout respectively. Similarly, in G_v we add two dummy vertices 0 and $n+1$ that represent the bottom-most and top-most boundary of the layout respectively. The area of the dummy vertices is 0. We have $x_0 = 0$ and $y_0 = 0$. Vertices 0 and $n+1$ are defined as the source and the sink in the graphs respectively. Thus, in both G_h and G_v, we add one edge from the source to each vertex that does not have any incoming edge, and add one edge from each vertex that does not have any outgoing edge to the sink.

In our problem formulation, we assume the constraint graphs G_h and G_v are given. Given an upper bound on the layout width as W, we want to minimize the layout height y_{n+1} by only shaping the soft blocks in the design, such that the layout width $x_{n+1} \leq W$. Such problem can be mathematically formulated as follows:

PROBLEM 1. **Height Minimization with Fixed Upper-Bound Width**

$$
\begin{array}{lll}
\text{Minimize} & y_{n+1} & \\
\text{subject to} & x_{n+1} \leq W & \\
& x_j \geq x_i + w_i, & \forall (i,j) \in G_h \\
& y_j \geq y_i + h_i, & \forall (i,j) \in G_v \\
& W_i^{min} \leq w_i \leq W_i^{max}, & 1 \leq i \leq n \\
& H_i^{min} \leq h_i \leq H_i^{max}, & 1 \leq i \leq n \\
& w_i h_i = A_i, & 1 \leq i \leq n \\
& x_0 = 0 & \\
& y_0 = 0 &
\end{array}
$$

It is clear that if W is set as the width of a predefined fixed outline, Problem 1 can be applied in fixed-outline floorplanning.

3. BASIC SLACK-DRIVEN SHAPING

In this section, we present the basic slack-driven shaping algorithm, which solves Problem 1 optimally for most cases.

First of all, we introduce some notations used in the discussion. Given the constraint graphs and the shape of the blocks, we can pack the blocks to four lines, i.e., the left (LL), right (RL), bottom (BL) and top (TL) lines. LL, RL, BL and TL are set as "$x = 0$", "$x = W$", "$y = 0$" and "$y = y_{n+1}$", respectively. Let Δ_{x_i} denote the difference of x_i when packing block i to RL and LL. Similarly, Δ_{y_i} denotes the difference of y_i when packing block i to TL and BL. For block i ($1 \leq i \leq n$), the horizontal slack s_i^h and vertical slack s_i^v are calculated as follows:

$$ s_i^h = max(0, \Delta_{x_i}), \quad s_i^v = max(0, \Delta_{y_i}) $$

In G_h, given any path [2] from the source to the sink, if for all blocks on this path, their horizontal slacks are equal to zero, then we define such path as a horizontal critical path (HCP). The length of one HCP is the summation of the width of blocks on this path. Similarly, we can define the vertical critical path (VCP) and the length of one VCP is the summation of the height of blocks on this path. Note that,

---

[2] By default, all paths in this paper are from the source to the sink in the constraint graph.

because we set RL as the "$x = W$" line, if $x_{n+1} < W$, then there is no HCP in G_h.

The algorithm flow of the basic *SDS* is simple and straightforward. The soft blocks are shaped iteratively. At each iteration, we apply the following two operations:

1. Shape the soft blocks on all VCPs by increasing the width and decreasing the height. This reduces the lengths of the VCPs.

2. Shape the soft blocks on all HCPs by decreasing the width and increasing the height. This reduces the lengths of the HCPs.

The purpose of the first operation is to minimize the layout height y_{n+1} by decreasing the lengths of all VCPs. As mentioned previously, if $x_{n+1} < W$ then there is no HCP. Thus, the second operation is applied only if $x_{n+1} = W$. This operation seems to be unnecessary, yet actually is critical for the proof of the optimality conditions. The purpose of this operation will be explained in Section 4. At each iteration, we first globally distribute the total amount of slack reduction to the soft blocks, and then locally shape each individual soft block on the critical paths based on the allocated amount of slack reduction. The algorithm stops when we cannot find any soft block to shape on the critical paths. During the whole shaping process, the layout height y_{n+1} is monotonically decreasing and thus the algorithm always converges.

In the following subsections, we first identify which soft blocks to be shaped (which we called *target soft blocks*) at each iteration. Secondly, we mathematically derive the shaping scheme on the target soft blocks. Finally, we present the algorithm flow of the basic *SDS*.

3.1 Target Soft Blocks

For a given shaping solution, the set of n blocks can be divided into the following seven disjoint subsets ($1 \le i \le n$).

$$
\left\{
\begin{array}{ll}
\text{Subset I} & = \{i \text{ is hard}\} \\
\text{Subset II} & = \{i \text{ is soft}\} \cap \{s_i^h \ne 0, s_i^v \ne 0\} \\
\text{Subset III} & = \{i \text{ is soft}\} \cap \{s_i^h = 0, s_i^v = 0\} \\
\text{Subset IV} & = \{i \text{ is soft}\} \cap \{s_i^h \ne 0, s_i^v = 0\} \cap \{w_i \ne W_i^{max}\} \\
\text{Subset V} & = \{i \text{ is soft}\} \cap \{s_i^h \ne 0, s_i^v = 0\} \cap \{w_i = W_i^{max}\} \\
\text{Subset VI} & = \{i \text{ is soft}\} \cap \{s_i^h = 0, s_i^v \ne 0\} \cap \{h_i \ne H_i^{max}\} \\
\text{Subset VII} & = \{i \text{ is soft}\} \cap \{s_i^h = 0, s_i^v \ne 0\} \cap \{h_i = H_i^{max}\}
\end{array}
\right.
$$

Based on the definitions of critical paths, we have the following observations [3].

OBSERVATION 1. *If block $i \in$ subset II, then i is not on any HCP nor VCP.*

OBSERVATION 2. *If block $i \in$ subset III, then i is on both HCP and VCP, i.e., at the intersection of some HCP and some VCP.*

OBSERVATION 3. *If block $i \in$ subset IV or V, then i is on some VCP but not on any HCP.*

OBSERVATION 4. *If block $i \in$ subset VI or VII, then i is on some HCP but not on any VCP.*

As mentioned previously, y_{n+1} can be minimized by reducing the height of the soft blocks on the vertical critical paths, and such block-height reduction will result in a decrease on the horizontal slacks of those soft blocks. From the above observations, only soft blocks in subsets III, IV and V are on the vertical critical paths. However, for block $i \in$ subset III, $s_i^h = 0$, which means its horizontal slack cannot be further reduced. And for block $i \in$ subset V, $w_i = W_i^{max}$, which means its height cannot be further reduced. As a result, to minimize y_{n+1} we can only shape blocks in subset IV. Similarly, we conclude

---
[3] Please refer to Theorem 1 in [14] for the proof of these observations.

that whenever we need to reduce x_{n+1} we can only shape blocks in subset VI. For the hard blocks in subset I, they cannot be shaped anyway.

Therefore, the target soft blocks are the blocks in subsets IV and VI.

3.2 Shaping Scheme

Let δ_i^h denote the amount of increase on w_i for block $i \in$ subset IV, and δ_i^v denote the amount of increase on h_i for block $i \in$ subset VI. In the remaining part of this subsection, we present the shaping scheme to shape the target soft block $i \in$ subset IV by setting δ_i^h. Similar shaping scheme is applied to shape the target soft block $i \in$ subset VI by setting δ_i^v. By default, all blocks mentioned in the following part are referring to the target soft blocks in subset IV.

We use "$i \in p$" to denote that block i is on a path p in G_h. Suppose the maximum horizontal slack over all blocks on p is s_{max}^p. Basically, s_{max}^p gives us a budget on the total amount of increase on the block width along this path. If $\sum_{i \in p} \delta_i^h > s_{max}^p$, then after shaping, we have $x_{n+1} > W$, which violates the constraint "$x_{n+1} \le W$". So we have to set δ_i^h accordingly, such that $\sum_{i \in p} \delta_i^h \le s_{max}^p$ for all p in G_h.

To determine the value of δ_i^h, we first define a distribution ratio α_i^p ($\alpha_i^p \ge 0$) for block $i \in p$. We assign the value of α_i^p, such that

$$\sum_{i \in p} \alpha_i^p = 1$$

LEMMA 1. *For any path p in G_h, we have*

$$\sum_{i \in p} \alpha_i^p s_i^h \le s_{max}^p$$

PROOF. Because $s_{max}^p = \text{MAX}_{i \in p}(s_i^h)$, this lemma can be proved as follows:

$$\sum_{i \in p} \alpha_i^p s_i^h \le \sum_{i \in p} \alpha_i^p s_{max}^p = s_{max}^p \sum_{i \in p} \alpha_i^p = s_{max}^p$$

□

Based on Lemma 1, for a single path p, it is obvious that if $\delta_i^h \le \alpha_i^p s_i^h$ ($i \in p$), then we can guarantee $\sum_{i \in p} \delta_i^h \le s_{max}^p$.

More generally, if there are multiple paths going through block i ($1 \le i \le n$), then δ_i^h needs to satisfy the following inequality:

$$\delta_i^h \le \alpha_i^p s_i^h, \forall p \in P_i^h \tag{1}$$

where P_i^h is the set of paths in G_h going through block i. Inequality 1 is equivalent to the following inequality.

$$\delta_i^h \le \underset{p \in P_i^h}{\text{MIN}}(\alpha_i^p) s_i^h \tag{2}$$

Essentially, Inequality 2 gives an upper bound on the amount of increase on w_i for block $i \in$ subset IV.

For block $i \in p$, the distribution ratio is set as follows:

$$
\alpha_i^p =
\begin{cases}
0 & i \text{ is the source or the sink} \\
\frac{W_i^{max} - w_i}{\sum_{k \in p}(W_k^{max} - w_k)} & otherwise
\end{cases}
\tag{3}
$$

The insight is that if we allocate more slack reduction to the blocks that have potentially more room to be shaped, the algorithm will converge faster. And we allocate zero amount of slack reduction to the dummy blocks at the source and the sink in G_h. Based on Equation 3, Inequality 2 can be rewritten as follows ($1 \le i \le n$):

$$\delta_i^h \le \frac{(W_i^{max} - w_i)s_i^h}{\underset{p \in P_i^h}{\text{MAX}}(\sum_{k \in p}(W_k^{max} - w_k))} \tag{4}$$

From the above inequality, to calculate the upper bound of δ_i^h, we need to obtain the value of three terms, $(W_i^{max} - w_i)$, s_i^h and $\text{MAX}_{p \in P_i^h}(\sum_{k \in p}(W_k^{max} - w_k))$. The first term can be obtained in constant time. Using the longest path algorithm, s_i^h for all i can be calculated in linear time. A trivial approach to calculate the third term is via traversing each path in G_h. This takes exponential time, which is not practical. Therefore, we propose a dynamic programming (DP) based approach that only takes linear time to calculate the third term.

In G_h, suppose vertex i ($0 \le i \le n+1$) has in-coming edges coming from the vertices in the set V_i^{in}, and out-going edges going to the vertices in the set V_i^{out}. Let P_i^{in} denote the set of paths that start at the source and end at vertex i in G_h, and P_i^{out} denote the set of paths that start at vertex i and end at the sink in G_h. For the source of G_h, we have $V_0^{in} = \phi$ and $P_0^{in} = \phi$. For the sink of G_h, we have $V_{n+1}^{out} = \phi$ and $P_{n+1}^{out} = \phi$. We notice that $\text{MAX}_{p \in P_i^h}(\sum_{k \in p}(W_k^{max} - w_k))$ can be calculated recursively by the following equations.

$$\text{MAX}_{p \in P_0^{in}}(\sum_{k \in p}(W_k^{max} - w_k)) = 0$$

$$\text{MAX}_{p \in P_{n+1}^{out}}(\sum_{k \in p}(W_k^{max} - w_k)) = 0$$

$$\text{MAX}_{p \in P_i^{in}}(\sum_{k \in p}(W_k^{max} - w_k)) = \text{MAX}_{j \in V_i^{in}}(\text{MAX}_{p \in P_j^{in}}(\sum_{k \in p}(W_k^{max} - w_k))) \\ + (W_i^{max} - w_i) \quad (5)$$

$$\text{MAX}_{p \in P_i^{out}}(\sum_{k \in p}(W_k^{max} - w_k)) = \text{MAX}_{j \in V_i^{out}}(\text{MAX}_{p \in P_j^{out}}(\sum_{k \in p}(W_k^{max} - w_k))) \\ + (W_i^{max} - w_i) \quad (6)$$

$$\text{MAX}_{p \in P_i^h}(\sum_{k \in p}(W_k^{max} - w_k)) = \text{MAX}_{p \in P_i^{in}}(\sum_{k \in p}(W_k^{max} - w_k)) \\ + \text{MAX}_{p \in P_i^{out}}(\sum_{k \in p}(W_k^{max} - w_k)) \\ - (W_i^{max} - w_i) \quad (7)$$

Based on the equations above, the DP-based approach can be applied step by step as follows ($1 \le i \le n$):

1. We apply topological sort algorithm on G_h.

2. We scan the sorted vertices from the source to the sink, and calculate $\text{MAX}_{p \in P_i^{in}}(\sum_{k \in p}(W_k^{max} - w_k))$ by Equation 5.

3. We scan the sorted vertices from the sink to the source, and calculate $\text{MAX}_{p \in P_i^{out}}(\sum_{k \in p}(W_k^{max} - w_k))$ by Equation 6.

4. $\text{MIN}_{p \in P_i^h}(\sum_{k \in p}(W_k^{max} - w_k))$ is obtained by Equation 7.

It is clear that by the DP-based approach, the whole process of calculating the upper bound of δ_i^h for all i takes linear time.

3.3 Flow of Basic Slack-Driven Shaping

The algorithm flow of basic slack-driven shaping is shown in Figure 1. In this flow, for each block i in the design, we set its initial width $w_i = W_i^{min}$ ($1 \le i \le n$). Based on the input G_h, G_v and initial block shape, we can calculate an initial value of x_{n+1}. If such initial value is already bigger than W, then Problem 1 is not feasible.

At each iteration we set $\delta_j^v = \beta \times \text{MIN}_{p \in P_j^v}(\alpha_j^p)s_j^v$ for target block $j \in$ subset VI. By default, $\beta = 100\%$, which means we set δ_j^v exactly at its upper bound. One potential problem with this strategy is that the layout height y_{n+1} may remain the same, i.e., never decreasing. This is because after one iteration of shaping, the length of some non-critical vertical path increases, and consequently its length may become equivalent to the length of the VCP in the previous iteration. Accidentally, such scenario may keep cycling forever, and

---

Basic Slack-Driven Shaping

Input: $w_i = W_i^{min}$ ($\forall 1 \le i \le n$); G_h and G_v; upper-bound width W.
Output: optimized y_{n+1}, w_i and h_i.
Begin
1. Set LL, BL and RL to "$x = 0$", "$y = 0$" and "$x = W$".
2. Pack blocks to LL and use longest path algorithm to get x_{n+1}.
3. If $x_{n+1} > W$,
4.   Return no feasible solution.
5. Else,
6.   Repeat
7.   Pack blocks to BL and use longest path algorithm to get y_{n+1}.
8.   Set TL to "$y = y_{n+1}$".
9.   Pack blocks to LL, RL and TL, respectively.
10.   Calculate s_i^h and s_i^v.
11.   Find target soft blocks.
12.   If there are target soft blocks,
13. $\forall j \in$ subset IV, increase w_j by $\delta_j^h = \text{MIN}_{p \in P_j^h}(\alpha_j^p)s_j^h$;
14. $\forall j \in$ subset VI, increase h_j by $\delta_j^v = \beta \times \text{MIN}_{p \in P_j^v}(\alpha_j^p)s_j^v$.
15.   Until there is no target soft block.
End

Figure 1: Flow of basic slack-driven shaping.

---

thus y_{n+1} would never decrease. This issue can be solved, as long as δ_j^v is set less than its upper bound. In this way, after one iteration of shaping we can guarantee that the length of the VCP will be shorter than the one in the previous iteration. Theoretically, any $\beta < 100\%$ can break the cycling scenario and guarantee the algorithm convergence. But because in SDS any amount of change that is less than 0.0001 would be masked by numerical error, we can actually calculate a lower bound of β, and obtain its range as follows.

$$\frac{0.01}{\text{MIN}_{p \in P_j^v}(\alpha_j^p)s_j^v}\% < \beta < 100\%$$

In the implementation, whenever we detect that y_{n+1} does not change for more than two iterations, we will set $\beta = 90\%$ for the next iteration. For δ_j^h, we always set it at its upper bound.

Because in each iteration the total increase on width or height of the target soft blocks would not exceed the budget, we can guarantee that the layout would not be outside of the four lines after shaping. As iteratively we set TL to the updated "$y = y_{n+1}$" line, y_{n+1} will be monotonically decreasing during the whole shaping process. Different from TL, because we set RL to the fixed "$x = W$" line, during the shaping process x_{n+1} may be bouncing i.e., sometimes increasing and sometimes decreasing, yet always no more than W. The shaping process stops when there is no target soft block.

4. OPTIMALITY CONDITIONS

For most cases, in the basic SDS the layout height y_{n+1} will converge to an optimal solution of Problem 1. However, sometimes the solution may be non-optimal as the one shown in Figure 2-(a). The layout in Figure 2-(a) contains four soft blocks 1, 2, 3 and 4, where $A_i = 4$, $W_i^{min} = 1$ and $W_i^{max} = 4$ ($1 \le i \le 4$). The given upper-bound width $W = 5$. In the layout, $w_1 = w_3 = 4$ and $w_2 = w_4 = 1$. There is no target soft block on any one of the four critical paths (i.e., two HCPs and two VCPs), so the basic SDS returns $y_{n+1} = 5$. But the optimal layout height should be 3.2, when $w_1 = w_2 = w_3 = w_4 = 2.5$ as shown in Figure 2-(b). In this section, we will look into this issue and present the optimality conditions for the shaping solution returned by the basic SDS.

Let L represent a shaping solution generated by the basic SDS in Figure 1. All proof in this section are established based on the fact that the *only* remaining soft blocks that could be shaped to possibly

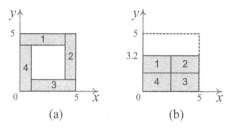

Figure 2: Example of a non-optimal solution from the basic *SDS*.

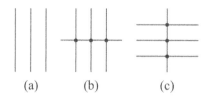

Figure 3: Examples of three optimal cases in L.

improve L are the ones in subset III. This is because L is the solution returned by the basic *SDS* and in L there is no soft block that belongs to subsets IV nor VI any more. This is also why we need apply the second shaping operation in the basic *SDS*. Its purpose is *not* reducing x_{n+1}, but eliminating the soft blocks in subset VI. From Observation 2, we know that any block in subset III is always at the intersection of some HCP and some VCP. Therefore, to improve L it is sufficient to just consider shaping the intersection soft blocks between the HCPs and VCPs.

Before we present the optimal conditions, we define two concepts.

- **Hard Critical Path:** If all intersection blocks on one critical path are hard blocks, then this path is a *hard* critical path.

- **Soft Critical Path:** A critical path, which is not hard, is a *soft* critical path.

LEMMA 2. *If there exists one hard VCP in L, then L is optimal.*

PROOF. Since all intersection blocks on this VCP are hard blocks, there is no soft block that can be shaped to possibly improve this VCP. Therefore, L is optimal. ☐

LEMMA 3. *If there exists at most one soft HCP or at most one soft VCP in L, then L is optimal.*

PROOF. As proved in Lemma 2, if there exists one hard VCP in L, then L is optimal. So in the following proof we assume there is no hard VCP in L. For any hard HCP, as all intersection blocks on it are hard blocks, we cannot change its length by shaping those intersection blocks anyway. So we can basically ignore all hard HCPs in this proof.

Suppose L is non-optimal. We should be able to identify some soft blocks and shape them to improve L. As mentioned previously, it is sufficient to just consider shaping the intersection soft blocks. If there is at most one soft HCP or at most one soft VCP, there are only three possible cases in L. (As we set TL as the "$y = y_{n+1}$" line, there is always at least one VCP in L.)

1. **There is no soft HCP, and there is one or multiple soft VCPs** (e.g., Figure 3-(a))
   In this case, L does not contain any intersection soft blocks.

2. **There is one soft HCP, and there is one or multiple soft VCPs** (e.g., Figure 3-(b))
   In this case, L has one or multiple intersection soft blocks. Given any one of such blocks, say i. To improve L, h_i has to be

reduced. But this increases the length of the soft HCP, which violates "$x_{n+1} \leq W$" constraint. So, none of the blocks can be shaped to improve L.

3. **There is one or multiple soft HCPs, and there is one soft VCP** (e.g., Figure 3-(c))
   In this case, L has one or multiple intersection soft blocks. Given any one of such blocks, say i. Similarly, it can be proved that "$x_{n+1} \leq W$" constraint will be violated, if h_i is reduced. So, none of the blocks can be shaped to improve L.

As a result, for all the above cases, we cannot find any soft block that could be shaped to possibly improve L. This means our assumption is not correct. Therefore, L is optimal. ☐

5. FLOW OF SLACK-DRIVEN SHAPING

Using the conditions presented in Lemmas 2 and 3, we can determine the optimality of the output solution from the basic *SDS*. Therefore, based on the algorithm flow in Figure 1, we propose the slack-driven shaping algorithm shown in Figure 4. *SDS* always returns an optimal solution for Problem 1.

Slack-Driven Shaping
Input: $w_i = W_i^{min}$ ($\forall 1 \leq i \leq n$); G_h and G_v; upper-bound width W.
Output: optimal y_{n+1}, w_i and h_i.
Begin
Lines 1 – 14 are the same as the ones in Figure 1.
15.   Else,
16.   If Lemma 2 or 3 is satisfied,
17. L is optimal.
18.   Else,
19.   Improve L by a single step of geometric programming.
20.   If no optimal solution is obtained,
21.   Go to Line 7.
22.   Else,
23. L is optimal.
24.   Until L is optimal.
End

Figure 4: Flow of slack-driven shaping.

The differences between *SDS* and the basic version are starting from line 15 in Figure 4. When there is not target soft block, instead of terminating the algorithm, *SDS* will first check the optimality of L, and if it is not optimal, L will be improved via geometric programming. The algorithm stops when an optimal solution is obtained.

As mentioned previously, if the solution L generated by the basic *SDS* is not optimal, we only need to shape the intersection soft blocks to improve L. In this way, the problem now becomes shaping the intersection blocks to minimize the layout height y_{n+1} subject to layout width constraint "$x_{n+1} \leq W$". In other words, it is basically the same as Problem 1, except that we only need to shape a smaller number of soft blocks (i.e., the intersection soft blocks). This problem is a geometric program. It can be transformed into a convex problem and solved optimally by any convex optimization technique. However, considering the runtime, we don't need to rely on geometric programming to converge to an optimal solution. We just run one step of some iterative convex optimization technique (e.g., deepest descent) to improve L. Then we can go back to line 7, and applied the basic *SDS* again. It is clear that *SDS* always converges to the optimal solution because as long as the solution is not optimal, the layout height will be improved.

In modern VLSI designs, the usage of Intellectual Property (IP) and embedded memory blocks becomes more and more popular. As a result, a design usually contains tens or even hundreds of big hard macros, i.e., hard blocks. Due to their big sizes, after applying the basic *SDS* most likely they are at the intersections of horizontal and vertical critical paths. Moreover, in our experiments we observe that there is always no more than one soft HCP or VCP in the solution returned by the basic *SDS*. Consequently, we never need to apply the geometric programming method in our experiments. Therefore, we believe that for most designs the basic slack-driven shaping algorithm is sufficient to achieve an optimal solution for Problem 1.

6. EXPERIMENTAL RESULTS

This section presents the experimental results. All experiments are run on a Linux server with AMD Opteron 2.59 GHz CPU and 16GB memory. We use two sets of benchmarks, MCNC [11] and HB [15]. For each circuit, the corresponding input G_h and G_v are provided by a floorplanner. The range of the aspect ratio for any soft block in the circuit is set to $[\frac{1}{3}, 3]$.

After the input data is read, *SDS* will set the initial width of each soft block at its minimum width. In *SDS*, if the amount of change on the width or height of any soft block is less than 0.0001, we would not shape such block because any change smaller than that would be masked by numerical error. Such numerical error, which is unavoidable, comes from the truncation of an infinite real number so as to make the computation possible and practical.

6.1 Experiments on MCNC Benchmarks

Using the MCNC benchmarks we compare *SDS* with the two shaping algorithms in [11] and [12]. All blocks in these circuits are soft blocks. The source code of [11] and [12] are obtained from the authors.

In fact, these three shaping algorithms *cannot* be directly compared, because their optimization objectives are all different:

- [11] is minimizing the layout area $x_{n+1}y_{n+1}$;

- [12] is minimizing the layout half perimeter $x_{n+1} + y_{n+1}$;

- *SDS* is minimizing the layout height y_{n+1}, s.t. $x_{n+1} \leq W$.

Still, to make some meaningful comparisons as best as we can, we setup the experiment in the following way.

- We conduct two groups of experiments: 1) *SDS v.s.* [11]; 2) *SDS v.s.* [12].

- As the circuit size are all very small, to do some meaningful comparison on the runtime, in each group we run both shaping algorithms 1000 times with the same input data.

- For group 1, we run [11] first, and use the returned final width from [11] as the input upper-bound width W for *SDS*. For group 2, similar procedure is applied.

- For groups 1 and 2, we compare the final results based on [11]'s and [12]'s objectives respectively.

Table 2 shows the results on group 1. The column "$ws(\%)$" gives the white space percentage over the total block area in the final layout. For all five circuits *SDS* achieves significantly better results on the floorplan area. On average, *SDS* achieves $394\times$ smaller white space and $23\times$ faster runtime than [11]. In the last column, we report the runtime *SDS* takes to converge to a solution that is better than [11]. To just get a slightly better solution than [11], on average *SDS* uses $253\times$ faster runtime. As pointed out by [12], [11] does not transform the problem into a convex problem before applying Lagrangian relaxation. Hence, algorithm [11] may not converge to an optimal solution.

Table 3 shows the results on group 2. The authors claims the shaping algorithm in [12] can find the optimal half perimeter on the floorplan layout. But, for all five circuits *SDS* gets consistently better half

Table 1: Comparison on runtime complexity.

Algorithm	Runtime Complexity
Young *et al.* [11]	$\mathcal{O}(m^3 + km^2)$
Lin *et al.* [12]	$\mathcal{O}(kn^2 m log(nC))$
Basic *SDS*	$\mathcal{O}(km)$

(k is the total number of iterations, n is the total number of blocks in the design, m is the total number of edges in G_h and G_v, and C is the biggest input cost.)

perimeter than [12], with on average $10\times$ faster runtime. Again, in the last column, we report the runtime *SDS* takes to converge to a solution that is better than [12]. To just get a slightly better solution than [12], on average *SDS* uses $33\times$ faster runtime. We believe algorithm [12] stops earlier, before it converges to an optimal solution.

From the runtime reported in Tables 2 and 3, it is clear that as the circuit size increases, *SDS* scales much better than both [11] and [12]. In Table 1, we list the runtime complexities among the three shaping algorithms. As in our experiments, it is never necessary to apply the geometric programming method in *SDS*, we list the runtime complexity of the basic *SDS* in Table 1. Obviously, the basic *SDS* has the best scalability.

6.2 Experiments on HB Benchmarks

This subsection presents the experimental results of *SDS* on HB benchmarks. As both algorithms [11] and [12] crashed on this set of circuits, we cannot compare *SDS* with them. The HB benchmarks contain both hard and soft blocks ranging from 500 to 2000 (see Table 4 for details).

For each test case, we set the upper-bound width W as the square root of 110% of the total block area in the corresponding circuit. Let Y denote the optimal y_{n+1} *SDS* converges to. The results are shown in Table 4. The "Convergence Time" column lists the total runtime of the whole convergence process. The "Total #.Iterations" column shows the total number of iteration *SDS* takes to converge to Y. For fixed-outline floorplanning, *SDS* can actually stop early as long as the solution is within the fixed outline. So in the subsequent four columns, we also report the number of iterations when $\frac{y_{n+1}-Y}{Y}$ starts to be less than 10%, 5%, 1% and 0.1%, respectively. The average total convergence time is 1.18 second. *SDS* takes average 1901 iterations to converge to Y. The four percentage numbers in the last row shows that on average after 6%, 10%, 22% and 47% of the total number of iterations, *SDS* converges to the layout height that is within 10%, 5%, 1% and 0.1% difference from Y, respectively. In order to show the convergence process more intuitively, we plot out the convergence graphs of y_{n+1} for four circuits in Figures 5(a)-5(d). In the figures, the four blue arrows point to the four points when y_{n+1} becomes less than 10%, 5%, 1% and 0.1% difference from Y, respectively.

Finally, we have four remarks on *SDS*.

1. As *SDS* sets the initial width of each soft block at its minimal width, such initial floorplan is actually considered as the *worse* start point for *SDS*. This means if any better initial shape is given, *SDS* will converge to Y even faster.

2. In our experiments, we never notice that the solution generated by the basic *SDS* contains more than one soft HCP or VCP. So if ignoring the numerical error mentioned previously, *SDS* obtains the optimal layout height for all circuits in the experiments simply by the basic *SDS*.

3. The experimental results show that after around $\frac{1}{5}$ of the total iterations, the difference between y_{n+1} and Y is already considered quite small, i.e., less than 1%. So in practice if it is not necessary to obtain an optimal solution, we can basically

Table 2: Comparison with [11] on MCNC Benchmarks († shows the total shaping time of 1000 runs and does not count I/O time).

Circuit	#. Soft Blocks	Young et al. [11]				SDS					SDS stops when result is better than [11]	
		ws (%)	Final Width	Final Height	Shaping Time† (s)	ws (%)	Final Width	Final Height	Upper-Bound Width W	Shaping Time† (s)	ws (%)	Time† (s)
apte	9	4.66	195.088	258.647	0.12	**0.00**	195.0880	246.6147	195.0880	0.26	2.85	0.01
xerox	10	7.69	173.323	120.945	0.08	**0.01**	173.3229	111.6599	173.3230	0.23	6.46	0.01
hp	11	10.94	83.951	120.604	0.08	**1.70**	83.9509	109.2605	83.9510	0.10	7.96	0.02
ami33a	33	8.70	126.391	100.202	22.13	**0.44**	126.3909	91.7830	126.3910	3.97	8.67	0.28
ami49a	49	10.42	144.821	273.19	203.80	**1.11**	144.8210	247.4727	144.8210	1.86	9.74	0.20
Normalized		393.919			23.351	**1.000**				**1.000**	313.980	0.092

Table 3: Comparison with [12] on MCNC Benchmarks († shows the total shaping time of 1000 runs and does not count I/O time).

Circuit	#. Soft Blocks	Lin et al. [12]				SDS					SDS stops when result is better than [12]	
		Half Perimeter	Final Width	Final Height	Shaping Time† (s)	Half Perimeter	Final Width	Final Height	Upper-Bound Width W	Shaping Time† (s)	Half Peri.	Time† (s)
apte	9	439.319	219.814	219.505	0.99	**439.3050**	219.8139	219.4911	219.8140	**0.59**	439.1794	0.01
xerox	10	278.502	138.034	140.468	1.24	**278.3197**	138.0339	140.2858	138.0340	**0.30**	278.4883	0.12
hp	11	190.3848	95.2213	95.1635	1.51	**190.2435**	95.2212	95.0223	95.2213	**0.17**	190.3826	0.10
ami33a	33	215.965	107.993	107.972	34.85	**215.7108**	107.9930	107.7178	107.9930	**1.45**	215.9577	0.46
ami49a	49	377.857	193.598	184.259	26.75	**377.5254**	193.5980	183.9274	193.5980	**2.20**	377.8242	0.44
Normalized		1.001			10.177	**1.000**				**1.000**	1.001	0.304

set a threshold value on the amount of change on y_{n+1} as the stopping criterion. For example, if the amount of change on y_{n+1} is less than 1% during the last 10 iterations, then *SDS* will stop.

4. Like all other shaping algorithms, *SDS* is *not* a floorplanning algorithm. To implement a fixed-outline floorplanner based on *SDS*, for example, we can simply integrate *SDS* into a similar annealing-based framework as the one in [14]. In each annealing loop, the input constraint graphs are sent to *SDS*, and *SDS* stops once the solution is within the fixed outline. The annealing process keeps refining the constraint graphs so as to optimize the various floorplanning objectives (e.g., wirelength, routability [16] [17], timing, etc.) in the cost function.

7. CONCLUSION AND FUTURE WORK

This work proposed an efficient, scalable and optimal slack-driven shaping algorithm for soft blocks in non-slicing floorplan. Unlike previous work, we formulate the problem in a way, such that it can be applied for fixed-outline floorplanning. For all cases in our experiments, the basic *SDS* is sufficient to obtain an optimal solution. Both the efficiency and effectiveness of *SDS* have been validated by comprehensive experimental results and rigorous theoretical analysis.

Due to the page limit, we have to reserve some problems on *SDS* as the motivation of future work, which includes: 1) To use the *duality gap* of Problem 1 as a better stopping criterion, because it indicates an upper-bound of the gap between the intermediate and optimal shaping solutions; 2) To propose a more scalable algorithm as a substitution of the geometric programing method in Figure 4; 3) To extend *SDS* to handle classical floorplanning. Also, because of the similarity between the *slack* in floorplanning and static timing analysis (STA), we believe *SDS* can be modified and applied on buffer/wire sizing for timing optimization.

Acknowledgment

The authors would like to thank Prof. H. Zhou from Northwestern University for providing us the source code of algorithms [11] and [12].

8. REFERENCES

[1] L. Stockmeyer. Optimal orientations of cells in slicing floorplan designs. *Information and Control*, 57:91–101, May/June 1983.

[2] G. Zimmermann. A new area and shape function estimation technique for VLSI layouts. In *Proc. DAC*, pages 60–65, 1988.

[3] J. Z. Yan and C. Chu. DeFer: Deferred decision making enabled fixed-outline floorplanning algorithm. *IEEE Trans. on Computer-Aided Design*, 43(3):367–381, March 2010.

[4] R. H. J. M. Otten. Efficient floorplan optimization. In *Proc. ICCD*, pages 499–502, 1983.

[5] P. Pan and C. L. Liu. Area minimization for floorplans. *IEEE Trans. on Computer-Aided Design*, 14(1):129–132, January 1995.

[6] T. C. Wang and D. F. Wong. Optimal floorplan area optimization. *IEEE Trans. on Computer-Aided Design*, 11(8):992–1001, August 1992.

[7] M. Kang and W. W. M. Dai. General floorplanning with L-shaped, T-shaped and soft blocks based on bounded slicing grid structure. In *Proc. ASP-DAC*, pages 265–270, 1997.

[8] S. Nakatake, K. Fujiyoshi, H. Murata, and Y. Kajitani. Module placement on BSG-structure and IC layout applications. In *Proc. ICCAD*, pages 484–491, 1996.

[9] T. S. Moh, T. S. Chang, and S. L. Hakimi. Globally optimal floorplanning for a layout problem. *IEEE Trans. on Circuits and Systems I*, 43:713–720, September 1996.

[10] H. Murata and E. S. Kuh. Sequence-pair based placement method for hard/soft/pre-placed modules. In *Proc. ISPD*, pages 167–172, 1998.

[11] F. Y. Young, C. C. N. Chu, W. S. Luk, and Y. C. Wong. Handling soft modules in general non-slicing floorplan using Lagrangian relaxation. *IEEE Trans. on Computer-Aided Design*, 20(5):687–692, May 2001.

[12] C. Lin, H. Zhou, and C. Chu. A revisit to floorplan optimization by Lagrangian relaxation. In *Proc. ICCAD*, pages 164–171, 2006.

[13] A. B. Kahng. Classical floorplanning harmful? In *Proc. ISPD*, pages 207–213, 2000.

[14] S. N. Adya and I. L. Markov. Fixed-outline floorplanning: Enabling hierarchical design. *IEEE Trans. on VLSI Systems*, 11(6):1120–1135, December 2003.

[15] J. Cong, M. Romesis, and J. R. Shinnerl. Fast floorplanning by look-ahead enabled recursive bipartitioning. In *Proc. ASP-DAC*, pages 1119–1122, 2005.

[16] Y. Zhang and C. Chu. CROP: Fast and effective congestion refinement of placement. In *Proc. ICCAD*, pages 344–350, 2009.

[17] Y. Zhang and C. Chu. RegularRoute: An efficient detailed router with regular routing patterns. In *Proc. ISPD*, pages 45–52, 2011.

Table 4: Experimental Results of *SDS* on HB Benchmarks.

Circuit	#.Soft Blocks / #.Hard Blocks	Upper-Bound Width W	Final Width	Final Height (Y)	Convergence Time (s)	Total #.Iterations	#.Iterations when $\frac{y_{n+1}-Y}{Y}$ becomes			
							< 10%	< 5%	< 1%	< 0.1%
ibm01	665 / 246	2161.9005	2161.9003	2150.3366	0.82	2336	54	85	225	629
ibm02	1200 / 271	3057.4816	3056.6026	3050.4862	0.40	485	65	102	230	431
ibm03	999 / 290	3298.2255	3298.2228	3305.6953	0.36	565	62	97	231	456
ibm04	1289 / 295	3204.7658	3204.7656	3179.9406	3.65	3564	53	87	271	1076
ibm05	564 / 0	2222.8426	2222.8424	2104.4136	0.29	1456	102	142	279	522
ibm06	571 / 178	3069.5289	3068.5232	2988.6851	0.14	500	58	105	265	419
ibm07	829 / 291	3615.5698	3615.5696	3599.6710	1.86	3966	63	114	269	1210
ibm08	968 / 301	3855.1451	3855.1449	3822.5919	0.42	690	75	111	232	545
ibm09	860 / 253	4401.0232	4401.0231	4317.0274	1.20	2512	50	82	234	687
ibm10	809 / 786	7247.6365	7246.7511	7221.0778	0.49	472	28	56	162	377
ibm11	1124 / 373	4844.2184	4844.2183	4820.8615	0.60	654	64	96	253	509
ibm12	582 / 651	6391.9946	6388.6978	6383.9537	0.10	157	26	47	91	138
ibm13	530 / 424	5262.6052	5262.6050	5204.0326	1.03	2695	52	78	244	753
ibm14	1021 / 614	5634.2142	5634.2140	5850.1577	2.88	2622	75	109	237	634
ibm15	1019 / 393	6353.8948	6353.8947	6328.6329	2.94	3770	100	152	331	1039
ibm16	633 / 458	7622.8724	7622.8723	7563.6297	0.95	2038	41	65	193	520
ibm17	682 / 760	6827.7756	6827.7754	6870.9049	1.78	2200	46	67	139	389
ibm18	658 / 285	6101.0694	6101.0692	6050.4116	1.35	3544	57	82	185	454
Average					**1.18**	**1901**	**5.9%**	**9.6%**	**22.3%**	**47.3%**

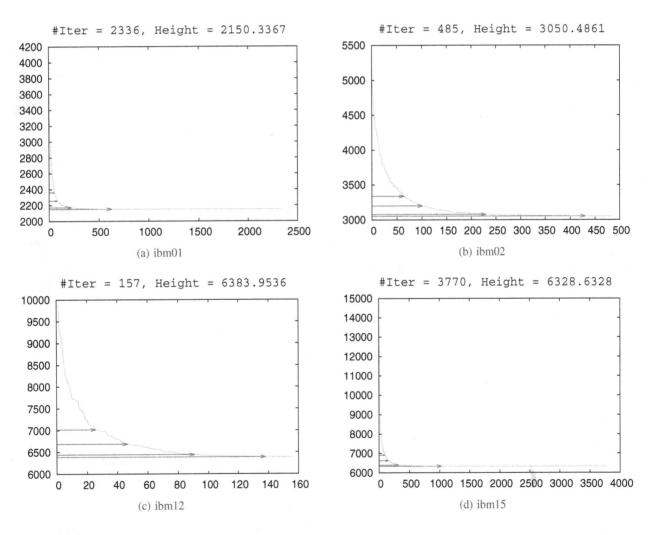

Figure 5: **Layout-height convergence graphs for circuits ibm01, ibm02, ibm12 and ibm15.** (*x-axis denotes the iteration number and y-axis denotes the layout height.*)

Scalable Hierarchical Floorplanning for Fast Physical Prototyping of Systems-on-Chip

Renshen Wang and Nimish Shah
Mentor Graphics Corporation
46871 Bayside Parkway, Fremont, CA 94538, USA
{renshen_wang, nimish_shah}@mentor.com

ABSTRACT

Floorplanning, as an early stage of the physical design flow, has been extensively studied in literature and developed into several branches. Recently, hierarchical floorplanning is regaining attention due to the rising scale of systems-on-chip, which necessarily requires divide-and-conquer strategies to handle the increasing complexity. This paper introduces a floorplanning scheme targeting hierarchical physical prototyping, answering some of the questions posed by Kahng [8] on classical floorplanning. Our scheme emphasizes practical requirements including runtime scalability, wire length and shape quality. We formulate a new hierarchical floorplanning problem with reduced computational complexity, but without weakening the problem as a global layout optimization. To achieve this goal, a placement seed is taken as input and converted into a slicing floorplan under the given constraints of region area and aspect ratio (region shape). We solve the problem by devising an efficient slicing algorithm with integrated dynamic programming. Implementation of the algorithm shows fast runtime and good quality of result.

Categories and Subject Descriptors

J.6 [**Computer Applications**]: Computer-aided design

General Terms

Algorithms, design

Keywords

Hierarchical, recursive slicing, dynamic programming

1. INTRODUCTION

The concept of floorplanning in IC design is generally perceived as area allocation for design blocks. Before placement-and-routing in the design flow, a floorplanning stage enables the blocks to be implemented independently in allocated areas, so that each block becomes a scaled-down sub-design, or even an existing reusable IP module. As feature sizes continue to scale down by Moore's law, complexity of

SoCs (system-on-chip) has risen to a level that a flat design flow is no longer viable or preferable. Thus, hierarchical approaches using divide-and-conquer strategies become necessary, where floorplanning is used for the "divide". In a hierarchical flow (e.g. as in figure 1), not only can we control the scale of each sub-system, we also enable the subsystems to be implemented in parallel to ensure a reasonable turn around cycle. Therefore, hierarchical design flows are increasingly being adopted, and a high quality floorplanner for large scale designs is required.

Floorplanning is not placement, although the two concepts are sometimes regarded as same and mixed together. To make it more confusing, typical design flows usually place hard macros before standard cell placement, where "floorplanning" actually means the placement of hard macros.

To clarify, we define "hierarchical floorplanning" as a step that decides the location and shape of each sub-system (design block). Standard cell placement decides the locations of standard cells (usually in a large number), and placement of hard macros is called "flat floorplanning". This paper will focus on hierarchical floorplanning. The objective is not the final location of certain cells or macros, but to provide a set of dissected chip areas for a set of smaller, mutually independent sub-designs.

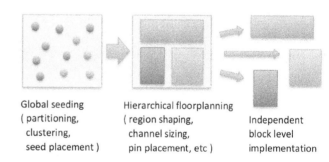

Global seeding   Hierarchical floorplanning
(partitioning,   (region shaping,   Independent
  clustering,   channel sizing,   block level
  seed placement)   pin placement, etc) implementation

Figure 1: Hierarchical flow for large scale designs

For the purpose of hierarchical floorplanning, the objectives at least include: adequate area for each sub-design with a good shape and short wire length (or estimated wire length) for interface nets among the sub-designs. Also, to be usable for designers, the solution should be obtained in a reasonably short amount of time. These requirements ask for a well formulated floorplanning problem and a fast scalable optimization algorithm.

1.1 Previous works

Flat floorplanning and placement are often done together

([1] [4] [11] [14]), and sometimes called mixed-size placement, or floorplacement. These tasks are generally categorized as placement problems, since the solutions require cell legalization, wire length reduction, and congestion alleviation throughout the entire circuit. Recent advancements like [6] [7] show that high quality placement necessarily requires sophisticated analytical solvers (e.g. conjugate gradient) applied to a minimization problem to achieve desired quality of results. Due to its modeling complexity and computation demand, the algorithm runtime of analytical placers will become a bottleneck as design sizes approach 100 million cells.

Hierarchical floorplanning is a convenient way to scale down the design size and parallelize the design process. Hierarchies exist in all modular designs, so we can pick a set of hierarchies $H = \{h_1, \cdots, h_n\}$ and assign a region r_i on the chip for each hierarchy h_i. As long as the regions $r_1, \cdots, r_n$ are non-overlapping, the implementations of these hierarchies are mutually independent, and thus can be processed on different machines, or even by different teams. Note that previous works like [3] also use hierarchy information to improve runtime or quality, but without a hard constraint of non-overlapping regions, the design is still worked on as a whole, making it much harder for effective parallelization.

Classical floorplanning generally addresses the hierarchical requirements. The task is usually formulated as packing a set of regions into a minimum area, with optional wire length objectives and shape constraints. The packing problem itself is NP-hard. Previous research has proposed a series of *floorplan representations* which can encode the packing topology into a string of numbers, including slicing tree [13], sequence-pair [12], o-tree [5], *etc.* Such a representation is usually employed in conjunction with simulated annealing to search for good solutions. Sometimes the region shapes are optimized by a separate routine, e.g. Lagrangian relaxation in [10], but the topology still needs to be obtained by searching among the representation encodings.

These approaches have been questioned in [8], which points out a set of drawbacks of the classical formulation and algorithms, mostly on runtime scalability. Even when we have merely dozens of regions, a good solution requires too much runtime because of the combinatorial explosion of the encodings. Furthermore, is it necessary, under the connectivity driven context, to consider the exponentially large solution space? When the regions have flexible shapes, it is easier to reduce the white space during packing, while the interconnect wire length, which affects performance and power consumption, becomes a major concern. Packing based representations generally have no direct modeling of connectivity, and focus instead on producing legalized packings. The optimization of wire length is therefore indirect and often diluted by other objectives and constraints.

1.2 Paper overview

We propose a hierarchical floorplanning scheme that addresses both runtime scalability and solution quality. First, the problem is formulated as region allocation/shaping on a fixed-die, given a set of regions and a global seed placement. The seed placement is optimized for wire length, timing, *etc.*, and it is generated by an algorithm like a standard cell placement, but largely simplified in the following two ways: (i) the design can be coarsened by cell clustering, and (ii) cells (or cell clusters) don't need to be strictly legalized or uniformly spread. The seed provides a guide on geometrical relations among the regions, which is the critical factor for reducing the complexity of our problem.

Although the packing flavor of the floorplanning problem cannot be eliminated, we are able to devise a fast and scalable slicing algorithm based on the seed placement to quickly find a good solution. And since the seed placement can be at a very coarse level, seed generation is fast as well. Overall, we are solving a hard, time consuming optimization problem by splitting it into two stages of less difficult problems, each solvable by a fast and scalable algorithm.

The remainder of this paper is organized as follows. Section 2 formulates the problem. Section 3 describes our algorithm and methodology to solve the problem, together with some analysis. Section 4 presents the algorithm's efficiency and solution quality by experimental results. Section 5 concludes and discusses future works.

2. PROBLEM FORMULATION

For the hierarchical design flow, we can assume that the available chip area is equal to or just slightly larger than the sum of regions' area. Regions shaped as rectangle with low aspect ratio are preferred for block implementation (illustrated in figure 2). Hence, the solution we need is a tight packing of a set of shape-constrained regions. Unlike the macro legalization in mixed-size placement [11] [14], where we always have standard cells to fill the fragmented area (white space) among macros, a hierarchical floorplan has very small amount or even zero white space to spare.

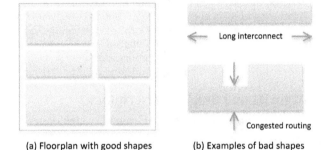

(a) Floorplan with good shapes   (b) Examples of bad shapes

Figure 2: Region shapes in floorplans

To lower the difficulty of the packing optimization, we introduce a seed placement as input, supported by the following observation. In global placement, most hierarchies have their cells closely placed, more often in a single continuous shape, because in modular designs, complex logic connections inside far outnumber interface nets outside hierarchies. Although these seed points cannot be directly used as region shapes, they provide a strong guidance on the topological orders among the regions. And since the placement can be optimized for timing, wire length and routing congestion, we can expect good floorplan quality if we can approximately shape each region around the seeds in its corresponding hierarchy. The problem now becomes one of seeking a floorplan of rectangular regions that matches up best to the seed distributions.

Thus, our hierarchical floorplanning is formulated as:

- Input: A rectangle die area D, an upper bound on region aspect ratio A_u, a set of n regions $r_1, \cdots, r_n$, each region r_i has m_i cells $e_{i,1}, \cdots, e_{i,m_i}$, and each cell $e_{i,j}$ is placed at a point $p(e_{i,j}) \in D$

- Solution: A set of rectangles $R_1, \cdots, R_n \subseteq D$ such that

$area(R_i) \geq \sum_j area(e_{i,j})$,   (adequate area)

$R_i \bigcap R_j = \phi$ for $i \neq j$,   (non-overlapping)

$max(\frac{l_1(R_i)}{l_2(R_i)}, \frac{l_2(R_i)}{l_1(R_i)}) \leq A_u$ (good shape)

- Objective: Maximize $\sum_i \{ \sum_j area(e_{i,j}) : p(e_{i,j}) \in R_i \}$

The idea of using a placement seed for generating a floorplan appears as early as in [13], where the seed is a set of points $\{ p_1, \cdots, p_n \}$ obtained by the Schoenberg embedding of a distance matrix D. However, the shape requirements are not guaranteed in the algorithm of [13], because it relies on a single sweep of recursive slicing without back tracking. Also, each point p_i only provides a hint of a good location for its region, without directly influencing its shape.

In our approach, the seed can be composed of real cells or cell clusters at any level of design coarsening (e.g. by cell clustering in [2]). A very coarse level of placement can still be useful, because the floorplanning solution does not require many details. Different placement algorithms ([1] [2] [4] [7] [9]) can be used to generate an initial seed placement.

3. ALGORITHM AND ANALYSIS

We use the basic slicing tree described in [13] because of its simplicity and convenience for area allocation. It is not the most powerful model as a floorplan representation as some floorplans are "non-slicable", but studies in [15] show that when the regions' aspect ratios are flexible, slicing floorplans can be very close to optimal in terms of space utilization. Industrial experience also suggests that slicing floorplans are effective in practical designs.

In the main flow of our algorithm, a slicing floorplan is constructed by recursively applying a slice-and-partition operation on the chip area and the set of regions. By the formulation in section 2, the slicing operation is straightforward: find a best cut line across the area, which maximizes the total area of cells that fall on the "correct" side of the cut line, or minimizes the area on the "wrong" side. The correct side of a region cell depends on other cells in the region and the cut line's position. Algorithm details are elaborated below.

3.1 Recursive slicing

For efficiency, we capture the placement seed in a grid map of size $m_c \times m_r$. Each grid unit g_{ij} is a small square in the chip area, and stores total cell area contribution $a_{ij}(r_k)$ from each region r_k to g_{ij}. We also pre-compute a cumulative cell area for each region in the map,

$$c_{ij}(r_k) = \sum_{u=1}^{i} \sum_{v=1}^{j} a_{uv}(r_k)$$

so that $c_{ij}(r_k)$ is the total cell area of region r_k in the rectangle from g_{11} to g_{ij}, and the cell area of region r_k in a rectangle $[x_0, x_1] \times [y_0, y_1]$ can be computed by:[1]

$c_{x_1 y_1}(r_k) - c_{x_1 (y_0-1)}(r_k) - c_{(x_0-1) y_1}(r_k) + c_{(x_0-1)(y_0-1)}(r_k)$

With the grid map defined, we use the grid lines as cut line candidates. The granularity of the grid will be sufficient if we pick m_c and m_r around a multiple of n (e.g. $10n$) since our hierarchical floorplanning is for area allocation, not for precise placement.

---

[1] Define $c_{ij}(r_k) = 0$ if $i \leq 0$ or $j \leq 0$.

Consider a vertical cut line cutting the area into left and right partitions. Allocation of each region r_k can be decided by comparing its cell area on the left (a_{left}) and right (a_{right}) sides of the cut line. The region is put on the side with the larger area, and the cost of this cut line is increased by $min(a_{left}, a_{right})$. Figure 3 illustrates the process.

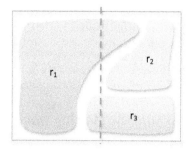

Slicing a floorplan by a vertical cut line, on the left side: r_1 on the right side: r_2, r_3

Cost of this cut line = $area_{right}(r_1) + area_{left}(r_3)$

Figure 3: A placement seed of 3 regions and the best cut line for the first slice-and-partition

If the seed cells are distributed evenly, we can apply the same algorithm recursively on the left and right partitions, until all regions are allocated. However, it is possible that seed distribution is uneven, and one side of the chip area may not be enough to hold the regions which are allocated in that partition. If it happens on the left side, we need to shift some regions on the left to the right side, and vice versa. The shifting heuristic is based on each region's center of gravity. We repeatedly pick the region with its center of gravity closest to the cut line, and move it to the right, until the total cell area on the left side is small enough. Region r_i's center of gravity $c_g(r_k)$ is computed as a vector:

$$\vec{c_g}(r_i) = \frac{\sum_{1 \leq j \leq m_i} area(e_{i,j}) \vec{p}(e_{i,j})}{\sum_{1 \leq j \leq m_i} area(e_{i,j})}$$

We pick each region by its position instead of its area to ensure that the relative positioning (topological order) among regions decided by the seed placement is not disrupted by the slice-and-partition operation. If the area of the picked region is too large to fit in the other side, it means the cut line is not favorable and will be discarded.

Table 1: Slice-and-partition

Create a slicing floorplan for a set of region S_r, within the grid map area $[x_{low}, x_{high}] \times [y_{low}, y_{high}]$.

if ($|S_r| = 1$)
      $D_m = \sqrt{\frac{\sum_{e_{i,j} \in r_i \in S_r} area(e_{i,j})}{A_u}}$ // for aspect ratio
     if ($min(x_{high} - x_{low}, y_{high} - y_{low}) \geq D_m$)
        return **success**; else return **failure**;
 foreach possible cut line l_i,
     slice the grid map area and compute $cost_i$;
 sort all pairs of $(l_i, cost_i)$ by $cost_i$ from low to high;
 for each $(l_i, cost_i)$ // assuming l_i is vertical
     slice-and-partition(left(S_r), $[x_{low}, l_i], [y_{low}, y_{high}]$)
     slice-and-partition(right(S_r), $[l_i, x_{high}], [y_{low}, y_{high}]$)
     if (both return success) return **success**;
 end_for_$(l_i, cost_i)$
 return **failure**;

After we gather all the feasible cut lines, they are sorted by cost (from low to high) and processed as follows: for each cut line, we recursively do the slice-and-partition operation on both sides. The recursion ends when there is only one region, and the algorithm finishes successfully if every region has been allocated a part of the chip area with an acceptable aspect ratio. The recursive algorithm is shown in table 1.

Note that in the recursive calls, region r_k's area on the left side can always be computed as $c_{im_r}(r_k)$. Although we assume in each iteration that the region cells on the "wrong" side of the cut line are shifted to the other side after a slice operation, the actual shifting can be omitted, and the total region area on the left side $c_{im_r}(r_k)$ is always equal to the shifted total area in a sub-rectangle. Therefore, the area distribution maps $a_{ij}(r_k)$ and $c_{ij}(r_k)$ can be pre-computed, and no update is needed.

3.2 Dynamic programming by caching intermediate solutions

The slicing algorithm above is essentially a search on the complete combinations of slicing trees, which in the worst case will need exponential runtime. This is because the packing nature of floorplanning is still present in our problem, and the problem cannot be solved by a series of straightforward decisions without backtracking. For each slicing operation, we can only guarantee enough area on both sides, but cannot predict if the regions with constrained shapes can actually be packed (or legalized) in the shape on each side. The slicing tree construction in [13] uses a deformation function as a heuristic to lower the aspect ratios of regions, but not under hard constraints. We believe the constrained solution can only be guaranteed by search and backtrack.

The runtime, however, can be reduced to a polynomial if we exploit the information in the seed placement. Because the topological order among the regions is largely decided by the seed, the total combinations on region subset S_r can be reduced to $O(n^4)$ during the recursive calls.

We use the center of gravity $(c_g(r_k))$ to represent the precise position of each region r_k. Assume our slicing algorithm preserves the relative positions among the regions. That is to say, for example, if r_i's cells are placed on the left side of r_j's cells, then in the floorplan there will never be a vertical cut line that puts r_j on left and r_i on right.[2] We observe that a set of regions S_r appearing in a slice-and-partition call can always be determined by at most four boundary regions, as illustrated by figure 4.

THEOREM 1. *Under preservation of relative positions, any subset S_r in the slice-and-partition calls can be contained in a rectangle R, such that $c_g(r_k) \in R$ for all $r_k \in S_r$, and $c_g(r_k) \notin R$ for all $r_k \notin S_r$.*

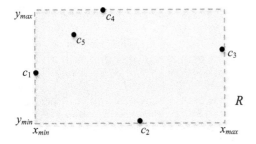

Figure 4: Center of gravity points in S_r

[2]Preservation of relative positions is not strictly guaranteed by the slicing scheme, but exceptions are rare in practice.

PROOF. In a recursive call with region subset S_r in a subpartition P, we first take all the center of gravity points $S_c = \{c_g(r_k) : r_k \in S_r\}$, and find the smallest rectangle covering all the points in S_c, denoted as R. Note that R contains S_c, so any point out of R is not in S_c.

Since R is smallest, each edge must be touching a point in S_c. Without losing generality, assume we have the situation in figure 4: c_1 at the left edge, c_2 at the lower edge, c_3 at the right edge, and c_4 at the upper edge. For a point $c_5 = c_g(r_k)$ in R, since c_5 is on the right side of c_1, it means r_k has never been partitioned to the left side of S_r by a vertical slicing. By the same reasoning, r_k has not been partitioned to the right side, lower side, or upper side of S_r. Therefore, r_k must be within S_r. □

Now we look at the routine of "slice-and-partition" in table 1, where the three input parameters are S_r, $[x_{low}, x_{high}]$ and $[y_{low}, y_{high}]$. The potential of combinatorial explosion is in $S_r \subseteq \{r_1, \cdots, r_n\}$, which has 2^n possible values. Because of theorem 1, it is reduced to at most n^4 in our algorithm.

Using dynamic programming techniques, for each input $(S_r, [x_{low}, x_{high}], [y_{low}, y_{high}])$, we cache the solution (including failure). The cache lookup helps avoid repeated search and backtrack for the same input. Additionally, the cache is built lazily, i.e., only visited solutions are stored. To further reduce complexity, the key for the cache entry can be chosen as $(S_r, x_{high} - x_{low}, y_{high} - y_{low})$, because it suffices to know how the regions in S_r can be packed in a rectangle of size $(x_{high} - x_{low} + 1) \times (y_{high} - y_{low} + 1)$, while the exact position of this rectangle is of less importance.

Table 2: Dynamic programming

Create a slicing floorplan for a set of region S_r, within the grid map area $[x_{low}, x_{high}] \times [y_{low}, y_{high}]$.

if (cache.find(S_r, $x_{high} - x_{low}$, $y_{high} - y_{low}$))
 return (cached solution)
 slice-and-partition(S_r, $[x_{low}, x_{high}]$, $[y_{low}, y_{high}]$)
 // recursive calls are on dynamic programming
 cache.insert(solution from above)

3.3 Runtime analysis

In the algorithm of table 1 and table 2, the computation time is mainly in sweeping the cut lines and in the recursive calls. Assume we have totally n regions, and in a certain iteration we have $m \leq m_c + m_r$ possible cut lines. The run time due to the cut line sweep is $O(mn)$. Note that each cut line corresponds to two recursive calls, so the number of recursive calls in this iteration is $2m$. Hence, the average runtime corresponding to each recursive call is bounded by $\frac{O(mn)}{2m} = O(n)$. Since this bound applies to any iteration, the final average runtime on each recursive call is $O(n)$.

As indicated by the dynamic programming, the total number of calls in the worst case is the total number of combinations on $(S_r, x_{high} - x_{low}, y_{high} - y_{low})$, which is bounded by $O(m_c m_r n^2)$, as shown by the following argument. When we count the combinations by the rectangle $[x_{low}, x_{high}] \times [y_{low}, y_{high}]$, there are $m_c m_r$ different rectangle sizes; for each size, we can decide an S_c (set of center points) by picking a pair of (lower, left) points, a pair of (lower, right) points, a pair of (upper, left) points, or a pair of (upper, right) points, and the number of choices is bounded by $4n^2$.

Thus, the time complexity of our algorithm is $O(m_c m_r n^3)$, and the space complexity is $O(m_c m_r n^2)$. If the grid map's

190

granularity is proportional to the number of regions, then we need $O(n^5)$ time and $O(n^4)$ space.

We see a great reduction in combinations, from super exponential to a polynomial. The reason is that we only do the search in a space where the order of regions is largely preserved from the placement seed, and thus avoid the combinatorial explosion. By preserving the relative positions, we skip a large number of packing configurations, which may contain some instances with better space utilization; but under wire length dominated design objectives, most of these instances are of little value. Only the floorplans with regions distributed closely to a good placement seed (which is optimized for connectivity) are good candidates for the search. Therefore, we believe that our floorplanning scheme enhances efficiency without sacrificing quality of results.

4. EXPERIMENTAL RESULTS

We implemented our floorplan algorithm in C++, and tested it on a set of industrial design cases. We ran the program on servers each with a quad-core 2.6GHz AMD Opteron processor and 128GB of memory. Our implementation runs on a single CPU in a single thread.

The test cases are from real industrial designs, ranging from 500,000 to 10 million cells with hierarchies. The subdesigns (regions) for hierarchical floorplanning are selected from the hierarchies with total area within a certain range.[3] We do not use the mixed-size placement benchmarks (e.g. IBM-MS [1]) in previous works, because: (i) we are not solving the same problem; (ii) logic hierarchies are not included in the benchmarks; and (iii) the design sizes are relatively small and therefore not suitable for hierarchical flows.

A floorplan example (Design$_1$) is shown in figure 5, where the regions are shaped in seconds of runtime. The final shape of each region is generated by a routine that shrinks the allocated sub-partition to a shape with the exact required area $\sum_j area(e_{i,j})$. The geometric heuristics in the routine are basic and therefore omitted.

4.1 Runtime

Our slice-and-partition algorithm is partially a greedy approach (as in table 1), because it starts from the best cut and returns as soon as a feasible solution is found. The backtrack only happens when a subset of regions cannot be packed in their sub-partition due to the constraint on region aspect ratios. To make a stress test on the algorithm, we imposed very tight constraints and measured the runtime.

The following table shows the results on three test cases. The first case is Design$_1$ with 20 regions, and total region

---

[3]Heuristics on generating multi-hierarchy regions is another topic of hierarchical floorplanning.

area at 75% of the die area. The other two cases are from Design$_2$ with 27 regions, with region area ratio at 67% and 90% (cell area bloated). The values of A_u are 1, 1.01, 1.05, 1.1, and then increase by 10% up to 1.95. For each A_u, we show the total run time (T in seconds) and total number of intermediate solutions (#Int.Sol) during the recursive calls.

Table 3: Runtimes and cache sizes

A_u	Design$_1$ (75%)		Design$_2$ (67%)		Design$_2$ (90%)	
	T(s)	#Int.Sol	T(s)	#Int.Sol	T(s)	#Int.Sol
1	125	1905666	4.5	64763	1753	9386249
1.01	99	1444349	4.0	60189	2055	9536067
1.05	3.2	22565	2.5	22288	2501	12985300
1.1	3.2	29244	1.6	8309	282	2554501
1.21	2.6	1313	1.5	326	57.6	756856
1.33	2.6	412	1.5	172	3.6	40680
1.46	2.6	936	1.5	145	1.6	9508
1.61	2.6	549	1.5	123	1.5	3551
1.77	2.6	38	1.5	109	1.5	1523
1.95	2.6	44	1.5	95	1.5	806

These results show that a large number of backtracks scales up significantly when the aspect ratios are bounded under 1.2 towards 1, except when constraints are very tight (e.g. Design$_2$(90%), no solution with $A_u \leq 1.05$ as the solution space shrinks when region shapes become less flexible). On Design$_1$, the program returns "no solution" with $A_u = 1$ with 2 million intermediate solutions. However, with A_u increased to 1.01, a solution is found (figure 6(a)). For cases with A_u close to 2 or above, very few backtracks are needed to find a good solution. This is possibly related to the conclusion in [15], stating that when $A_u \geq 2$ and a small fraction of white space is available, a packing solution is guaranteed.

Practical designs usually have much looser constraints on aspect ratios. So although the algorithm's time complexity $O(n^5)$ is a high order polynomial, the computation in most real cases will be only a small fraction of the upper bound.

We also tried another stress test on Design$_1$ where the placement seed is degenerate, i.e., all the cells in the seed are placed on the center point of the chip area. Such input may exist when the seed is provided on a very coarse level. The result in figure 6(b) shows the robustness of our algorithm: it can exploit the information in the seed placement, but does not depend on its quality or correctness.

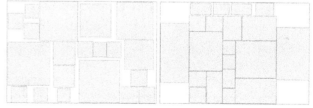

(a) Aspect ratio ≤ 1.01   (b) Degenerate seed

Figure 6: Stress test results on Design$_1$

4.2 Quality of results

Besides the shape quality of each region, the wire length of interface nets among the regions is also an index of quality. We use a placement engine[4] to perform global placement on

---

[4]The analytical placer available in Olympus-SoC, Mentor Graphics' place-and-route platform.

Figure 5: Design$_1$'s seed placement and its floorplan with aspect ratio ≤ 4

the entire design before and after floorplanning, and use the wire length difference to measure the floorplan's impact.

Before floorplanning, the placement is on a flat design, and we tune the placer to minimize the total wire length (WL_{flat}). After floorplaning, since the regions add a constraint on cell locations (each cell must be inside its region), we expect the total wire length to be higher than WL_{flat}. On the other hand, our floorplan is using the flat or clustered placement result as seed input, the regions are likely to be shaped close to their cells, which means the increase in wire length should be small despite the additional constraint.

The results are listed in table 4. We find that the regions' shapes are generally similar between results from coarse and refined seed placement. There are differences in wire lengths, but both positive and negative differences on higher level of clustering are observed, and the percentages based on total wire lengths are usually small. In our opinion, the reason seems to be that our floorplan only takes a coarse indication of positions from the input seed, so the difference in refinement levels do not have a large effect. The impact on final wire length after floorplanning depends on details in the region shapes as well as the placement engine.

5. CONCLUSION AND FUTURE WORK

We have shown a preliminary validation of a hierarchical floorplanning flow for large scale designs. Floorplanning based on a coarse placement can quickly provide us a set of shaped regions. Next, we can place interface pins (by pin placement algorithms) for inter-region connections. Now, each region becomes an independent sub-system design, and the whole chip can be implemented in a "divide-and-conquer" mode. For highly modular system-on-chip designs, we believe this is a promising approach.

Future works will study other steps in the hierarchical flow, e.g., fast and high quality seed placement, region generation based on hierarchies and/or placement, enhancement of region shaping, etc. A lot of new challenges and opportunities lie in the class of hierarchical methodologies.

6. REFERENCES

[1] S. N. Adya and I. L. Markov. Combinatorial techniques for mixed-size placement. *ACM Transactions on Design Automation of Electronic Systems*, 10:58–90, January 2005.

[2] C. Alpert, A. B. Kahng, G.-J. Nam, S. Reda, and P. Villarrubia. A semi-persistent clustering technique for VLSI circuit placement. *Proceedings of the international symposium on Physical design*, pages 200 – 207, 2005.

[3] Y.-L. Chuang, G.-J. Nam, C. Alpert, Y.-W. Chang, J. Roy, and N. Viswanathan. Design-hierarchy aware mixed-size placement for routability optimization. *Proceedings of the international conference on Computer-aided design*, pages 663 – 668, 2010.

[4] H. Eisenmann and F. M. Johannes. Generic global placement and floorplanning. *Proceedings of the Design automation conference*, pages 269 – 274, 1998.

[5] P.-N. Guo, C.-K. Cheng, and T. Yoshimura. An O-tree representation of non-slicing floorplan and its applications. *Proceedings of the Design automation conference*, pages 268 – 273, 1999.

Table 4: Wire length results (unit: Å)

Test case	WL_{flat}	Placement seed	$WL_{floorplan}$
Design$_1$	3.04×10^{11}	flat	3.91×10^{11}
		cluster level 1	3.58×10^{11}
		cluster level 2	3.49×10^{11}
Design$_2$	1.39×10^{11}	flat	1.52×10^{11}
		cluster level 1	1.52×10^{11}
		cluster level 2	1.44×10^{11}
Design$_3$	7.58×10^{11}	flat	7.72×10^{11}
		cluster level 1	8.59×10^{11}
		cluster level 2	8.60×10^{11}
Design$_4$	1.55×10^{12}	flat	1.66×10^{12}
		cluster level 1	1.74×10^{12}
		cluster level 2	1.74×10^{12}
Design$_5$	5.15×10^{11}	flat	5.21×10^{11}
Design$_6$	2.05×10^{11}	flat	2.23×10^{11}
		cluster level 1	2.32×10^{11}
Design$_7$	1.42×10^{11}	flat	1.44×10^{11}
		cluster level 1	1.50×10^{11}
		cluster level 2	1.55×10^{11}
		cluster level 3	1.63×10^{11}
Design$_8$	2.65×10^{12}	cluster level 1	3.03×10^{12}

[6] M.-K. Hsu, Y.-W. Chang, and V. Balabanov. TSV-aware analytical placement for 3-D IC designs. *Proceedings of the Design automation conference*, pages 664 – 669, 2011.

[7] Z.-W. Jiang, B.-Y. Su, and Y.-W. Chang. Routability-driven analytical placement by net overlapping removal for large-scale mixed-size designs. *Proceedings of the 45th annual Design Automation Conference*, pages 167 – 172, 2008.

[8] A. B. Kahng. Classical floorplanning harmful? *Proceedings of the international symposium on Physical design*, pages 207 – 213, 2000.

[9] M.-C. Kim, D.-J. Lee, and I. Markov. SimPL: an effective placement algorithm. *Proceedings of the international conference on Computer-aided design*, pages 649 – 656, 2010.

[10] C. Lin, H. Zhou, and C. Chu. A revisit to floorplan optimization by Lagrangian relaxation. *Proceedings of the international conference on Computer-aided design*, pages 164 – 171, 2006.

[11] M. D. Moffitt, J. A. Roy, I. L. Markov, and M. E. Pollack. Constraint-driven floorplan repair. *ACM Transactions on Design Automation of Electronic Systems*, 13:67:1–67:13, October 2008.

[12] H. Murata, K. Fujiyoshi, S. Nakatake, and Y. Kajitani. Rectangle-packing-based module placement. *Proceedings of the international conference on Computer-aided design*, pages 472 – 479, 1995.

[13] R. H. J. M. Otten. Automatic floorplan design. *Proceedings of the 19th Design automation conference*, pages 261 – 267, 1982.

[14] J. Roy, S. Adya, D. Papa, and I. L. Markov. Min-cut floorplacement. *IEEE Transactions on Computer-Aided Design of Integrated Circuits and Systems*, 25:1313 – 1326, 2006.

[15] E. F. Y. Young and M. D. F. Wong. How good are slicing floorplans? *Proceedings of the international symposium on Physical design*, pages 144 – 149, 1997.

MAPLE: Multilevel Adaptive PLacEment for Mixed-Size Designs

Myung-Chul Kim[†‡], Natarajan Viswanathan[‡], Charles J. Alpert[‡], Igor L. Markov[†], Shyam Ramji[§]

[†]University of Michigan, EECS Department, Ann Arbor, MI 48109

[‡]IBM Corporation, Austin, TX 78758 / [§]IBM Corporation, Hopewell Junction, NY 12533

mckima@umich.edu, {nviswan, alpert}@us.ibm.com, imarkov@umich.edu, ramji@us.ibm.com

ABSTRACT

We propose a new multilevel framework for large-scale placement called MAPLE that respects utilization constraints, handles movable macros and guides the transition between global and detailed placement. In this framework, optimization is adaptive to current placement conditions through a new density metric. As a baseline, we leverage a recently developed flat quadratic optimization that is comparable to prior multilevel frameworks in quality and runtime. A novel component called Progressive Local Refinement (ProLR) helps mitigate disruptions in wirelength that we observed in leading placers. Our placer MAPLE outperforms published empirical results — RQL, SimPL, mPL6, NTUPlace3, FastPlace3, Kraftwerk and APlace3 — across the ISPD 2005 and ISPD 2006 benchmarks, in terms of official metrics of the respective contests.

Categories and Subject Descriptors

B.7.2 [**Hardware, Integrated Circuits**]: Design Aids—*Placement and routing*

General Terms

Algorithms, Design, Performance

1. INTRODUCTION

Large-scale placement remains one of the most influential optimizations in interconnect-driven physical design and physical synthesis [3]. Despite the long history of research, three ISPD contests on placement have shown that recent algorithms achieve sizable gains over prior state of art [22]. The ISPD 2011 routability-driven placement contest [30] has demonstrated that the choice of the wirelength-driven global placement engine is paramount even in multi-objective placement — two of the top three teams relied on the high-quality SimPL framework [18], including the contest winners, who reimplemented SimPL without having access to the original source code [12]. Yet, no placer dominated across the entire benchmark set, indicating possible improvements. Such improvements are described in this paper, although our work is orthogonal to and compatible with the innovations developed for the ISPD 2011 contest [12, 13, 17].

In this work, we develop MAPLE — a multilevel force-directed placement algorithm that pioneers key algorithmic components and a more effective way of combining individual components into a reliable multi-objective optimization. MAPLE generates the coarsest-level placement by a variant of the SimPL algorithm [18] but also employs multilevel extensions reinforced by our new Progressive Local Refinement (ProLR).[1] This combination enhances trade-offs between wirelength and module density. Compared to recent literature, our implementation produces superior solution quality with reasonable runtimes.

The improvement on ISPD 2006 benchmarks is particularly encouraging because it demonstrates that MAPLE not only reduces the wirelength but also avoids highly concentrated placements, thus promoting routability and providing greater flexibility for timing optimization transforms. Note that the original SimPL algorithm was not evaluated with utilization constraints of the ISPD 2006 benchmark suite and could not handle movable macros present in those benchmarks. At a more conceptual level, our work explores limits to optimization imposed by noise inherent in analytic placement algorithms. After studying sources of this noise, we develop techniques to avoid noise or suppress it, which consistently improve end results beyond the best reported in the literature.

Our key contributions include:

- A study of obstacles to extending analytic placement with multilevel techniques. We observe that straightforward extensions cause disruptions between successive optimizations during global placement.

- A key insight to combine unclustering with two-tier Progressive Local Refinement (ProLR) so as to ensure graceful transitions between optimizations at different cluster levels. Optimization adapts to current wirelength/density trade-offs, which we track by a newly developed metric — ABU_γ.

- A placement algorithm (MAPLE) that relies on SimPL iterations, but augments them with two-level clustering and ProLR. MAPLE guides the transition from global to detailed placement to avoid unnecessary disruptions. This guidance allows MAPLE to derive the final placement from the lower- rather than the upper-bound placement as in the original SimPL, enhancing solution quality.

- Extensions of the MAPLE algorithm to handle movable macros. This includes extending the SimPL algorithm and dealing with macros during refinement.

- Empirical evaluation against best published results on ISPD 2005 and ISPD 2006 benchmarks using official metrics. MAPLE consistently outperforms all leading-edge placers described in the literature.

---

[1]The implementation used in this work was written from scratch.

The remainder of this paper is structured as follows. Section 2 presents background and prior art. Section 3 analyzes disruptions during multilevel placement optimization that undermine solution quality. In Sections 4 and 5, we present the MAPLE algorithm and specific techniques to ensure graceful transitions between successive optimizations. Section 6 describes extensions of the MAPLE algorithm to handle movable macros. Section 7 empirically validates our ideas and algorithms. Section 8 concludes our paper.

2. BACKGROUND AND PRIOR ART

Given a netlist $\mathcal{N} = (E, V)$ with nets E and nodes (cells) V, *global placement* seeks node locations (x_i, y_i) such that the area of nodes within any rectangular region does not exceed the area of (cell sites in) that region. Some locations of cells may be given initially and fixed. The interconnect objective optimized by global placement is the Half-Perimeter WireLength (HPWL). For node locations $\vec{x} = \{x_i\}$ and $\vec{y} = \{y_i\}$, $\text{HPWL}_{\mathcal{N}}(\vec{x}, \vec{y}) = \text{HPWL}_{\mathcal{N}}(\vec{x}) + \text{HPWL}_{\mathcal{N}}(\vec{y})$, where

$$HPWL_{\mathcal{N}}(\vec{x}) = \Sigma_{e \in E} [\max_{i \in e} x_i - \min_{i \in e} x_i] \quad (1)$$

A consistent 2% HPWL improvement is considered significant and can affect routability, timing and power. For optimization, HPWL can be approximated by differentiable functions [7, 10, 16].

Quadratic optimization represents the netlist by a weighted graph $\mathcal{G} = (E_{\mathcal{G}}, V)$, using the star, clique or Bound2Bound net model [26]. Here we denote vertices by V and edges by $E_{\mathcal{G}}$. Edge weights $w_{ij} > 0$ for all edges $e_{ij} \in E_{\mathcal{G}}$. The *quadratic objective* $\Phi_{\mathcal{G}}$ is defined as

$$\Phi_{\mathcal{G}}(\vec{x}, \vec{y}) = \Sigma_{i,j} w_{i,j} [(x_i - x_j)^2 + (y_i - y_j)^2] \quad (2)$$

$$\Phi_{\mathcal{G}}(\vec{x}, \vec{y}) = \frac{1}{2} \vec{x}^T Q_x \vec{x} + \vec{c}_x^T \vec{x} + \frac{1}{2} \vec{y}^T Q_y \vec{y} + \vec{c}_y^T \vec{y} + \text{const} \quad (3)$$

The connectivity matrix Q_x captures connections between pairs of movable vertices, while vector $\vec{c}_x$ captures connections between movable and fixed vertices. Since Q_x is positive semi-definite, $\Phi_{\mathcal{G}}(\vec{x})$ is a convex function with a unique minimum, which can be found by solving the system of linear equations $Q_x \vec{x} = -\vec{c}_x$ using preconditioned Conjugate Gradient (CG) as in FastPlace, RQL and SimPL.

FastPlace-Global [28] is a force-directed quadratic placer with two-level Best-choice clustering [2]. It relies on a hybrid (star-clique) net model[2] and employs *cell shifting* to spread the modules during the early stages of placement flow. The *Iterative Local Refinement (ILR)* technique is applied after quadratic optimization to reduce HPWL and spread the modules (see Section 5). **RQL [29]** extends FastPlace-Global by limiting spreading forces (*force-vector modulation*). **FastPlace-DP [24]** is a wirelength-driven detailed placer based on (i) single segment cell clustering, (ii) global cell swapping, (iii) vertical cell swapping, and (iv) local reordering. **SimPL [18]** is a flat, force-directed global placer. It maintains a lower-bound and an upper-bound placement and progressively narrows the displacement between the two. The final solution is derived from the upper-bound placement when the two bounds converge. The upper-bound placement is generated by *lookahead legalization (LAL)*, which is based on top-down geometric partitioning and non-linear scaling. Applying the upper-bound placement as fixed-points, the lower-bound placement is generated by minimizing the quadratic objective using the CG method. Unlike FastPlace-Global and RQL, the SimPL algorithm relies on the Bound2Bound net model [26].

---

[2]The numerical equivalence of the clique model and the star model with a star node was pointed out in [20] and proven in [19].

3. ANALYSIS OF DISRUPTIONS DURING ANALYTIC OPTIMIZATION

State-of-the-art algorithms for placement integrate multiple optimization steps, which sometimes target different objectives. Poor coordination between successive steps may cause radical changes in intermediate placements. These changes become disruptive when they reverse improvement obtained by previous steps, increasing overall runtime and undermining final solution quality. We now investigate the sources of disruptive changes between successive stages of analytic placement.

Unclustering. In multilevel global placement algorithms, placement iterations after unclustering often include changes to the optimization objective as well as the netlist. This may abruptly increase wirelength as illustrated in [15, Figure 4] for APlace. The authors state that "*Clustering helps to spread cells more quickly, but wirelength is impaired during cell expansion. It is clearly seen from the figures that when wirelength weight is decreased and the conjugate gradient optimizer restarts, discrepancy drops sharply and wirelength is often increased at first and then refined during the optimization*". However, in contrast to our observation in Section 5, the authors claim that when both discrepancy (overflow) and wirelength change slowly, they obtained a near stable suboptimal solution, in which additional iterations did not further reduce discrepancy and wirelength without a major change to the parameters.

Transition to the HPWL objective. FastPlace [28] and RQL [29] use ILR iterations to recover HPWL after quadratic optimization and before detailed placement. ILR iterations include bin resizing over wide ranges to allow large moves across the placement region [22, Chapter 8]. Moreover, each bin maintains a bin-specific utilization weight $0 \leq \theta \leq 1$, which changes depending upon the current bin's utilization. As history accumulates on dense bins over iterations, ILR increasingly penalizes such bins and allows abrupt moves to decrease local density (Figure 1). The density metric ABU_{10} is defined in Section 4.2.

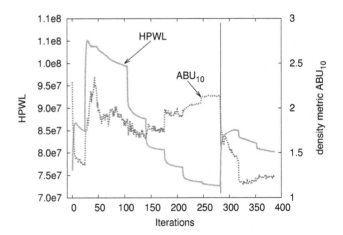

Figure 1: Progressions of wirelength and the density metric ABU_{10} over ILR iterations on ADAPTEC1. Unclustering is marked with a vertical line. ILR disruptively improves ABU_{10} and increases the wirelength. Each ILR iteration traverses all movable modules once.

Hand-off to detailed placement. Recall that the SimPL algorithm maintains two placements throughout its iterations, and legalization is invoked on the *upper-bound* placement, when the lower- and upper-bound placements are reasonably close. The lower-bound placement within SimPL is analogous to module locations maintained by other algorithms. Instead of using the upper-bound, invoking (full) legalization on the lower-bound placement should be potentially better in preserving wirelength optimized by the linear system solver. However, these placements typically exceed target utilization and undergo significant changes during full legalization (Figure 2). Despite local improvement in wirelength during detailed placement, such abrupt changes are detrimental to solution quality in terms of wirelength, routing congestion and timing.

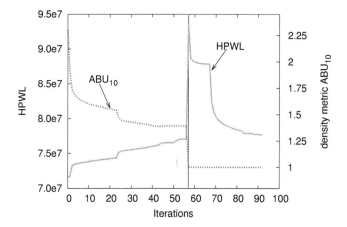

Figure 2: Progressions of wirelength and the density metric ABU_{10} over FastPlace-DP iterations on ADAPTEC1. The start of detailed placement is marked with a vertical line. Placements with high utilization undergo significant changes as full legalization completes.

Strategies for mitigating disruptions. Disruptions during analytic optimization can be mitigated by ensuring gradual transitions between successive optimizations. With this in mind, we develop a new use of placement metrics to make these transitions more adaptive to the actual module distribution and interconnect characteristics. **(1)** the overall placement flow is modified at the points where the objective function abruptly changes, as identified in the above analysis — before/after unclustering, and before detailed placement. We introduce a new intermediate stage that optimizes a linear combination of the preceding and succeeding objective functions, while gradually modifying parameters to ensure smooth transition between the objectives. **(2)** At each substage, we seek near-monotone improvement of either wirelength or module density in a predictable manner without disrupting the other objective. **(3)** Specifically, each intermediate stage prohibits abrupt cell movement and significant changes in key objective functions. Small moves are encouraged instead, as this smoothens changes in wirelength and module density. **(4)** Weighting is adaptively updated according to a new placement metric. These ideas are developed in Progressive Local Refinement (ProLR) in Section 5.

4. MULTILEVEL ADAPTIVE PLACEMENT

We developed our global placement algorithm to address or circumvent the pitfalls in prior art discussed above. This technique consists of three phases: clustering, top-level (coarsest-level) placement iterations, and Progressive Local Refinement (ProLR) used in conjunction with unclustering (Algorithm 1). We apply Best-choice clustering [2] until the number of clusters is reduced to half the size of the flat netlist. Top-level placement iterations perform quadratic optimization on a coarsened netlist and globally regulate module densities over the placement region while moderating wire-

lengh increase. We adopt a variant of the SimPL algorithm [18] for this phase. The ProLR technique discussed in Section 5 improves both wirelength and module density before/after unclustering. Section 7.3 gives an outlook for using more than 2 levels of clustering.

Algorithm 1 Multilevel Adaptive PlacEment (MAPLE)

1: **Phase 0: Clustering of Standard Cells**
2:   N_0 = number_of_modules in flat netlist
3:   **while** number_of_clusters $> N_0$ / 2.0 **do**
4:     cluster netlist using the Best-choice clustering algorithm
5:   **end while**
6:
7: **Phase 1: Top-level Placement Iterations (SimPL extended)**
8:   initial HPWL optimization
9:   **while** ABU_{10} of lower-bound placement $>$ threshold **do**
10:     transform the lower-bound placement into an upper-bound
—   placement by Extended Lookahead Legalization (E-LAL)
11:     fix movable macros upon stabilization (Section 6)
12:     update *pseudopin* locations and *pseudonet* weights
—   in the linear system [18]
13:     solve the updated linear system using
—   the preconditioned CG method
14:   **end while**
15:
16: **Phase 2: Refinement for Mixed-size Netlists**
17:   determine parameters for ProLR
18:   perform ProLR-w and ProLR-d optimizations
19:   legalize and fix all movable macros   **// the end of Phase2a**
20:   **while** number_of_modules $< N_0$ **do**
21:     uncluster the netlist
22:     place unclustered cells side by side
23:   **end while**
24:   recalculate parameters for ProLR
25:   perform ProLR-w and ProLR-d   **// the end of Phase2b**

4.1 Top-level placement iterations

Top-level placement for the coarsest netlist is performed by the SimPL force-directed placement. It generates lower- and upper-bound placements at each iteration and reduces the displacement gap between the two upon convergence. In contrast to the original SimPL algorithm, MAPLE chooses the last lower-bound placement as a final solution of quadratic placement iterations. This choice is based on our observation that our implementation of SimPL in MAPLE does not completely close the gap between lower and upper bounds. Also, given that *lookahead legalization* [18] is unaware of wirelength objectives, the upper-bound placements are likely to suffer suboptimality. On ISPD 2005 benchmarks, MAPLE typically exhibits a gap of 5.63% to 13.89% between lower and upper bounds at its final iterations. However, even with superior wirelength, lower-bound placements typically exhibit worse module density than upper-bound placements. To address this challenge, we improve lower-bound placements using local-search techniques, as described in Section 5.

4.2 A placement density metric - ABU_γ

We now explore density metrics during global placement, which provide insights into the quality of module spreading in intermediate placements and estimate wirelength impact of legality enforcement. Based on such a metric, the global placer can adaptively adjust its parameters depending on how concentrated the placement is, as described in Section 5.[3] To this end, we propose a new den-

---

[3]Little is published on density metrics for global placement. Metrics based on *averaged overflow* (including *scaled-overflow per bin* in the ISPD 2006 contest) often fail to capture uneven module distribution. The *maximum utilization* metric leads to pessimistic estimation in the presence of many fixed modules.

sity metric, ABU_γ — average bin utilization of the top $\gamma\%$ densest bins excluding bins fully occupied by fixed macros. Given that the top $\gamma\%$ densest bin are averaged,[4] this metric reflects the non-uniformity of module distribution (Figures 1 and 2). Compared to overflow-based metrics, ABU_γ provides a more intuitive, cross-design perspective into the quality of module spreading.[5] Monitoring density along with wirelength during placement enables comparisons of different parameter settings and even different placers (Figure 3). Such comparisons speed up algorithm development.

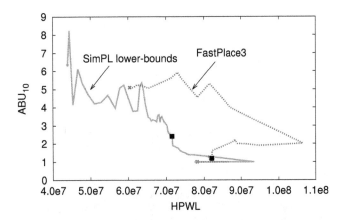

Figure 3: Progression of the density metric ABU_{10} versus wirelength, comparing SimPL lower-bounds (w/ FastPlace-DP) and FastPlace3 on ADAPTEC1. **Steeper slope and datapoints closer to the origin indicate better trade-offs. Each square box indicates the beginning of detailed placement.**

5. A METHODOLOGY FOR GRACEFUL OPTIMIZATION IN PLACEMENT

After quadratic optimization, placements typically exceed the target utilization in many regions, and their HPWL can be improved without increasing max module density. Furthermore, unclustering traditionally counts on subsequent quadratic placement and can be simple-minded in placing modules within clusters. MAPLE improves this situation by using ProLR — a two-tier technique to reduce wirelength and max module density. ProLR adopts single iterations of ILR [28, 29] — **L**ocal **R**efinement (LR) — as a baseline and a vehicle for placement modification. While ILR tends to be disruptive, ProLR promotes gradual transitions via (1) limited bin resizing, (2) *Explicit Bin-Blocking (EBB)*, (3) careful scheduling of utilization weights (θ) between wirelength and module density, and (4) optimizing one objective at a time, while limiting changes to other objectives; such optimizations are alternated.

Bin sizing. ILR and ProLR use regular bin structures and greedily move modules between adjacent bins based on Formula 4. Unlike in ILR, the bins in ProLR are small and remain unchanged during each invocation of LR. Each bin is 5 times the average movable-module area (bins shrink after unclustering). This restricts moves in ProLR.

Explicit Bin-Blocking (EBB) makes local-refinement moves less disruptive. The technique consists of two components: EBB^+ and EBB^-. EBB^+ stops the inflow of modules to some bins (when

[4]In our experiments $\gamma = 10\%$ and the equal-sized square bins in the grid have 6 standard-cell heights on the side.

[5]Empirical validation of the ABU_γ metric is not reported due to page limitations.

such moves are expected to be harmful), while EBB^- stops the outflow of modules from some bins and encourages the inflow of modules into these bins. Therefore, EBB^+ is applied to a handful of bins to limit density, while EBB^- is applied to a larger set of bins to attract modules from remaining bins (the density of these bins may decrease).

Joint optimization of density and wirelength. Local refinement moves individual modules based on the linear combination of improvements in HPWL and density.

$$\text{Score}(m) = \alpha \cdot \Delta_{HPWL} + \beta \cdot \theta \cdot \Delta_{density} \qquad (4)$$

where θ is the *utilization weight*, and α and β are normalizing coefficients [22, Chapter 8]. In FastPlace and RQL, bin-specific θ_b values are managed after they are reset to values $0.4 \leq \theta \leq 0.6$ when ILR iterations start at each level.

Existing move-based algorithms for optimizing (*i*) max density and (*ii*) HPWL use effective techniques for finding highest-gain moves. Yet, no known algorithms are currently known for directly finding the best moves with respect to Formula 4. ProLR inspects best moves for each objective and select those that do not harm the other objective. ProLR performs two simpler optimizations ProLR-w and ProLR-d, which optimize wirelength and module density, respectively. To smoothen placement changes, utilization weight (θ) starts from a small value $\theta_w^0 = 0.1$ for ProLR-w with a coarsened netlist, and θ_{step}^0 is found via a monotonic function

$$\theta_{step}^0 = f(\Upsilon_{target} - \Upsilon_{design}) \qquad (5)$$

When the difference between *design utilization* (Υ_{design}) and *target utilization* (Υ_{target}) is small, placement iterations should aggressively reduce density, which is achieved by using a large θ_{step}^0 (greater emphasis on spreading in LR). On the other hand, a wider gap between the two justifies a greater weight for wirelength, and the best wirelength is often achieved by using a small θ_{step}^0 (greater emphasis on wirelength in LR). Details can be found in the Appendix. The utilization weight for ProLR-w with a flat netlist, θ_w^1 is determined as $\theta_w^1 = \theta_d^{M-1}$ where M is the number of ProLR-d invocations performed for the coarsened netlist. The θ_d^k values in the k-th invocation of ProLR-d are determined by

$$\theta_{step}^k = \theta_{step}^{k-1} \cdot (1 + \frac{ABU_{10}}{100\Upsilon_{target}}) \qquad (6)$$

$$\theta_d^k = \theta_w^{k/M} + \theta_{step}^k \quad \forall k \in \{0, M\} \qquad (7)$$

$$\theta_d^k = \theta_d^{k-1} + \theta_{step}^k \quad \forall k \notin \{0, M\} \qquad (8)$$

ProLR-w improves placement wirelength while maintaining the initial module density distribution. As ProLR-w begins, bin-specific θ_b are reset to θ_w^0 for the clustered netlist and to θ_w^1 for the flat netlist. These values are updated throughout the LR iterations of ProLR-w. Given that ProLR-w maintains θ_b over the entire 300 LR iterations, it closely resembles the use of ILR in FastPlace [28]. However, ProLR-w prohibits abrupt cell movement and significant changes in placement by (1) EBB^+ for bins whose utilization exceeds ABU_{10} and (2) keeping small bin sizes. ProLR-w terminates when ABU_{10} of the current placement exceeds the initial ABU_{10}. Otherwise, ProLR-w continues until there is no improvement in wirelength.

ProLR-d reduces module density of a given placement while keeping wirelength low. The changes in wirelength and density are nearly monotonic. Unlike ProLR-w, ProLR-d consists of up to 15 LR iterations, and bin-specific θ are reset to θ_d^k of each ProLR-d invocation. ProLR-d initially rejects abrupt moves that greatly impact wirelength, and increasing θ_d^k progressively puts a greater emphasis on spreading over multiple invocations. In contrast to ProLR-w,

EBB^- is applied to bins with below-target utilization, attracting modules to sparse bins. We repeat ProLR-d up to 12 times until ABU_{10} stabilizes.

Refinement. When a cluster is broken down, constituent modules are placed side by side. The placement is refined by ProLR.[6] Note in Figures 1 and 2 that during disruptions, wirelength increases sharply and density decreases. Therefore, we schedule ProLR-d before the disruption and ProLR-w after the disruption. Figure 4 shows that this schedule smoothens disruptions in both objectives.

Hand-off to detailed placement. Preprocessing lower-bound placements by ProLR gives better trade-offs between wirelength and density than passing either upper-bound or lower-bound placements to detailed placement algorithms as in original SimPL [18].

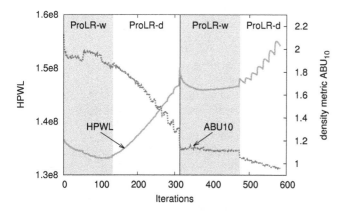

Figure 4: Progressions of wirelength and the density metric ABU_{10} over ProLR iterations (BIGBLUE2). Unclustering is marked with a vertical line. ProLR alternates ProLR-w (shaded) and ProLR-d phases.

6. PLACING MACRO BLOCKS

In placers based on nonconvex optimization, the handling of pre-placed macro blocks requires dedicated techniques (sigmoid functions, level smoothing, etc). In MAPLE, the handling of pre-placed macro blocks is inherited from the SimPL algorithm [18] and LR. To handle movable macros, we extend *lookahead legalization (LAL)* of SimPL, and call the resulting step *E-LAL*. With E-LAL, upper-bound placements are generated in two steps: macro positions are determined first, followed by standard-cell placement [23]. As in original SimPL, roughly legalized placements generated by E-LAL produce fixed pseudopins for subsequent quadratic optimization. Movable macros are legalized by a variant of the *cell shifting* algorithm in FastPlace2 [27]. Our variant uses larger regular bins at 6 times the row height, and employs a 3×3 Laplacian [28] to smoothen bin utilization. A broader view of utilization allows E-LAL to move macros further than FastPlace-Global can and find an almost-legal placement. In the early top-level placement iterations, MAPLE simultaneously places movable macros and standard cells. Upon stabilization (when the gap between the upper- and lower-bounds reduces below 50% from the gap at the 10th iteration), we fix only movable macros with heights $> 2\times$ the row height. Further iterations optimize locations of standard and double-height cells (Figure 5). Recent macro placement literature [8,11] points out that naive force-directed methods do not reliably find overlap-free placements and that a poor macro placement

---

[6]Unclustering is followed by *interpolation* in [6,9] to improve ordering, but ProLR explicitly optimizes HPWL and module density.

may cause large overlaps and substantial disruption when removing those overlaps. To address this problem, unlike other force-directed placers, MAPLE fixes macro positions from the upper-bound placement, which tend to have little overlap among macros (Figure 5). Local refinement (LR) moves double-height and standard cells. For double-height cells, bin-specific θ_b and the utilization weights are averaged over all relevant bins. Following the contest protocol, flipping and rotation of macro blocks were disallowed in this work. While macro placement [8, 11, 23] is not a *primary* focus of this work, our techniques produce competitive results on ISPD 2006 benchmarks. Ongoing work indicates that our algorithms for mixed-size placement can be improved further.

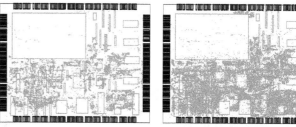
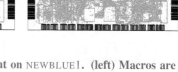

Figure 5: Macro placement on NEWBLUE1. (left) Macros are fixed at top-level placement iteration 30. (right) Further iterations optimize cell locations.

7. EMPIRICAL VALIDATION

The MAPLE algorithm is implemented in C/C++ within an industry infrastructure for placement optimization, including a variant of FastPlace-DP [24] for final legalization and detailed placement. We compared MAPLE to other state-of-the-art academic and industry placers on the ISPD 2005 and ISPD 2006 placement contest benchmark suites. For placers available to us, benchmark runs were performed on an Intel Core i7 860 Linux workstation running at 2.8GHz with 8GB RAM, using only one CPU core. For other placers (marked with asterisks), results were quoted from respective publications. To ensure the reproducibility of our empirical results, Formula 9 reports specific constants used in our experiments. All benchmarks were placed with identical parameter settings. HPWL of solutions produced by each placer was computed by the GSRC Bookshelf Evaluator [1].

7.1 ProLR versus ILR

Figure 6 illustrates the use of ProLR and ILR in MAPLE through snapshots of placements at different phases of Algorithm 1, starting with identical placements at Phase1. The use of ILR in Phase2a relocates many cells over great distances across fixed macros, as seen in the upper left regions of ILR plots on the left. These moves decrease maximal density, but change the placement abruptly and increase HPWL. After Phase2b, the difference in HPWL between ILR and ProLR decreases, but ILR results remain inferior. One can also see that ILR placements on the left are more clustered than the ProLR placements on the right *and* deviate more from the top-level placements. Table 1 compares MAPLE with ProLR to MAPLE with ILR on ISPD 2005 benchmarks in terms of final HPWL. The results confirm the superiority of ProLR. On the two largest benchmarks — BIGBLUE3 and BIGBLUE4, ProLR was on average, $1.5\times$ slower than ILR.

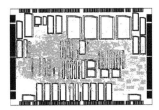

Phase1 (BestChoice+SimPL),HPWL=6.81e7

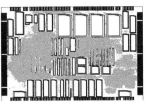

Phase2a (ILR),HPWL=7.99e7

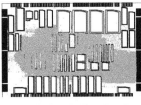

Phase2a (ProLR),HPWL=7.33e7

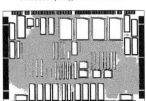

Phase2b (ILR),HPWL=8.25e7

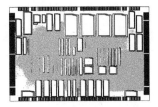

Phase2b (ProLR),HPWL=7.94e7

Figure 6: Snapshots of global placement (ADAPTEC1) after each phase of Algorithm 1 for MAPLE with ILR (left) and MAPLE with ProLR (right). Phase1 is top-level placement (BestChoice+SimPL). Phase2a and Phase2b perform LR placement of the coarsened and flat netlist, respectively.

7.2 Comparisons on ISPD 2005 testcases

As shown in Table 3, MAPLE found placements with the lowest HPWL for seven out of eight circuits in the ISPD 2005 benchmarks (no parameter tuning to specific benchmarks was employed). On average, MAPLE improves wirelength by 9.50%, 6.24%, 6.53%, 7.10%, 8.06%, 4.72%, 2.73% and 2.09% versus APlace2 [16], NTU-Place3 (V7.05.30) [10], FastPlace3 [28], Kraftwerk2 [26], mFAR [14], mPL6 [7], SimPL [18] and RQL [29], respectively.

Table 2 compares the runtime of MAPLE with mPL6, APlace2, NTUPlace3, FastPlace3 and SimPL. On average, MAPLE is $1.13\times$, $2.68\times$ faster than mPL6, APlace2, and $2.32\times$, $6.25\times$, and $7.14\times$ slower than NTUPlace3, FastPlace3 and SimPL, resp. On BIG-BLUE4, **top-level placement iterations** consume 26.3% of total runtime: 64.1% is in CG, and 18.3% in building sparse matrices for CG. **ProLR iterations** consume 65.4% split almost evenly between ProLR-w and ProLR-d. **Best-choice clustering and unclustering** consume 0.2% of the runtime. **Detailed placement** takes 5.5%.

Ckts	MAPLE W/ ILR	MAPLE W/ PROLR	IMPROV.
AD1	77.41	76.36	1.37%
AD2	89.07	86.95	2.38%
AD3	210.13	209.78	0.17%
AD4	190.07	179.91	5.35%
BB1	95.25	93.74	1.59%
BB2	149.84	144.55	3.53%
BB3	345.20	323.05	6.42%
BB4	792.20	775.71	2.08%
Avg	$1.03\times$	$1.00\times$	2.86%

Table 1: HPWL ($\times$10e6) produced by ProLR and ILR on ISPD 2005 benchmarks "ADAPTEC (AD)" and "BIGBLUE (BB)".

7.3 Runtime considerations

As MAPLE is currently slower than some of its competitors, we note that industry implementations like ours tend to be handicapped (versus standalone academic implementations) by the use of a multipurpose design database. Because such a database stores information unnecessary to placement, the decreased cache locality increases runtime. Other relevant legacy infrastructures in our database include netlist-query support for accurate timing analysis and physical synthesis. In contrast to academic placers, our industry-strength implementation can work with a netlist that is dynamically changed during physical synthesis.

Unlike the original SimPL, our implementation does not use SSE instructions and is almost twice as slow (so far, we focused on solution quality and not runtime). Also, ProLR should parallelize well on multicore CPUs. Another consideration deals with the role of placement in physical synthesis, where it is invoked several times [3]. Fast execution is particularly important for early runs that estimate interconnect before netlist optimization. The top-level placement step from MAPLE produces good estimates because the final placement result does not look very different (Figure 6). Top-level placement consumes only $25 - 30\%$ of MAPLE runtime and can be accelerated as outlined above. As timing analysis and optimizations dominate the runtime of physical synthesis, greater effort in placement can be justified by improved results.

Runtime can sometimes be reduced by *deeper clustering* (more levels). To estimate its potential impact in MAPLE, we note that top-level placement takes 26% and ProLR takes 65% of MAPLE runtime on BIGBLUE4 (195.52 min. / 91% total). ProLR runtime is split 1:2 between the coarse and flat netlists. For *three levels* of clustering, top-level placement will take 13%, and ProLR will take $11\% + 22\% + 43\% = 76\%$ runtime. The total (191.23 min. / 89%) is only a 2% reduction versus *two levels*.

7.4 Comparisons on ISPD 2006 testcases

We compared MAPLE to other state-of-the-art academic and industry placers on the ISPD 2006 benchmark suite. Table 4 reports scaled HPWL and overflow penalty for several placers. Following the contest protocol, scaled HPWL is calculated as $HPWL \cdot (1 + 0.01 \cdot overflow\_penalty)$. On average, MAPLE achieved 11.28%, 5.59%, 13.58%, 6.63%, 11.57%, 4.37%, 3.13% scaled HPWL improvements versus APlace3 [22], NTUPlace3 (V7.05.30) [10], FastPlace3 [28], Kraftwerk2 [26], mFAR [14], mPL6 [7], and RQL [29], respectively. MAPLE obtains the best scaled HPWL results on seven out of eight circuits. Furthermore, compared to the other two best-performing placers on the benchmarks — RQL and NTUPlace3, MAPLE achieves lower overflow penalty on average. Thus, MAPLE not only reduces the wirelength but also avoids highly concentrated placements. Recall that the original implementation of SimPL [18] does not support density constraints of ISPD 2006 benchmarks and does not perform mixed-size placement.

Ckts	AP2	NTU3	MPL6	FP3	SIMPL	MP
AD1	46.29	7.92	21.45	2.36	2.48	17.48
AD2	65.49	7.28	21.87	3.58	3.46	24.30
AD3	144.27	14.98	67.14	7.56	6.43	47.34
AD4	158.30	15.47	57.70	6.69	5.44	44.32
BB1	56.68	12.67	24.56	3.67	3.53	24.31
BB2	110.96	25.18	65.44	6.51	6.36	43.96
BB3	233.70	49.70	88.87	19.85	13.25	94.36
BB4	516.37	109.82	199.74	32.27	29.50	214.86
Avg	$2.68\times$	$0.43\times$	$1.13\times$	$0.16\times$	$0.14\times$	$1.00\times$

Table 2: Runtime comparison (minutes) on ISPD 2005 benchmarks for APlace2 (AP2), NTUPlace3 (NTU3), mPL6, FastPlace3 (FP3), SimPL and MAPLE (MP).

Benchmarks	APLACE2 [16]	NTUPLACE3 [10]	FASTPLACE3 [28]	KRAFTWERK2* [26]	MFAR* [22]	MPL6 [7]	SIMPL [18]	RQL* [29]	MAPLE
ADAPTEC1	78.35	81.82	78.66	82.43	82.50	77.93	78.58	77.82	**76.36**
ADAPTEC2	95.70	88.79	94.06	92.85	92.79	92.04	91.24	88.51	**86.95**
ADAPTEC3	218.52	214.83	214.13	227.22	217.56	214.16	**208.90**	210.96	209.78
ADAPTEC4	209.28	195.93	197.50	199.43	197.90	193.89	185.39	188.86	**179.91**
BIGBLUE1	100.02	98.41	96.67	97.67	98.80	96.80	97.54	94.98	**93.74**
BIGBLUE2	153.75	151.55	155.74	154.74	160.40	152.34	145.28	150.03	**144.55**
BIGBLUE3	411.59	360.66	365.16	343.32	368.70	344.10	340.24	323.09	**323.05**
BIGBLUE4	871.29	866.43	836.20	852.40	865.40	829.44	801.35	797.66	**775.71**
Geomean	**1.10×**	**1.07×**	**1.07×**	**1.08×**	**1.09×**	**1.05×**	**1.03×**	**1.02×**	**1.00×**

Table 3: Legal HPWL (×10e6) comparison on the ISPD 2005 benchmark suite. The previous best wirelengths are marked with gray. The placers marked by asterisks were unavailable to us in binary, and we reproduce HPWL from respective publications.

Benchmarks (Υ_{target})	APLACE3* [22]	NTUPLACE3 [10]	FASTPLACE3 [28]	KRAFTWERK2* [26]	MFAR* [22]	MPL6 [7]	RQL* [29]	MAPLE
ADAPTEC5 (0.5)	520.97 (15.9)	430.73 (12.2)	541.22 (36.5)	449.84 (3.69)	476.28 (6.21)	431.27 (1.09)	443.28 (9.25)	**407.33** **(4.76)**
NEWBLUE1 (0.8)	73.31 (0.14)	**62.39** **(0.76)**	76.56 (1.02)	65.95 (0.05)	77.54 (0.23)	68.08 (0.14)	64.43 (0.34)	69.25 (1.05)
NEWBLUE2 (0.9)	198.24 (0.42)	211.77 (3.21)	240.56 (1.97)	206.53 (1.28)	212.90 (0.59)	201.85 (1.52)	199.60 (1.45)	**191.66** **(1.01)**
NEWBLUE3 (0.8)	273.64 (0.00)	280.19 (0.01)	301.72 (0.78)	279.58 (0.38)	303.91 (0.11)	284.11 (0.59)	269.33 (0.07)	**268.07** **(0.77)**
NEWBLUE4 (0.5)	384.12 (1.74)	302.25 (9.22)	306.07 (7.74)	309.44 (1.71)	324.40 (5.42)	300.58 (1.63)	308.75 (15.2)	**282.49** **(5.86)**
NEWBLUE5 (0.5)	613.86 (12.5)	547.20 (20.82)	633.72 (28.31)	563.15 (2.69)	601.27 (5.92)	537.14 (1.42)	537.49 (13.6)	**515.04** **(4.05)**
NEWBLUE6 (0.8)	522.73 (0.03)	518.25 (6.08)	531.56 (1.26)	537.59 (1.70)	535.96 (1.63)	522.54 (1.40)	515.69 (4.33)	**494.82** **(1.08)**
NEWBLUE7 (0.8)	1098.9 (0.06)	1114.2 (5.19)	1116.7 (1.33)	1162.1 (3.15)	1153.8 (1.58)	1084.4 (1.14)	1057.8 (2.57)	**1032.6** **(1.70)**
Geomean	**1.13×** **(0.32)**	**1.04×** **(2.55)**	**1.16×** **(3.47)**	**1.07×** **(1.09)**	**1.13×** **(1.29)**	**1.06×** **(1.22)**	**1.03×** **(2.30)**	**1.00×** **(1.90)**

Table 4: Comparison of scaled HPWL (×10e6) which includes overflow penalty w.r.t the given target utilization on the ISPD 2006 benchmark suite. Overflow penalty values computed by the contest script are reported in parentheses. The placers marked by asterisks were unavailable to us in binary, and we reproduce results from respective publications. This hinders runtime comparisons.

8. CONCLUSIONS AND FUTURE WORK

The significance of large-scale placement in IC physical design is well-documented in recent literature [3] and is continuing to grow with the amount of on-chip random logic and current trends in interconnect scaling. Placement algorithms in the industry and academia were initially developed with the HPWL objective in mind [22] and later extended [3] to account for other objectives and concerns [12, 13, 17]. Despite known pitfalls, the HPWL objective appears to be a good performance predictor for various extensions of core placement algorithms. Focusing on the HPWL objective and module density, our research (*i*) contributes the discovery of essential deficiencies in prior techniques and (*ii*) advances the state of the art by developing algorithms that improve the quality of benchmark layouts beyond all published results. A full list of our contributions can be found in Section 1. For results on the ISPD 2011 routability-driven placement contest benchmark suite, see our related publication [17].

8.1 Perspectives

Our results bear some relevance to three recurring themes in physical design and physical synthesis. **One** is *the comparisons and trade-offs between linear and quadratic wirelength functions.* Since the 1960s, it was known that quadratic optimization was computationally efficient, but did not adequately track the demand for routing resources, which is much closer to the HPWL objective and its weighted variants [4]. Seminal work by Sigl, Doll

and Johannes in the early 1990s developed a *linearization* technique that represents the linear wirelength objective on graphs by a dynamically-weighted quadratic objective [25]. However, the modeling of multi-pin nets remained inaccurate, and the research community has largely replaced quadratic optimization by much more cumbersome and slow non-convex optimization techniques ten years later [7, 10, 16]. In the mid-2000s, Spindler and Johannes developed the Bound2Bound model [26], which considerably improved the modeling accuracy for multi-pin nets in quadratic placement by employing a dynamic (placement-dependent) graph topology. With additional improvements to flat quadratic placement, this technique has recently outperformed prior art in both runtime and quality of results, both in terms of HPWL and in routability-driven placement [12, 17, 18]. This development raised several key research questions:

- Is there a tangible gap between the Bound2Bound model and the HPWL objective in practice ?

- Can global quadratic optimization with the Bound2Bound model be effectively improved on multi-million gate netlists (with respect to HPWL) ?

- Is multilevel placement optimization compatible with Bound2Bound and competitive in performance ?

Our work answers these three questions in the affirmative. The gap between Bound2Bound and HPWL is illustrated by the SimPL line in Figure 3 — note the *return* to smaller HPWL when detailed placement is invoked. Global quadratic placement of multi-million gate netlists can be improved by using the ProLR technique proposed in Section 5. MAPLE demonstrates that multilevel placement is compatible with the Bound2Bound model and is competitive with state of the art, as long as abrupt changes to placement are avoided before/after clustering. However, Section 7.3 shows that only two levels of clustering are useful for current benchmarks. Larger netlists may justify deeper clustering.

The second theme addressed in our work is relatively new to physical design, but no less fundamental — *methodology for module spreading and handling of whitespace*. These considerations are essential not only to global placement, but also to buffer insertion, gate sizing and other physical synthesis transformations, as well as to congestion-driven placement. Until the late 1990s, whitespace was rare in IC layouts, but now can reach over 60% by area [22]. We develop efficient techniques for spreading modules during placement, while satisfying density constraints and optimizing HPWL beyond the accuracy of the Bound2Bound model.

The third fundamental theme explored in our work has not received as much recognition, but may deserve it — we study *the composition of multiple optimizations into a high-precision, reliable multi-objective optimization process*. Our key discovery is that transitions between multiple objective functions and optimization techniques in placement often lead to major disruptions. In particular, adding netlist clustering or ILR to the SimPL algorithm for quadratic placement with the Bound2Bound model does not directly improve quality of results because the disruptions overshadow the benefits of such integration. To this end, we developed new techniques, such as two-tier Progressive Local Refinement (ProLR), to facilitate graceful transitions between multiple optimizations. In placement, these techniques are applied before and after unclustering, during the transition from a quadratic objective to HPWL, and before detailed placement. Many more applications exist in physical synthesis.

8.2 Further directions for future work

Empirical results in Tables 3 and 4 indicate a trend — quadratic placers RQL, SimPL and MAPLE produce overall better solutions than placers APlace3, NTUPlace3 and mPL6 based on non-convex optimization, which also tend to be slower. This is due, in part, to the greater amount of recent research on quadratic placement, including the development of successful industry tools [5,29]. Yet, many of our contributions, such as ProLR, can be adapted for use in non-convex placers. Whether this will make non-convex placers competitive again, remains a curious direction for future work.

The SimPL placer used by MAPLE was recently extended to routability-driven placement [17] and power-driven placement with integrated clock-network synthesis [21]. Precision-handling of net weights demonstrated in [21] enables timing optimization. Opportunities remain for improving mixed-size placement in MAPLE.

Appendix - Computation of Initial θ_{step}

To implement Formula 5, MAPLE uses a step function that distinguishes three different cases: (i) emphasis on wirelength optimization, (ii) no bias, and (iii) emphasis on spreading. Given that Υ_{design} is fixed, the step function only depends on Υ_{target}, which is typically chosen by the designer. Assuming fixed-outline placement ($\Upsilon_{target} \geq \Upsilon_{design}$),

$$\theta_{step}^0 = \begin{cases} 0.0250, & \text{if } \Upsilon_{target} - \Upsilon_{design} \geq 0.5 \\ 0.0275, & \text{if } \Upsilon_{target} - \Upsilon_{design} \geq 0.05 \\ 0.0375, & \text{if } \Upsilon_{target} - \Upsilon_{design} < 0.05 \end{cases} \quad (9)$$

9. REFERENCES

[1] S. N. Adya, I. L. Markov, "Executable Placement Utilities," http://vlsicad.eecs.umich.edu/BK/PlaceUtils/

[2] C. J. Alpert et al., "A Semi-persistent Clustering Technique for VLSI Circuit Placement," *ISPD* 2005, pp. 200-207.

[3] C. J. Alpert et al., "Techniques for Fast Physical Synthesis," *Proc. IEEE* 95(3), 2007, pp. 573-599.

[4] A. E. Caldwell et al., "On Wirelength Estimations for Row-based Placement," *TCAD* 18(9), 1999, pp. 1265-1278.

[5] U. Brenner, M. Struzyna, J. Vygen, "BonnPlace: Placement of Leading-Edge Chips by Advanced Combinatorial Algorithms," *IEEE TCAD* 27(9) 2008, pp.1607-20.

[6] T. F. Chan, J. Cong, K. Sze, "Multilevel Generalized Force-directed Method for Circuit Placement," *ISPD* 2005, pp. 185-192.

[7] T. F. Chan et al., "mPL6: Enhanced Multilevel Mixed-Size Placement," *ISPD* 2006, pp. 212-214.

[8] H.-C. Chen et al., "Constraint Graph-based Macro Placement for Modern Mixed-size Circuit Designs," *ICCAD* 2008, pp. 218-223.

[9] H. Chen et al., "An Algebraic Multigrid Solver for Analytical Placement with Layout Based Clustering," *DAC* 2003, pp. 794-799.

[10] T.-C. Chen et al.,"NTUPlace3: An Analytical Placer for Large-Scale Mixed-Size Designs With Preplaced Blocks and Density Constraints," *IEEE TCAD* 27(7) 2008, pp.1228-1240.

[11] T.-C. Chen et al.,"MP-trees: A Packing-based Macro Placement Algorithm for Mixed-size Designs," *TCAD* 27(9) 2008, pp. 657-662.

[12] X. He et al., "Ripple: An Effective Routability-Driven Placer by Iterative Cell Movement," *ICCAD* 2011, pp. 74-79.

[13] M.-K. Hsu et al., "Routability-Driven Analytical Placement for Mixed-Size Circuit Designs," *ICCAD* 2011, pp. 80-84.

[14] B. Hu, M. Marek-Sadowska, "mFAR: Fixed-Points-Addition-based VLSI Placement Algorithm," *ISPD* 2005, pp. 239-241.

[15] A. B. Kahng, Q. Wang, "Implementation and Extensibility of an Analytic Placer," *IEEE TCAD* 2005, pp. 734-747.

[16] A. B. Kahng, Q. Wang, "A Faster Implementation of APlace," *ISPD* 2006, pp. 218-220.

[17] M.-C. Kim, J. Hu, I. L. Markov, "A SimPLR Method for Routability-driven Placement," *ICCAD* 2011, pp. 67-73.

[18] M.-C. Kim, D.-J. Lee, I. L. Markov, "SimPL: An Effective Placement Algorithm," *IEEE TCAD* 31(1), 2012, pp. 50-60.

[19] A. A. Kennings, I. L. Markov, "Smoothening Max-terms and Analytical Minimization of Half Perimeter Wirelength," *VLSI Design* 14(3), 2002, pp. 229-237.

[20] J. J. Kleinhans et al., "GORDIAN: VLSI Placement by Quadratic Programming and Slicing Optimization," *IEEE TCAD* 10(3), 1991, pp. 356-365.

[21] D.-J. Lee, I. L. Markov, "Obstacle-aware Clock-tree Shaping during Placement" to appear in *IEEE TCAD* 31(2), 2012.

[22] G.-J. Nam, J. Cong, "Modern Circuit Placement: Best Practices and Results," *Springer*, 2007.

[23] A. N. Ng et al., "Solving Hard Instances of Floorplacement," *ISPD* 2006, pp. 170-177.

[24] M. Pan, N. Viswanathan, C. Chu, "An Efficient & Effective Detailed Placement Algorithm," *ICCAD* 2005, pp. 48-55.

[25] G. Sigl, K. Doll, F. M. Johannes,"Analytical Placement: A Linear or a Quadratic Objective Function?" *DAC* 1991, pp.427-432.

[26] P. Spindler, U. Schlichtmann, F. M. Johannes, "Kraftwerk2 - A Fast Force-Directed Quadratic Placement Approach Using an Accurate Net Model," *IEEE TCAD* 27(8) 2008, pp. 1398-1411.

[27] N. Viswanathan, M. Pan, C. Chu, "FastPlace2.0: An Efficient Analytical Placer for Fixed-mode Designs," *ASPDAC* 2006, pp. 195-200.

[28] N. Viswanathan, M. Pan, C. Chu, "FastPlace3.0: A Fast Multilevel Quadratic Placement Algorithm with Placement Congestion Control," *ASPDAC* 2007, pp. 135-140.

[29] N. Viswanathan et al., "RQL: Global Placement via Relaxed Quadratic Spreading and Linearization," *DAC* 2007, pp. 453-458.

[30] N. Viswanathan et al., "Routability-Driven Placement Contest and Benchmark Suite," *ISPD* 2011, pp. 141-146.

A Size Scaling Approach for Mixed-size Placement

Kalliopi Tsota, Cheng-Kok Koh and Venkataramanan Balakrishnan
School of Electrical and Computer Engineering
Purdue University, West Lafayette, IN 47907
{ktsota,chengkok,ragu}@ecn.purdue.edu

ABSTRACT

We propose a global placement algorithm that employs size scaling of circuit components to provide continuity during placement. In the context of mixed-size placement, size scaling is utilized to handle significant variations among the sizes of the components, thereby avoiding additional complexity that is often associated with multiple levels of smoothing. By using the optimal region approach to first determine an initial placement, the size scaling approach allows the global placement algorithm to converge to better placement solutions.

Categories and Subject Descriptors

J.6 [**Computer-aided Engineering**]: Computer-aided design (CAD)

General Terms

Algorithms, Design

1. INTRODUCTION

As the sizes of transistors become smaller, the number of transistors integrated on a chip increases. Currently, the number of transistors in modern ICs has reached the billions and is increasing rapidly. Moreover, advances in the VLSI technology have brought to the forefront a series of issues related to power dissipation, noise, and interconnect delay.

The complexity of physical design problems imposes the need for elaborate optimization algorithms. Such optimization algorithms are utilized during all phases of the physical design flow. They have to consider stringent design constraints and to handle large problem sizes. For instance, the placement tools have to generate solutions that satisfy the cell density constraint. Moreover, satisfying the cell density constraint may conflict with the objective of minimizing interconnect length. The objective of minimizing wirelength may also adversely affect routability. CAD tools that are capable of handling these conflicting objective functions and constraints are very important to the performance of VLSI systems.

The majority of placement algorithms utilize clustering and execute placement in a multi-level scheme. This approach has been adopted by NTUplace3 [4], APlace2.0 [7], mPL6 [2], and FastPlace3.0 [18]. In NTUplace3 [4], the authors proposed an analytical multi-level placement algorithm that uses multiple levels of smoothing to handle cells and macros. APlace2.0 [7] used a multi-level analytical placement framework. A force-directed approach was used in mPL6 [2], while FastPlace3.0 [18] made use of iterative local refinement in a multi-level framework. In [17], the authors proposed a two-level congestion-aware placement framework. In [15], the authors proposed an incremental technique for congestion reduction by iterated spreading during placement. In the more recent algorithm simPL [9], the authors made use of a rough legalization scheme to determine the spreading forces during placement.

This paper focuses on preserving continuity in placement by proposing size scaling for large-scale mixed-size designs. Our placer performs size scaling of cells and macros to discourage abrupt changes during placement, thus enabling a continuous placement flow. The remainder of the paper is structured as follows: Section 2 discusses the placement algorithm. Section 3 reports our experimental results. Section 4 concludes the paper.

2. GLOBAL PLACER

The input to the global placer is a circuit netlist that represents a hypergraph with vertices corresponding to the circuit components and nets corresponding to the connections among the components. The algorithm has the objective of minimizing the wirelength on the chip, while simultaneously eliminating cell overlapping. To determine a placement solution, we take into account the sizes of the components and the locations of fixed macros on the chip. However, only approximate locations for movable macros and cells are determined at this stage. The final locations are assigned during the legalization and detailed placement stage.

In order to spread the components on the placement area, and to evenly distribute the cell density throughout the entire chip, the placement area is divided into uniform bins. Let $(\mathbf{x}, \mathbf{y})$ denote the vector of cell and movable macro coordinates and $WL(\mathbf{x}, \mathbf{y})$ denote the wirelength of the placement. Also, let $SD_b(\mathbf{x}, \mathbf{y})$ denote the cell potential function that is the total area of movable macros and cells inside bin b. Finally, let $SD_{b,fixed}(\mathbf{x}, \mathbf{y})$ denote the total area of fixed macros inside b, w_b the width of b, h_b the height of b, and

T_d the pre-assigned target density value for each bin. Using the approach of [4], we formulate the constrained wirelength-driven placement problem as

$$\min WL(\mathbf{x}, \mathbf{y})$$
$$\text{s.t.} \qquad (1)$$
$$SD_b(\mathbf{x}, \mathbf{y}) \leq T_d(w_b h_b - SD_{b,fixed}(\mathbf{x}, \mathbf{y})).$$

The fixed macros on the chip contribute to the cell potential of their overlapping bins. Thus, the area of the bin that is not occupied by pre-placed macros is set to be $T_d(w_b h_b - SD_{b,fixed}(\mathbf{x}, \mathbf{y}))$. The solution approaches wirelength-driven placement as a constrained optimization problem and transforms this formulation to an unconstrained problem. The cell potential is incorporated into the wirelength objective as a weighted penalty and $\frac{1}{2u}$ is the weight of the penalty function. Define

$$A(\mathbf{x}, \mathbf{y}) =$$
$$\max(SD_b(\mathbf{x}, \mathbf{y}) - T_d(w_b h_b - SD_{b,fixed}(\mathbf{x}, \mathbf{y})), 0)^2. \qquad (2)$$

Using (2), the unconstrained optimization problem for (1) is expressed as

$$\min \ WL(\mathbf{x}, \mathbf{y}) + \frac{1}{2u} \sum_b A(\mathbf{x}, \mathbf{y}). \qquad (3)$$

When the cell potential $SD_b(\mathbf{x}, \mathbf{y})$ exceeds the area of the bin that is not occupied by pre-placed macros $T_d(w_b h_b - SD_{b,fixed}(\mathbf{x}, \mathbf{y}))$, we add to the objective function the difference of the two terms squared, and multiplied by the factor $\frac{1}{2u}$. We solve the placement problem as a sequence of at most $maxnu = 50$ unconstrained optimization problems (3), each problem leading to a solution that satisfies the placement constraints. The wirelength of a net is calculated as the half-perimeter wirelength ($HPWL$) of the smallest bounding box of a net. Then, for net $e \in E$, the $HPWL$ is expressed as

$$HPWL_e = \max_{i \in e}\{x_i\} - \min_{i \in e}\{x_i\} + \max_{i \in e}\{y_i\} - \min_{i \in e}\{y_i\}.$$

The total $HPWL$ of the placement is expressed as

$$HPWL = \sum_{e \in E} HPWL_e.$$

Our smoothing approach is based on the two-variable Chen-Harker-Kanzow-Smale ($CHKS$) function that was initially proposed in [3, 8, 16]. The $CHKS$ function [10, 11] is used to smooth the two-variable maximum function

$$CHKS(x_1, x_2) = \frac{\sqrt{(x_1 - x_2)^2 + \alpha^2} + x_1 + x_2}{2}.$$

As shown in [1], to obtain the multi-variable $CHKS$ approximation of the multi-variable minimum and maximum functions, the two-variable $CHKS$ function is called recursively. The multi-variable $CHKS$ approximation was applied in the context of VLSI placement in [11]. To obtain an even distribution of macros and cells across the chip, the cell density has to be uniform. Following the approach of APlace 2.0 [7], the cell potential function is approximated

using a bell-shaped function. Let w_v be the width and h_v be the height of cell v. Also, let v be located at a horizontal distance d_x from the center of b. As in [7], an approximation for the cell potential function in the horizontal direction is expressed as

$$p_x(b, v) = \begin{cases} 1 - \frac{4d_x^2}{(w_v + 2w_b)(w_v + 4w_b)}, & \text{if } d_x \in [0, \frac{w_v + 2w_b}{2}], \\[2ex] \frac{2(d_x - \frac{w_v + 4w_b}{2})^2}{w_b(w_v + 4w_b)}, & \text{if } d_x \in [\frac{w_v + 2w_b}{2}, \frac{w_v + 4w_b}{2}], \\[2ex] 0, & \text{otherwise.} \end{cases}$$

Similarly, an approximation for the cell potential function in the vertical direction is $p_y(b, v)$. Then, the cell potential inside bin b is determined by the product of the horizontal cell potential and the vertical cell potential as

$$SD_b(\mathbf{x}, \mathbf{y}) = \sum_v C_v p_x(b, v) p_y(b, v).$$

2.1 Size scaling

In a multi-level placement framework, the algorithm places a sequence of hypergraphs from coarsest to finest granularity, obtaining the initial placement of a finer granularity hypergraph from the placement of a coarser granularity hypergraph. Despite being fast and scalable, the multi-level approach has the inherent drawback of being dependent of the clustering algorithm.

In addition, large macros of the design often create discontinuities on the placement region. How the placer handles these discontinuities is a non-trivial task. Regarding large macros, whether pre-placed and fixed or movable, they pose barriers and prevent cells from migrating to their natural locations. As shown in Fig. 1, the natural location for the cell on the top-right side of the chip is to be placed closer to its neighboring cells. However, the existence of the large macro next to the cell prevents the cell from migrating to its natural location. Thus, unevenness of the placement region essentially makes it hard for cells to move to their natural locations.

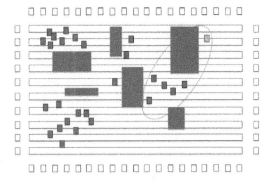

Figure 1: Cell located at the top-right of the placement area is blocked by large macro. Size scaling of the macro handles this problem. Thus, the cell is allowed to move to its natural location (closer to its neighboring cells) more easily.

Furthermore, for large-scale mixed-size designs, the multi-level approach often employs multiple smoothing techniques to control the different sizes of the various placed components. For a typical mixed-size design, the circuit netlist contains a large number of cells, combined with macros of sizes significantly larger than cells. For this reason, most placers apply clustering and various smoothing approaches to simultaneously place macros and cells on the chip area. For instance, NTUPlace 3.0 [4] performs global placement in a multi-level framework and applies two types of smoothing during global placement to reduce the cell potential levels.

To enhance continuity in placement and provide the means for a smoother placement flow, we propose handling placement discontinuities using size scaling. In this approach, we first scale down the dimensions of the placement components. Then, it gradually adjusts the dimensions of the components until they have reached their original sizes. Overall, our proposed approach targets a series of issues for mixed-size placement:

- Placement quality affected by the clustering algorithm

- Abrupt changes to cell locations during the multi-level placement framework

- Large macros acting as obstacles and preventing cells from migrating to their natural placement locations

- Existence of discontinuities during the placement flow and unevenness of the placement region

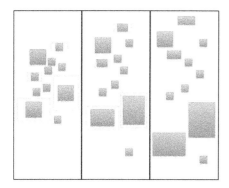

Figure 2: Size scaling for a design that includes macros and cells with large variations in size. To facilitate simultaneous placement of macros and cells, the placement components gradually scale up to the original sizes. The placement flow is presented in three steps, with the initial step corresponding to the placement of the scaled design (left) and the final step to the placement of the original design (right).

Size scaling in mixed-size placement is illustrated in Fig. 2. We effectively use size scaling to achieve smoothing. Let w_{cell} be the width and h_{cell} be the height of the cell in the modified circuit. Also, let w_{macro} be the width and h_{macro} be the height of the macro in the modified circuit. These values are determined using size scaling. Once they get updated, a modified circuit is placed. Also, let $cell_{width,min}$ ($cell_{height,min}$) be the width (height) of the smallest cell in the original circuit and $cell_{width,orig}$ ($cell_{height,orig}$) be the

width (height) of a cell in the original circuit. Finally, let $macro_{width,min}$ ($macro_{height,min}$) be the width (height) of the smallest macro in the original circuit and $macro_{width,orig}$ ($macro_{height,orig}$) be the width (height) of a macro in the original circuit.

During the final placement all components have reached their original sizes. The value of $NumStep$ is proportional to the ratio between the area of the largest macro and the area of the smallest macro of the design. For $0 \leq Step < NumStep$, the pattern for size scaling is

$$
\begin{aligned}
w_{cell} &= cell_{width,min}+ \\
&+(cell_{width,orig} - cell_{width,min}) * \frac{1 + Step}{NumStep}, \\
h_{cell} &= cell_{height,min}+ \\
&+(cell_{height,orig} - cell_{height,min}) * \frac{1 + Step}{NumStep} \\
w_{macro} &= macro_{width,min}+ \\
&+(macro_{width,orig} - macro_{width,min}) * \frac{1 + Step}{NumStep}, \\
h_{macro} &= macro_{height,min}+ \\
&+(macro_{height,orig} - macro_{height,min}) * \frac{1 + Step}{NumStep}.
\end{aligned} \tag{4}
$$

The algorithm makes use of an unconstrained optimization flow to converge to a non-overlapping placement solution. However, while the placement framework for multi-level algorithms is determined by performing clustering of the circuit netlist, our algorithm follows a different approach and employs size scaling of macros and cells to preserve smoothness during placement. For each step in $0 \leq Step < NumStep$, we solve a sequence of at most $maxnu$ unconstrained optimization problems (3).

Let $gDP_{total}(\mathbf{x}, \mathbf{y})$ be the gradient of the cell potential function and $gWL_{total}(\mathbf{x}, \mathbf{y})$ be the gradient of the wire-length function. Each unconstrained optimization problem is solved iteratively. For the first unconstrained problem u is set to be $\frac{gDP_{total}(\mathbf{x},\mathbf{y})}{gWL_{total}(\mathbf{x},\mathbf{y})}$. When the convergence criteria for the first problem have been satisfied, the algorithm uses an update scheme that halves u. Then, the next unconstrained optimization problem of the sequence is solved with an updated u, starting from the placement solution of the previous problem.

In this way, there are two loops in the placement algorithm. The outer-most loop iterates for $NumStep$ times. The inner-most loop is a sequence of at most $maxnu$ unconstrained optimization problems. Then, for each iteration of the outer-most loop, we update the sizes of the placement components using (4) to place a modified circuit, and initialize u as $\frac{gDP_{total}(\mathbf{x},\mathbf{y})}{gWL_{total}(\mathbf{x},\mathbf{y})}$. In each inner loop iteration, we initialize $u = \frac{gDP_{total}(\mathbf{x},\mathbf{y})}{gWL_{total}(\mathbf{x},\mathbf{y})}$ to formulate the first unconstrained optimization problem. The update scheme halves u every time the convergence criteria for the unconstrained optimization problem have been satisfied and a new problem of the sequence is solved.

2.2 Optimal region approach

The optimal region approach is proposed in the context of global placement in order to determine target regions for cells [5, 6] and determine an initial placement. Previously, the authors of [14] made use of the target region approach in the context of detailed placement. The optimal region approach examines the hyperedges corresponding to movable macros and cells, and makes use of fixed macros to determine their initial locations. For each cell or movable macro, we construct bounding boxes based on the coordinates of the pins of fixed macros that belong to the same hyperedge and excluding the location of that particular cell or movable macro. Then, the optimal region is defined by the intersection of the median of the horizontal coordinates and the median of the vertical coordinates of the bounding boxes associated with the cell or movable macro.

Let $(left_{reg}, bot_{reg})$ be the coordinates of the bottom-left of the optimal region and w_{reg} (h_{reg}) be the width (height) of the optimal region corresponding to v. Also, let $RAND_{MAX}$ be the maximum value returned by a pseudo-random integer $rand()$. Then $rand()$ is in the range 0 to $RAND_{MAX}$ and the location of the cell or movable macro inside its optimal region is determined using

$$x_v = (w_{reg} - \frac{w_v}{2})\frac{rand()}{RAND_{MAX}} + left_{reg} + \frac{w_v}{2},$$
$$y_v = (h_{reg} - \frac{h_v}{2})\frac{rand()}{RAND_{MAX}} + bot_{reg} + \frac{h_v}{2}.$$

3. EXPERIMENTAL RESULTS

The global placement flow involved an initial step of optimal region identification for movable macros and cells, and a secondary step of applying size scaling. The algorithm employed size scaling of the components to discourage abrupt changes to cell placements during spreading, thus enabling a smoother placement flow. We determined the number of steps and scaled the sizes of macros and cells as in (4). After the global placement step, the designs were legalized using [4] and the detailed placement step was performed using [18]. The initial locations of movable macros and cells were determined using the optimal region approach, whereas preplaced macros remained at fixed locations. For the circuits of the ISPD 2006 [12] mixed-size placement suite, we performed an additional step of fast legalization for movable macros. All experiments were performed on an Intel Core2 DUO at 2.66GHz with 2GB RAM.

For the circuits of the ISPD 2005 [13] benchmark suite, the results are reported in Table 1. We compare our placer with mPL6 [2], NTUPlace 3.0 [4], APlace 2.0 [7], and simPL [9]. For each placement tool, the first column reports the wirelength and the second column the total CPU time. The last two rows report the normalized wirelength and average reduction in wirelength of our placer compared against each placement tool. For each design, we calculate the reduction in wirelength that was obtained from our placer. Then, we calculate the average reduction among the individual improvements in wirelength. Also, we calculate the normalized wirelength to be the normalized value of the average reduction for each placement tool. The results reported for simPL were obtained directly using [9] and these are indicated by a "*". As reported in [9], the run-time for simPL is approximately half of that of NTUPlace 3.0 [4]. Our placer obtained an average wirelength reduction by 2.50%, 4.36%, 5.18%, and 9.02% compared against simPL [9], mPL6 [2], NTUPlace 3.0 [4], and APlace 2.0 [7], respectively. In terms of run-time, our placer was comparable to mPL6 [2] and faster than APlace 2.0 [7], but slower than simPL [9] and NTUPlace 3.0 [4].

For the circuits of the ISPD 2006 [12] benchmark suite, the results are reported in Table 2. As in Table 1, for each placement tool, the first column reports the wirelength and the second column reports the total CPU time for global placement, legalization, and detailed placement. SimPL was developed only for the ISPD 2005 [13] benchmarks and is not included in the comparison. The last two rows report the normalized wirelength and average reduction in wirelength of our placer compared against each placement tool. Our placer obtained an average wirelength reduction by 2.51%, 3.95%, and 4.81% compared against mPL6 [2], NTUPlace 3.0 [4], and APlace 2.0 [7], respectively. In terms of runtime, our placer was faster than mPL6 [2] and APlace 2.0 [7], but slower than NTUPlace 3.0 [4].

4. CONCLUSION

The algorithm was proposed to perform effective flat placement for large-scale mixed-size designs. The placer combined size scaling with the optimal region approach as an alternative to multi-level circuit placement. The placement algorithm gradually increased the dimensions of the placement components to preserve smoothness during the placement process. Overall, our algorithm produced high quality placement solutions for the circuits of the ISPD 2005 [13] and ISPD 2006 [12] mixed-size placement benchmark suites.

5. ACKNOWLEDGMENTS

The authors would like to acknowledge the support of SRC (Task 1822.001). We are also grateful to many reviewers for their constructive comments and suggestions.

6. REFERENCES

[1] S. I. Birbil, S. C. Fang, H. Frenk, and S. Zhang. Recursive Approximation of the High-dimensional Max Function. Technical Report SEEM 2002-12, Department of Systems Engineering and Engineering Management SEEM, The Chinese University of Hong Kong, Hong Kong, Dec. 2002.

[2] T. F. Chan, J. Cong, M. Romesis, J. R. Shinnerl, K. Sze, and M. Xie. mPL6: A Robust Multilevel Mixed-size Placement Engine. In *Proc. ACM/SIGDA International Symposium on Physical Design*, pages 227–229, San Francisco, California, Apr. 2005.

[3] B. Chen and P. T. Harker. A Non-interior Point Continuation Method for Linear Complementarity Problems. *SIAM Journal on Matrix Analysis and Applications*, 14:1168–1190, 1993.

[4] T.-C. Chen, Z.-W. Jiang, T.-C. Hsu, H.-C. Chen, and Y.-W. Chang. NTUplace3: An Analytical Placer for Large-Scale Mixed-Size Designs With Preplaced Blocks and Density Constraints. *IEEE Transactions on Integrated Circuits and Systems*, 27(7):1228–1240, 2008.

[5] S. Goto. An Efficient Algorithm for the Two-dimensional Placement Problem in Electrical Circuit Layout. *Proc. IEEE/ACM International Symposium on Circuits and Systems*, 28(1):12–18, Jan. 1981.

[6] S. Hakimi. Optimum Locations of Switching Centers and the Absolute Centers and Medians of a Graph. *Journal of the Operational Research Society*, 12.

Table 1: Wirelength after legalization and detailed placement on ISPD 2005 [13] mixed-size placement benchmark suite.

Circuit	Ours		mPL6 [2]		NTUPlace 3.0 [4]		APlace 2.0 [7]		simPL [9]	
	dWL (x E8)	pCPU (s)	dWL (x E8)	pCPU (s)	dWL (x E8)	pCPU (s)	dWL (x E8)	pCPU (s)	dWL (x E8)	pCPU (s)
adaptec1	0.79	1380	0.80	2041	0.81	689	0.78	3164	0.78	136*
adaptec2	0.90	1560	0.92	2095	0.90	685	0.96	4598	0.90	209*
adaptec3	1.93	3534	2.14	6196	2.15	1719	2.19	11167	2.09	422*
adaptec4	1.80	2280	1.94	5705	1.94	2047	2.09	12726	1.87	318*
bigblue1	9.42	2100	9.68	2531	9.74	1387	1.00	4124	9.74	241*
bigblue2	1.41	4980	1.52	6338	1.52	4457	1.54	9545	1.46	497*
bigblue3	3.30	6300	3.44	8720	3.61	5213	4.12	19299	3.40	827*
bigblue4	8.10	13920	8.29	20062	8.29	10200	8.71	49572	8.08	2148*
normalized wirelength	1.00		1.04		1.05		1.09		1.03	
average reduction	–		4.36%		5.19%		9.02%		2.50%	

Table 2: Wirelength after legalization and detailed placement on ISPD 2006 [12] mixed-size placement benchmark suite.

Circuit	Ours		mPL6 [2]		NTUPlace 3.0 [4]		APlace 2.0 [7]	
	dWL (x E8)	pCPU (s)	dWL (x E8)	pCPU (s)	dWL (x E8)	pCPU (s)	dWL (x E8)	pCPU (s)
adaptec5	3.31	5100	3.33	10688	3.71	3335	3.54	17592
newblue1	6.20	1925	6.32	2552	6.05	887	6.55	6764
newblue2	1.92	4420	1.99	6253	2.03	2144	1.96	12543
newblue3	2.83	4970	2.84	6701	2.80	1142	2.78	16779
newblue4	2.45	7564	2.47	7430	2.51	3977	2.64	12559
newblue5	4.20	8964	4.23	13693	4.34	8529	4.40	19494
newblue6	4.52	9132	4.99	12591	4.98	7111	5.03	21485
normalized wirelength	1.00		1.03		1.04		1.05	
average reduction	–		2.51%		3.95%		4.81%	

[7] A. B. Kahng, S. Reda, and Q. Wang. Architecture and Details of a High-quality, Large-scale Analytical Placer. In *Proc. IEEE/ACM International Conference on Computer-Aided Design*, pages 891–898, San Jose, California, Nov. 2005.

[8] C. Kanzow. Some Non-interior Continuation Methods for Linear Complementarity Problems. *SIAM Journal on Matrix Analysis and Applications*, 17:851–868, 1996.

[9] M.-C. Kim, D.-J. Lee, and I. L. Markov. SimPL: An Effective Placement Algorithm. In *Proc. IEEE/ACM International Conference on Computer-Aided Design*, pages 649–656, San Jose, California, Nov. 2010.

[10] C. Li. *Placement of VLSI Circuits*. PhD thesis, Purdue University, West Lafayette, 2006.

[11] C. Li and C.-K. Koh. Recursive Function Smoothing of Half-perimeter Wirelength for Analytical Placement. In *Proc. International Symposium on Quality Electronic Design*, pages 829–834, San Jose, California, Mar. 2007.

[12] G.-J. Nam, C. J. Alpert, and P. Villarrubia. ISPD 2006 Placement Contest Benchmark Suite. In *http://www.sigda.org/ispd2006/contest.htm*, 2006.

[13] G.-J. Nam, C. J. Alpert, P. Villarrubia, B. Winter, and M. Yildiz. ISPD 2005 Placement Contest

Benchmark Suite. In *http://www.sigda.org/ispd2005/contest.htm*, 2005.

[14] M. Pan, N. Viswanathan, and C. Chu. An Efficient and Effective Detailed Placement Algorithm. In *Proc. IEEE/ACM International Conference on Computer-Aided Design*, pages 48–55, San Jose, California, Nov. 2005.

[15] J. A. Roy, N. Viswanathan, G.-J. Nam, C. J. Alpert, and I. L. Markov. CRISP: Congestion Reduction by Iterated Spreading during Placement. In *Proc. IEEE/ACM International Conference on Computer-Aided Design*, pages 357–362, San Jose, California, Nov. 2009.

[16] S. Smale. Algorithms for Solving Equations. In *Proc. International Congress of Mathematicians*, pages 172–195, Providence, Rhode Island, 1987.

[17] K. Tsota, C.-K. Koh, and V. Balakrishnan. Guiding Global Placement with Wire Density. In *Proc. IEEE/ACM International Conference on Computer-Aided Design*, pages 212–217, San Jose, California, Nov. 2008.

[18] N. Viswanathan, M. Pan, and C. C. Chu. FastPlace 3.0: A Fast Multilevel Quadratic Placement Algorithm with Placement Congestion Control. In *Proc. Asia and South Pacific Design Automation Conference*, pages 135–140, 2007.

Author Index

NOTES

NOTES